Statistical Tables

Statistical Tables
SECOND EDITION

F. James Rohlf

Robert R. Sokal

State University of New York
at Stony Brook

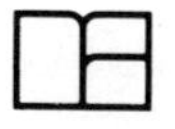

W. H. Freeman and Company
New York

Library of Congress Cataloging in Publication Data

Rolhf, F. James, 1936-
Statistical tables.

"Each table is accompanied by . . . a reference to the section or sections in our textbook Biometry giving explanations and applications of the table."—Notes on the 2nd ed.
1. Biometry—Tables, etc. 2. Mathematical statistics —Tables, etc. I. Sokal, Robert R.
II. Sokal, Robert R. Biometry. 2nd ed. III. Title.
QH323.5.R63 1981 519.5'0212 81-2576
ISBN 0-7167-1257-1 AACR2
ISBN 0-7167-1258-X (pbk.)

5 6 7 8 9 10 MP 4 3 2 1 0 8 9 8 7 6

Contents

Preface *vii*
Notes on the Second Edition *ix*
Introduction: Interpolation *xi*
1 Common logarithms of factorials *1*
2 $f \ln f$ as a function of f *5*
3 $(f + ½) \ln (f + ½)$ as a function of f *47*
4 Orthogonal polynomials *50*
5 The angular transformation *53*
6 Proportions corresponding to angle θ *57*
7 The z-transformation of correlation coefficient r *61*
8 Correlation coefficient r as a function of transform z *65*
9 C_n—Gurland and Tripathi's correction for the standard deviation *69*
10 Ten thousand random digits *71*
11 Areas of the normal curve *77*
12 Critical values of Student's t-distribution *80*
13 Critical values of Student's t-distribution based on Šidák's multiplicative inequality *83*
14 Critical values of the chi-square distribution *97*
15 Critical values of the chi-square distribution based on Šidák's multiplicative inequality *101*
16 Critical values of the F-distribution *108*
17 Critical values of F_{max} *131*
18 Critical values of the studentized range *133*
19 Critical values for Welsch's step-up procedure *139*
20 Critical values of the studentized augmented range *145*
21 Critical values of the studentized maximum modulus distribution *147*
22 Shortest unbiased confidence limits for the variance *154*
23 Confidence limits for percentages *156*
24 Relative expected frequencies for individual terms of the Poisson distribution *163*
25 Critical values for correlation coefficients *166*
26 Mean ranges of samples from a normal distribution *169*
27 Rankits (normal order statistics) *171*
28 Critical values of the number of runs *175*
29 Critical values of U, the Mann–Whitney statistic *185*
30 Critical values of the Wilcoxon rank sum *191*

31 Critical values of the two-sample Kolmogorov–Smirnov statistic *195*
32 Critical values of the one-sample Kolmogorov–Smirnov statistic *203*
33 Critical values of the one-sample Kolmogorov–Smirnov statistic for intrinsic hypotheses *205*
34 Critical values for Page's test *207*
35 Critical values of Olmstead and Tukey's test criterion *210*
36 Critical values for testing outliers (according to Dixon) *211*
37 Critical values for testing outliers (according to Grubbs) *213*
38 Critical values for Kendall's rank correlation coefficient τ *215*
39 Critical values for runs up and down *217*
40 Some mathematical constants *219*

Preface

This set of tables grew out of our dissatisfaction with the customary placement of statistical tables at the end of textbooks of biometry and statistics. Serious users of these books and tables are constantly inconvenienced by having to turn back and forth between the text material on a certain method and the table necessary for the test of significance or for some other computational step. Occasionally, the tables are interspersed throughout a textbook at sites of their initial application; they are then difficult to locate, and turning back and forth in the book is not avoided. Frequent users of statistics, therefore, generally use one or more sets of statistical tables, not only because these usually contain more complete and diverse statistical tables than the textbooks, but also to avoid the constant turning of pages in the latter.

When we first planned to write our textbook on biometry (R. R. Sokal and F. J. Rohlf, *Biometry,* second edition, W. H. Freeman and Company, San Francisco, 1981) we thought to eliminate tables altogether, asking readers to furnish their own statistical tables from those available. However, for pedagogical reasons, it was found desirable to refer to a standard set of tables, and we consequently undertook to furnish such tables to be bound separately from the text. Once embarked upon the task of preparing these tables, we gave considerable thought to making them as useful as we could for statistical work in the biological and social sciences.

Several of the tables would have been very complicated and tedious to recompute. These have been copied with permission of authors and publishers, whose courtesy is here acknowledged collectively. We are indebted to the Literary Executor of the late Sir Ronald A. Fisher, F.R.S., Cambridge, to Dr. Frank Yates, F.R.S., Rothamsted, and to Messrs. Oliver and Boyd, Limited, Edinburgh, for permission to reprint tables III and XX from their book *Statistical Tables for Biological, Agricultural and Medical Research.* Other specific acknowledgements are found beneath the particular table. We appreciate the constructive comments of Professor K. R. Gabriel (University of Rochester), who read a draft of the introductory material.

We hope that users of statistics will find our tables as useful as we have already found them to be in our work. We shall be grateful for any suggestions about changes, additions, or deletions, as well as for any corrections.

F. James Rohlf

Robert R. Sokal

Notes on the Second Edition

This new edition of our Statistical Tables reflects developments in statistics over the last decade. Some recently developed or adopted methods require tables for significance testing. Thirteen new tables have been added on this account. They include five new tables for multiple comparisons tests; a table for Gurland and Tripathi's correction for the standard deviation; tables for critical values of the two-sample Kolmogorov–Smirnov statistic and of the one-sample Kolmogorov–Smirnov statistic for intrinsic hypotheses; tables for Page's test, for testing outliers by Dixon's method, for testing outliers by Grubb's method, for Kendall's rank correlation coefficient τ, and for runs up and down. Tables of simple mathematical functions such as square roots, logarithms, and trigonometric functions that are freely available on even simple pocket calculators have been eliminated.

An introductory section on interpolation precedes the tables. Each table is accompanied by a brief explanation of its nature, a demonstration of how to look up a value, and a reference to the section or sections in our textbook *Biometry* giving explanations and applications of the table. All references to section, table, or box numbers unaccompanied by a citation pertain to this textbook. (Those who use the set of tables but not the textbook should simply disregard these references.) Each table is accompanied by a short note on the method of its generation. Most of the tables were generated using simple FORTRAN programs on a variety of computers (IBM 7040, IBM 370/168, GE 625, UNIVAC 1110). In such cases equations are given to explain how the tables were prepared.

F. James Rohlf
Robert R. Sokal

Stony Brook, New York
January 1981

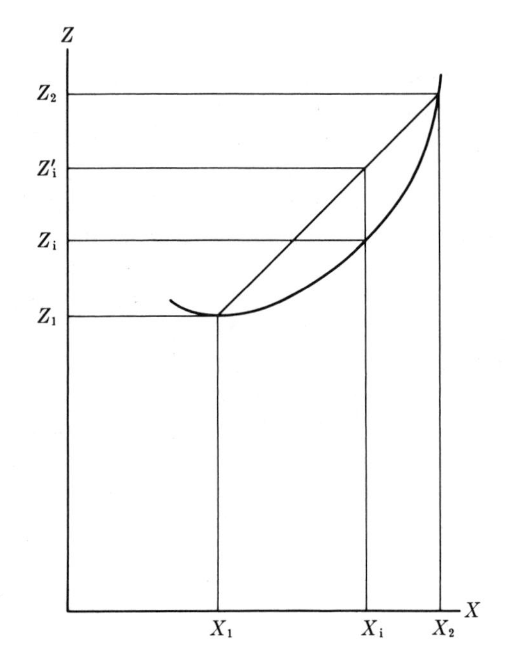
Z
Z₂
Z′ᵢ
Zᵢ
Z₁
X
X₁
Xᵢ
X₂

Introduction: *Interpolation*

Finding a value of a function for an argument that is intermediate between two arguments in a table requires *interpolation*. In some tables published earlier, aids to mental interpolation (proportional parts) are furnished. Since the present tables are oriented toward use with calculators, we furnish several formulas especially adapted for their use.

We shall employ the following symbolism. The tabled arguments to each side of the desired argument X_i are identified as X_1 and X_2, respectively. Argument X_i must lie between the tabled arguments, $X_1 < X_i < X_2$ or $X_1 > X_i > X_2$. The functions shown in the table are Z_1 and Z_2 corresponding to X_1 and X_2 and the desired function corresponding to argument X_i is labeled Z_i.

The simplest method is *linear interpolation*. It assumes that the function $Z = f(X)$ is approximately linear over the interval from X_1 to X_2. It serves as an adequate method where the interval over which one needs to interpolate is not very wide, or where the function is either truly or approximately linear in that interval. The effect of linear interpolation is seen in the accompanying figure, which illustrates a linear function approximating a curvilinear function over the interval from X_1 to X_2. The true function Z_i corresponding to argument X_i is approximated by the linear interpolate Z_i'.

To carry out a linear interpolation with a calculator, first compute $p = (X_i - X_1)/(X_2 - X_1)$. Then substitute the given values of the function and p into the following equation:

$$Z_i' = pZ_2 + (1 - p)Z_1$$

The coefficients p and $1 - p$ represent complementary proportions of the distance from the tabled arguments to the intermediate value. When, as frequently happens, the length of the interval from X_1 to X_2 is 1, the computation is especially simple, since $p = X_i - X_1$. Some examples will show the use of this equation. Suppose we wish to find the value of the z-transformation of the correlation coefficient $r = 0.9913$. In Table **7** we find arguments $X_1 = 0.991$, $X_2 = 0.992$ and corresponding functions $Z_1 = 2.6996$, and $Z_2 = 2.7587$. We then compute proportion $p = (0.9913 - 0.991)/\ (0.992 - 0.991) = 0.0003/0.001 = 0.3$. Then

$$\begin{aligned} Z_i' &= (0.3)(2.7587) + (0.7)(2.6996) \\ &= 2.71733 \end{aligned}$$

This compares with 2.71661 given by direct computation of $\tanh^{-1} 0.9913$.

As an example of interpolation when the interval from X_1 to X_2 is 1, and p is simply $X_i - X_1$, we interpolate for $f \ln f$ in Table **2** when $f = 103.5$.

$$Z_i' = (0.5)(483.017) + (0.5)(477.377) = 480.197$$

The correct value, shown in Table **3**, is 480.196.

Inverse interpolation is employed to evaluate an argument given a value of a function intermediate between two tabled values. Using the same symbolism as above, one approximates the desired argument X_i by X'_i as follows:

$$X'_i = X_1 + \frac{(Z_i - Z_1)(X_2 - X_1)}{Z_2 - Z_1}$$

By way of illustration, interpolate for the argument in the earlier example from Table **7** where 2.71733 was obtained as the linearly interpolated value for $r = 0.9913$. Bracketing this value Z_i are functions $Z_1 = 2.6996$ and $Z_2 = 2.7587$. The corresponding arguments in the table are $X_1 = 0.991$ and $X_2 = 0.992$. On substitution in the inverse interpolation formula, one obtains

$$\begin{aligned} X'_i &= 0.991 + \frac{(2.71733 - 2.6996)(0.992 - 0.991)}{2.7587 - 2.6996} \\ &= 0.991 + \frac{(0.01773)(0.001)}{0.0591} \\ &= 0.991 + 0.0003 \\ &= 0.9913 \end{aligned}$$

Four-point interpolation may provide more exact results than linear interpolation. The symbolism is as before, except that Z_0 is the tabled function corresponding to X_0, the argument before X_1, and Z_3 is the function corresponding to X_3, the argument after X_2. It is assumed that the X's are equally spaced.

$$Z'_i = \tfrac{1}{2}\{[2 + (p - p^2)][pZ_2 + (1 - p)Z_1] - \frac{(p - p^2)}{3}[(1 + p)Z_3 + (2 - p)Z_0]\}$$

Applying this formula to the problem of finding z when $r = 0.9913$, computed earlier by linear interpolation, one obtains from Table **7**: $X_1 = 0.991$, $X_2 = 0.992$, $Z_0 = 2.6467$, $Z_1 = 2.6996$, $Z_2 = 2.7587$, $Z_3 = 2.8257$. Therefore proportion $p = (0.9913 - 0.991)/(0.992-0.991) = 0.3$. Solving for Z'_i one obtains

$$\begin{aligned} Z'_i &= \tfrac{1}{2}\{[2 + (0.3 - 0.3^2)][(0.3)(2.7587) + (1 - 0.3)(2.6996)] \\ &\quad - \frac{(0.3 - 0.3^2)}{3}[(1 + 0.3)(2.8257) + (2 - 0.3)(2.6467)]\} \\ &= \tfrac{1}{2}\{[2.21][2.71733] - \frac{0.21}{3}[8.17280]\} \\ &= \tfrac{1}{2}\{5.4332033\} \\ &= 2.71660 \end{aligned}$$

which is very close to the correct value 2.71661.

Many tables, such as Table **12**, are arranged for *harmonic interpolation*. For the upper range of the arguments, functions in these tables will be approximately linearly related to the reciprocal of the arguments. Usually the arguments are degrees of freedom spaced as follows: 30, 40, 60, 120, ∞. For purposes of convenience in interpolation these are changed by dividing them into the last finite value of the argument, yielding 120/30, 120/40, 120/60, 120/120, 120/∞, or 4, 3, 2, 1, 0. These integral values are the new arguments; functions for any argument between these tabled values are linearly interpolated between these transformed values. One advantage of this method is that it permits interpolation between a finite and an infinite argument. An example will illustrate the method. Find the value of $t_{.001[200]}$. In Table **12** can be found $Z_1 = t_{.001[120]} = 3.373$ and $Z_2 = t_{.001[\infty]} = 3.291$. Transform the arguments $X_1 = 120$ and $X_2 = \infty$ to $X_1 = 120/120 = 1$ and $X_2 = 120/\infty = 0$. Transform X_i to $120/200 = 0.6$, and apply the formula for linear interpolation:

$$p = \frac{0.6 - 1}{0 - 1} = 0.4$$

$$\begin{aligned} Z_i' &= (0.4)(3.291) + (0.6)(3.373) \\ &= 3.3402 \end{aligned}$$

which is rounded back to 3.340.

Statistical Tables

TABLE 1 Common logarithms of factorials

This table gives the common logarithms of factorials of integers from 0 to 499. The mantissa is accurate to four decimal places.

For example, log 5! = log 120 = 2.0792; or log 281! = 567.6733.

Logarithms of factorials are needed whenever much computation is done with factorials, as, for example, in working out Fisher's exact test of independence in a 2 × 2 contingency table (Section 17.4 and Box 17.7).

This table was generated by summing logarithms from the FORTRAN IV library function, using double precision arithmetic.

TABLE 1 Common logarithms of factorials

	0	1	2	3	4
0	0.0000	0.0000	0.3010	0.7782	1.3802
10	6.5598	7.6012	8.6803	9.7943	10.9404
20	18.3861	19.7083	21.0508	22.4125	23.7927
30	32.4237	33.9150	35.4202	36.9387	38.4702
40	47.9116	49.5244	51.1477	52.7811	54.4246
50	64.4831	66.1906	67.9066	69.6309	71.3633
60	81.9202	83.7055	85.4979	87.2972	89.1034
70	100.0784	101.9297	103.7870	105.6503	107.5196
80	118.8547	120.7632	122.6770	124.5961	126.5204
90	138.1719	140.1310	142.0948	144.0632	146.0364
100	157.9700	159.9743	161.9829	163.9958	166.0128
110	178.2009	180.2462	182.2955	184.3485	186.4054
120	198.8254	200.9082	202.9945	205.0844	207.1779
130	219.8107	221.9280	224.0485	226.1724	228.2995
140	241.1291	243.2783	245.4306	247.5860	249.7443
150	262.7569	264.9359	267.1177	269.3024	271.4899
160	284.6735	286.8803	289.0898	291.3020	293.5168
170	306.8608	309.0938	311.3293	313.5674	315.8079
180	329.3030	331.5606	333.8207	336.0832	338.3480
190	351.9859	354.2669	356.5502	358.8358	361.1236
200	374.8969	377.2001	379.5054	381.8129	384.1226
210	398.0246	400.3489	402.6752	405.0036	407.3340
220	421.3587	423.7031	426.0494	428.3977	430.7480
230	444.8898	447.2534	449.6189	451.9862	454.3555
240	468.6094	470.9914	473.3752	475.7608	478.1482
250	492.5096	494.9093	497.3107	499.7138	502.1186
260	516.5832	518.9999	521.4182	523.8381	526.2597
270	540.8236	543.2566	545.6912	548.1273	550.5651
280	565.2246	567.6733	570.1235	572.5753	575.0287
290	589.7804	592.2443	594.7097	597.1766	599.6449
300	614.4858	616.9644	619.4444	621.9258	624.4087
310	639.3357	641.8285	644.3226	646.8182	649.3151
320	664.3255	666.8320	669.3399	671.8491	674.3596
330	689.4509	691.9707	694.4918	697.0143	699.5380
340	714.7076	717.2404	719.7744	722.3097	724.8463
350	740.0920	742.6373	745.1838	747.7316	750.2806
360	765.6002	768.1577	770.7164	773.2764	775.8375
370	791.2290	793.7983	796.3689	798.9406	801.5135
380	816.9749	819.5559	822.1379	824.7211	827.3055
390	842.8351	845.4272	848.0205	850.6149	853.2104
400	868.8064	871.4096	874.0138	876.6191	879.2255
410	894.8862	897.5001	900.1150	902.7309	905.3479
420	921.0718	923.6961	926.3214	928.9478	931.5751
430	947.3607	949.9952	952.6307	955.2672	957.9047
440	973.7505	976.3949	979.0404	981.6868	984.3342
450	1000.2389	1002.8931	1005.5482	1008.2043	1010.8614
460	1026.8237	1029.4874	1032.1520	1034.8176	1037.4841
470	1053.5028	1056.1758	1058.8498	1061.5246	1064.2004
480	1080.2742	1082.9564	1085.6394	1088.3234	1091.0082
490	1107.1360	1109.8271	1112.5191	1115.2119	1117.9057

5	6	7	8	9	
2.0792	2.8573	3.7024	4.6055	5.5598	0
12.1165	13.3206	14.5511	15.8063	17.0851	10
25.1906	26.6056	28.0370	29.4841	30.9465	20
40.0142	41.5705	43.1387	44.7185	46.3096	30
56.0778	57.7406	59.4127	61.0939	62.7841	40
73.1037	74.8519	76.6077	78.3712	80.1420	50
90.9163	92.7359	94.5619	96.3945	98.2333	60
109.3946	111.2754	113.1619	115.0540	116.9516	70
128.4498	130.3843	132.3238	134.2683	136.2177	80
148.0141	149.9964	151.9831	153.9744	155.9700	90
168.0340	170.0593	172.0887	174.1221	176.1595	100
188.4661	190.5306	192.5988	194.6707	196.7462	110
209.2748	211.3751	213.4790	215.5862	217.6967	120
230.4298	232.5634	234.7001	236.8400	238.9830	130
251.9057	254.0700	256.2374	258.4076	260.5808	140
273.6803	275.8734	278.0693	280.2679	282.4693	150
295.7343	297.9544	300.1771	302.4024	304.6303	160
318.0509	320.2965	322.5444	324.7948	327.0477	170
340.6152	342.8847	345.1565	347.4307	349.7071	180
363.4136	365.7059	368.0003	370.2970	372.5959	190
386.4343	388.7482	391.0642	393.3822	395.7024	200
409.6664	412.0009	414.3373	416.6758	419.0162	220
433.1002	435.4543	437.8103	440.1682	442.5281	220
456.7265	459.0994	461.4742	463.8508	466.2292	230
480.5374	482.9283	485.3210	487.7154	490.1116	240
504.5252	506.9334	509.3433	511.7549	514.1682	250
528.6830	531.1078	533.5344	535.9625	538.3922	260
553.0044	555.4453	557.8878	560.3318	562.7774	270
577.4835	579.9399	582.3977	584.8571	587.3180	280
602.1147	604.5860	607.0588	609.5330	612.0087	290
626.8930	629.3787	631.8659	634.3544	636.8444	300
651.8134	654.3131	656.8142	659.3166	661.8204	310
676.8715	679.3847	681.8993	684.4152	686.9324	320
702.0631	704.5894	707.1170	709.6460	712.1762	330
727.3841	729.9232	732.4635	735.0051	737.5479	340
752.8308	755.3823	757.9349	760.4888	763.0439	350
778.3997	780.9632	783.5279	786.0937	788.6608	360
804.0875	806.6627	809.2390	811.8165	814.3952	370
829.8909	832.4775	835.0652	837.6540	840.2440	380
855.8070	858.4047	861.0035	863.6034	866.2044	390
881.8329	884.4415	887.0510	889.6617	892.2734	400
907.9660	910.5850	913.2052	915.8264	918.4486	410
934.2035	936.8329	939.4633	942.0948	944.7272	420
960.5431	963.1826	965.8231	968.4646	971.1071	430
986.9825	989.6318	992.2822	994.9334	997.5857	440
1013.5194	1016.1783	1018.8383	1021.4991	1024.1609	450
1040.1516	1042.8200	1045.4893	1048.1595	1050.8307	460
1066.8771	1069.5547	1072.2332	1074.9127	1077.5930	470
1093.6940	1096.3806	1099.0681	1101.7565	1104.4458	480
1120.6003	1123.2958	1125.9921	1128.6893	1131.3874	490

TABLE 2 $f \ln f$ as a function of f

Table of $f \ln f$ for integers ranging from 0 to 9999. Three-decimal-place accuracy is furnished.

Looking up $f \ln f$ for any number is quite straightforward. Since f is an integer, no interpolation is required. Thus, for example, when $f = 1287$, the function in the table is $1287 \ln 1287 = 9215.009$.

This table is used especially for the log likelihood ratio test (G-test) for goodness of fit and independence in contingency tables. See Section 17.1 for an explanation of G-tests and later sections in Chapter 17 for examples.

When $f > 9999$, and can be factored $f = ab$, where both a and $b < 9999$, the following identity is useful for computing $f \ln f$: $ab \ln ab = a(b \ln b) + b(a \ln a)$.

This table was computed from the FORTRAN IV library function for natural logarithms, using double precision arithmetic.

0–490

TABLE 2 $f \ln f$ as a function of f

f	0	1	2	3	4
0	0.000	0.000	1.386	3.296	5.545
10	23.026	26.377	29.819	33.344	36.947
20	59.915	63.935	68.003	72.116	76.273
30	102.036	106.454	110.904	115.385	119.896
40	147.555	152.256	156.982	161.732	166.504
50	195.601	200.523	205.465	210.425	215.405
60	245.661	250.763	255.882	261.017	266.169
70	297.395	302.650	307.920	313.204	318.501
80	350.562	355.950	361.351	366.764	372.189
90	404.983	410.488	416.005	421.532	427.070
100	460.517	466.127	471.747	477.377	483.017
110	517.053	522.758	528.472	534.195	539.927
120	574.499	580.291	586.091	591.899	597.715
130	632.779	638.651	644.530	650.416	656.311
140	691.830	697.775	703.727	709.687	715.653
150	751.595	757.609	763.630	769.657	775.691
160	812.028	818.106	824.191	830.281	836.378
170	873.086	879.224	885.369	891.519	897.676
180	934.732	940.928	947.129	953.336	959.548
190	996.935	1003.184	1009.439	1015.699	1021.964
200	1059.663	1065.964	1072.270	1078.581	1084.896
210	1122.893	1129.242	1135.596	1141.955	1148.319
220	1186.598	1192.994	1199.394	1205.799	1212.209
230	1250.758	1257.198	1263.643	1270.092	1276.545
240	1315.353	1321.836	1328.323	1334.814	1341.309
250	1380.365	1386.889	1393.416	1399.948	1406.483
260	1445.777	1452.340	1458.906	1465.477	1472.051
270	1511.574	1518.174	1524.778	1531.386	1537.997
280	1577.741	1584.378	1591.018	1597.661	1604.309
290	1644.265	1650.937	1657.612	1664.291	1670.972
300	1711.135	1717.840	1724.549	1731.261	1737.976
310	1778.337	1785.076	1791.817	1798.562	1805.309
320	1845.863	1852.633	1859.406	1866.182	1872.961
330	1913.701	1920.501	1927.305	1934.111	1940.921
340	1981.842	1988.672	1995.505	2002.342	2009.181
350	2050.277	2057.136	2063.998	2070.863	2077.731
360	2118.997	2125.885	2132.775	2139.668	2146.564
370	2187.996	2194.911	2201.829	2208.749	2215.672
380	2257.265	2264.207	2271.151	2278.097	2285.047
390	2326.797	2333.765	2340.735	2347.707	2354.682
400	2396.586	2403.579	2410.574	2417.571	2424.572
410	2466.624	2473.642	2480.662	2487.684	2494.709
420	2536.907	2543.948	2550.992	2558.038	2565.087
430	2607.428	2614.493	2621.560	2628.629	2635.701
440	2678.181	2685.269	2692.359	2699.451	2706.546
450	2749.161	2756.272	2763.384	2770.499	2777.616
460	2820.364	2827.496	2834.631	2841.768	2848.906
470	2891.784	2898.938	2906.094	2913.252	2920.412
480	2963.417	2970.592	2977.769	2984.948	2992.129
490	3035.259	3042.454	3049.652	3056.851	3064.053

5	6	7	8	9	f
8.047	10.751	13.621	16.636	19.775	0
40.621	44.361	48.165	52.027	55.944	10
80.472	84.711	88.988	93.302	97.652	20
124.437	129.007	133.604	138.228	142.879	30
171.300	176.118	180.957	185.818	190.699	40
220.403	225.420	230.454	235.506	240.575	50
271.335	276.517	281.714	286.927	292.153	60
323.812	329.136	334.473	339.823	345.186	70
377.625	383.074	388.534	394.006	399.489	80
432.618	438.177	443.747	449.327	454.917	90
488.666	494.325	499.993	505.670	511.357	100
545.667	551.416	557.174	562.941	568.716	110
603.539	609.372	615.212	621.060	626.916	120
662.212	668.121	674.037	679.961	685.892	130
721.626	727.607	733.594	739.587	745.588	140
781.731	787.778	793.831	799.890	805.956	150
842.481	848.590	854.705	860.826	866.953	160
903.838	910.005	916.179	922.357	928.542	170
965.766	971.989	978.217	984.451	990.690	180
1028.235	1034.510	1040.791	1047.077	1053.368	190
1091.217	1097.542	1103.873	1110.208	1116.548	200
1154.687	1161.060	1167.438	1173.820	1180.207	210
1218.623	1225.041	1231.464	1237.891	1244.322	220
1283.003	1289.464	1295.930	1302.400	1308.875	230
1347.808	1354.312	1360.819	1367.330	1373.846	240
1413.022	1419.565	1426.113	1432.664	1439.218	250
1478.628	1485.210	1491.795	1498.385	1504.977	260
1544.612	1551.231	1557.853	1564.479	1571.108	270
1610.959	1617.614	1624.271	1630.933	1637.597	280
1677.658	1684.346	1691.038	1697.734	1704.433	290
1744.695	1751.417	1758.142	1764.871	1771.602	300
1812.060	1818.815	1825.572	1832.332	1839.096	310
1879.743	1886.529	1893.317	1900.108	1906.903	320
1947.734	1954.549	1961.368	1968.190	1975.014	330
2016.023	2022.868	2029.716	2036.566	2043.420	340
2084.602	2091.475	2098.352	2105.231	2112.113	350
2153.463	2160.364	2167.268	2174.175	2181.084	360
2222.597	2229.526	2236.456	2243.390	2250.326	370
2291.999	2298.953	2305.910	2312.870	2319.832	380
2361.660	2368.640	2375.623	2382.608	2389.596	390
2431.574	2438.579	2445.587	2452.597	2459.609	400
2501.736	2508.765	2515.797	2522.831	2529.868	410
2572.138	2579.191	2586.247	2593.305	2600.365	420
2642.776	2649.852	2656.931	2664.012	2671.095	430
2713.643	2720.742	2727.844	2734.947	2742.053	440
2784.735	2791.857	2798.980	2806.106	2813.234	450
2856.047	2863.191	2870.336	2877.483	2884.633	460
2927.575	2934.739	2941.905	2949.074	2956.245	470
2999.312	3006.497	3013.685	3020.874	3028.065	480
3071.256	3078.462	3085.669	3092.879	3100.090	490

TABLE 2 $f \ln f$ as a function of f

f	0	1	2	3	4
500	3107.304	3114.520	3121.737	3128.957	3136.178
510	3179.549	3186.785	3194.022	3201.262	3208.503
520	3251.991	3259.246	3266.502	3273.761	3281.022
530	3324.625	3331.899	3339.174	3346.452	3353.731
540	3397.447	3404.740	3412.034	3419.330	3426.628
550	3470.455	3477.766	3485.079	3492.393	3499.709
560	3543.645	3550.973	3558.304	3565.636	3572.971
570	3617.013	3624.359	3631.708	3639.058	3646.409
580	3690.556	3697.920	3705.286	3712.653	3720.022
590	3764.272	3771.653	3779.036	3786.420	3793.806
600	3838.158	3845.556	3852.955	3860.356	3867.759
610	3912.210	3919.624	3927.040	3934.458	3941.877
620	3986.426	3993.857	4001.289	4008.722	4016.158
630	4060.803	4068.250	4075.698	4083.148	4090.599
640	4135.340	4142.802	4150.266	4157.731	4165.198
650	4210.032	4217.510	4224.989	4232.470	4239.952
660	4284.878	4292.371	4299.866	4307.362	4314.859
670	4359.876	4367.384	4374.894	4382.405	4389.917
680	4435.023	4442.546	4450.070	4457.596	4465.123
690	4510.317	4517.855	4525.393	4532.934	4540.476
700	4585.756	4593.308	4600.861	4608.416	4615.972
710	4661.338	4668.904	4676.471	4684.040	4691.610
720	4737.061	4744.641	4752.222	4759.805	4767.389
730	4812.923	4820.516	4828.111	4835.708	4843.306
740	4888.921	4896.528	4904.137	4911.747	4919.359
750	4965.055	4972.676	4980.298	4987.921	4995.546
760	5041.322	5048.956	5056.591	5064.228	5071.866
770	5117.721	5125.368	5133.016	5140.666	5148.317
780	5194.249	5201.909	5209.570	5217.233	5224.897
790	5270.906	5278.579	5286.253	5293.928	5301.604
800	5347.689	5355.375	5363.061	5370.749	5378.438
810	5424.598	5432.295	5439.994	5447.694	5455.396
820	5501.630	5509.339	5517.051	5524.763	5532.477
830	5578.783	5586.505	5594.229	5601.953	5609.679
840	5656.058	5663.792	5671.527	5679.263	5687.001
850	5733.451	5741.197	5748.944	5756.692	5764.441
860	5810.962	5818.719	5826.478	5834.238	5841.999
870	5888.589	5896.358	5904.128	5911.900	5919.672
880	5966.331	5974.112	5981.893	5989.676	5997.460
890	6044.187	6051.979	6059.772	6067.566	6075.361
900	6122.155	6129.958	6137.762	6145.567	6153.374
910	6200.235	6208.049	6215.864	6223.680	6231.497
920	6278.424	6286.249	6294.075	6301.902	6309.730
930	6356.722	6364.557	6372.394	6380.232	6388.071
940	6435.127	6442.973	6450.821	6458.670	6466.519
950	6513.639	6521.496	6529.354	6537.213	6545.073
960	6592.256	6600.123	6607.992	6615.861	6623.732
970	6670.977	6678.855	6686.734	6694.614	6702.495
980	6749.802	6757.690	6765.579	6773.469	6781.360
990	6828.728	6836.626	6844.525	6852.426	6860.327

5	6	7	8	9	f
3143.402	3150.628	3157.855	3165.085	3172.316	500
3215.746	3222.991	3230.238	3237.487	3244.738	510
3288.284	3295.548	3302.815	3310.083	3317.353	520
3361.013	3368.296	3375.581	3382.868	3390.157	530
3433.928	3441.230	3448.533	3455.839	3463.146	540
3507.027	3514.347	3521.669	3528.992	3536.318	550
3580.307	3587.644	3594.984	3602.325	3609.668	560
3653.763	3661.118	3668.475	3675.834	3683.194	570
3727.393	3734.765	3742.140	3749.515	3756.893	580
3801.194	3808.583	3815.975	3823.367	3830.762	590
3875.163	3882.569	3889.977	3897.386	3904.797	600
3949.298	3956.720	3964.144	3971.570	3978.997	610
4023.595	4031.033	4038.473	4045.915	4053.359	620
4098.052	4105.506	4112.962	4120.420	4127.879	630
4172.666	4180.136	4187.608	4195.081	4202.556	640
4247.436	4254.921	4262.408	4269.897	4277.387	650
4322.358	4329.859	4337.361	4344.864	4352.370	660
4397.431	4404.947	4412.463	4419.982	4427.502	670
4472.652	4480.182	4487.714	4495.247	4502.781	680
4548.019	4555.563	4563.109	4570.657	4578.206	690
4623.529	4631.088	4638.649	4646.210	4653.774	700
4699.182	4706.755	4714.329	4721.905	4729.482	710
4774.974	4782.561	4790.150	4797.739	4805.330	720
4850.905	4858.505	4866.107	4873.711	4881.315	730
4926.971	4934.585	4942.201	4949.817	4957.435	740
5003.172	5010.799	5018.428	5026.058	5033.689	750
5079.505	5087.146	5094.787	5102.431	5110.075	760
5155.969	5163.622	5171.277	5178.933	5186.591	770
5232.562	5240.228	5247.896	5255.564	5263.235	780
5309.282	5316.961	5324.641	5332.323	5340.005	790
5386.128	5393.819	5401.512	5409.206	5416.901	800
5463.098	5470.802	5478.507	5486.213	5493.921	810
5540.191	5547.907	5555.624	5563.343	5571.063	820
5617.405	5625.134	5632.863	5640.593	5648.325	830
5694.739	5702.479	5710.220	5717.963	5725.706	840
5772.192	5779.943	5787.696	5795.450	5803.206	850
5849.761	5857.524	5865.289	5873.054	5880.821	860
5927.446	5935.221	5942.997	5950.774	5958.552	870
6005.245	6013.031	6020.818	6028.607	6036.396	880
6083.157	6090.955	6098.753	6106.553	6114.353	890
6161.181	6168.990	6176.799	6184.610	6192.422	900
6239.316	6247.135	6254.956	6262.777	6270.600	910
6317.559	6325.390	6333.221	6341.053	6348.887	920
6395.911	6403.752	6411.594	6419.437	6427.282	930
6474.370	6482.221	6490.074	6497.928	6505.783	940
6552.934	6560.797	6568.660	6576.524	6584.390	950
6631.604	6639.476	6647.350	6655.225	6663.100	960
6710.377	6718.259	6726.143	6734.028	6741.914	970
6789.252	6797.145	6805.039	6812.935	6820.831	980
6868.229	6876.132	6884.037	6891.942	6899.848	990

TABLE 2 $f \ln f$ as a function of f

f	0	1	2	3	4
1000	6907.755	6915.664	6923.573	6931.483	6939.394
1010	6986.883	6994.801	7002.720	7010.640	7018.561
1020	7066.109	7074.037	7081.966	7089.896	7097.827
1030	7145.434	7153.371	7161.310	7169.250	7177.191
1040	7224.855	7232.802	7240.751	7248.700	7256.651
1050	7304.373	7312.330	7320.288	7328.247	7336.207
1060	7383.986	7391.952	7399.920	7407.888	7415.857
1070	7463.693	7471.669	7479.646	7487.623	7495.602
1080	7543.494	7551.479	7559.465	7567.452	7575.440
1090	7623.387	7631.381	7639.377	7647.373	7655.370
1100	7703.372	7711.376	7719.380	7727.385	7735.392
1110	7783.448	7791.461	7799.474	7807.488	7815.504
1120	7863.614	7871.636	7879.658	7887.681	7895.706
1130	7943.869	7951.900	7959.931	7967.963	7975.996
1140	8024.213	8032.252	8040.293	8048.334	8056.375
1150	8104.645	8112.693	8120.742	8128.791	8136.842
1160	8185.163	8193.220	8201.277	8209.336	8217.395
1170	8265.768	8273.833	8281.899	8289.966	8298.034
1180	8346.458	8354.532	8362.606	8370.682	8378.758
1190	8427.233	8435.315	8443.398	8451.482	8459.567
1200	8508.092	8516.183	8524.274	8532.366	8540.459
1210	8589.035	8597.133	8605.233	8613.333	8621.435
1220	8670.059	8678.167	8686.274	8694.383	8702.492
1230	8751.166	8759.282	8767.398	8775.514	8783.632
1240	8832.355	8840.478	8848.602	8856.727	8864.853
1250	8913.624	8921.755	8929.887	8938.020	8946.154
1260	8994.972	9003.112	9011.252	9019.393	9027.534
1270	9076.401	9084.548	9092.696	9100.845	9108.994
1280	9157.908	9166.063	9174.218	9182.375	9190.532
1290	9239.493	9247.656	9255.819	9263.983	9272.149
1300	9321.155	9329.326	9337.497	9345.669	9353.842
1310	9402.895	9411.073	9419.252	9427.432	9435.612
1320	9484.711	9492.897	9501.083	9509.270	9517.458
1330	9566.603	9574.796	9582.990	9591.185	9599.380
1340	9648.569	9656.770	9664.972	9673.174	9681.377
1350	9730.611	9738.819	9747.028	9755.238	9763.448
1360	9812.726	9820.942	9829.158	9837.375	9845.593
1370	9894.915	9903.138	9911.362	9919.586	9927.812
1380	9977.178	9985.408	9993.639	10001.870	10010.103
1390	10059.512	10067.749	10075.988	10084.226	10092.466
1400	10141.919	10150.163	10158.408	10166.654	10174.901
1410	10224.396	10232.648	10240.901	10249.154	10257.407
1420	10306.945	10315.204	10323.463	10331.724	10339.985
1430	10389.565	10397.830	10406.097	10414.364	10422.632
1440	10472.254	10480.526	10488.800	10497.074	10505.349
1450	10555.012	10563.292	10571.572	10579.853	10588.135
1460	10637.840	10646.126	10654.414	10662.702	10670.990
1470	10720.736	10729.029	10737.323	10745.618	10753.913
1480	10803.700	10812.000	10820.301	10828.603	10836.905
1490	10886.732	10895.039	10903.346	10911.654	10919.963

5	6	7	8	9	f
6947.307	6955.220	6963.134	6971.049	6978.965	1000
7026.484	7034.407	7042.331	7050.256	7058.182	1010
7105.759	7113.692	7121.626	7129.561	7137.497	1020
7185.132	7193.075	7201.018	7208.963	7216.909	1030
7264.602	7272.554	7280.507	7288.462	7296.417	1040
7344.167	7352.129	7360.092	7368.055	7376.020	1050
7423.828	7431.799	7439.771	7447.744	7455.718	1060
7503.582	7511.562	7519.544	7527.526	7535.509	1070
7583.429	7591.419	7599.409	7607.401	7615.393	1080
7663.368	7671.367	7679.367	7687.368	7695.369	1090
7743.399	7751.407	7759.416	7767.426	7775.436	1100
7823.520	7831.537	7839.555	7847.574	7855.593	1110
7903.731	7911.757	7919.783	7927.811	7935.840	1120
7984.030	7992.065	8000.101	8008.137	8016.175	1130
8064.418	8072.462	8080.506	8088.552	8096.598	1140
8144.893	8152.946	8160.999	8169.053	8177.108	1150
8225.455	8233.516	8241.578	8249.640	8257.704	1160
8306.103	8314.172	8322.242	8330.313	8338.385	1170
8386.835	8394.913	8402.992	8411.071	8419.152	1180
8467.652	8475.739	8483.826	8491.914	8500.003	1190
8548.553	8556.648	8564.743	8572.839	8580.937	1200
8629.537	8637.640	8645.743	8653.848	8661.953	1210
8710.603	8718.714	8726.826	8734.939	8743.052	1220
8791.750	8799.870	8807.990	8816.111	8824.232	1230
8872.979	8881.106	8889.234	8897.363	8905.493	1240
8954.288	8962.423	8970.559	8978.696	8986.834	1250
9035.677	9043.820	9051.964	9060.109	9068.254	1260
9117.144	9125.295	9133.447	9141.600	9149.753	1270
9198.690	9206.849	9215.009	9223.170	9231.331	1280
9280.314	9288.481	9296.649	9304.817	9312.986	1290
9362.016	9370.190	9378.365	9386.541	9394.718	1300
9443.793	9451.975	9460.158	9468.342	9476.526	1310
9525.647	9533.837	9542.027	9550.218	9558.410	1320
9607.577	9615.774	9623.971	9632.170	9640.369	1330
9689.581	9697.785	9705.991	9714.197	9722.403	1340
9771.659	9779.871	9788.084	9796.297	9804.511	1350
9853.812	9862.031	9870.251	9878.472	9886.693	1360
9936.037	9944.264	9952.491	9960.719	9968.948	1370
10018.336	10026.570	10034.804	10043.039	10051.275	1380
10100.706	10108.947	10117.189	10125.431	10133.675	1390
10183.149	10191.397	10199.646	10207.895	10216.145	1400
10265.662	10273.917	10282.173	10290.430	10298.687	1410
10348.246	10356.508	10364.771	10373.035	10381.299	1420
10430.900	10439.170	10447.440	10455.710	10463.982	1430
10513.624	10521.901	10530.177	10538.455	10546.733	1440
10596.418	10604.701	10612.984	10621.269	10629.554	1450
10679.279	10687.569	10695.860	10704.151	10712.443	1460
10762.210	10770.506	10778.804	10787.102	10795.401	1470
10845.208	10853.511	10861.815	10870.120	10878.426	1480
10928.273	10936.583	10944.894	10953.205	10961.518	1490

TABLE 2 $f \ln f$ as a function of f

f	0	1	2	3	4
1500	10969.831	10978.144	10986.458	10994.773	11003.089
1510	11052.996	11061.316	11069.637	11077.959	11086.281
1520	11136.228	11144.555	11152.882	11161.210	11169.539
1530	11219.525	11227.859	11236.193	11244.527	11252.863
1540	11302.888	11311.228	11319.568	11327.910	11336.251
1550	11386.316	11394.662	11403.009	11411.357	11419.705
1560	11469.808	11478.161	11486.514	11494.868	11503.223
1570	11553.365	11561.724	11570.083	11578.444	11586.805
1580	11636.985	11645.350	11653.716	11662.083	11670.450
1590	11720.668	11729.040	11737.412	11745.785	11754.159
1600	11804.414	11812.792	11821.171	11829.550	11837.930
1610	11888.223	11896.607	11904.992	11913.378	11921.764
1620	11972.094	11980.484	11988.876	11997.267	12005.660
1630	12056.027	12064.423	12072.820	12081.218	12089.617
1640	12140.020	12148.423	12156.827	12165.231	12173.635
1650	12224.075	12232.484	12240.894	12249.304	12257.714
1660	12308.191	12316.606	12325.021	12333.437	12341.854
1670	12392.367	12400.788	12409.209	12417.631	12426.054
1680	12476.602	12485.029	12493.457	12501.885	12510.313
1690	12560.898	12569.330	12577.764	12586.198	12594.632
1700	12645.252	12653.691	12662.130	12670.570	12679.010
1710	12729.665	12738.110	12746.555	12755.001	12763.447
1720	12814.137	12822.587	12831.038	12839.490	12847.942
1730	12898.667	12907.123	12915.580	12924.037	12932.495
1740	12983.254	12991.716	13000.179	13008.642	13017.105
1750	13067.899	13076.367	13084.835	13093.304	13101.773
1760	13152.602	13161.075	13169.549	13178.023	13186.498
1770	13237.361	13245.840	13254.319	13262.799	13271.280
1780	13322.176	13330.661	13339.146	13347.632	13356.118
1790	13407.048	13415.538	13424.029	13432.520	13441.012
1800	13491.975	13500.471	13508.968	13517.465	13525.962
1810	13576.959	13585.460	13593.962	13602.464	13610.967
1820	13661.997	13670.504	13679.011	13687.519	13696.028
1830	13747.090	13755.603	13764.116	13772.629	13781.143
1840	13832.238	13840.756	13849.274	13857.793	13866.313
1850	13917.441	13925.964	13934.488	13943.012	13951.537
1860	14002.697	14011.226	14019.755	14028.284	14036.815
1870	14088.007	14096.541	14105.076	14113.611	14122.146
1880	14173.371	14181.910	14190.450	14198.990	14207.531
1890	14258.788	14267.332	14275.877	14284.423	14292.969
1900	14344.257	14352.807	14361.358	14369.909	14378.460
1910	14429.780	14438.335	14446.891	14455.447	14464.003
1920	14515.354	14523.915	14532.476	14541.037	14549.599
1930	14600.981	14609.547	14618.113	14626.679	14635.247
1940	14686.660	14695.231	14703.802	14712.374	14720.946
1950	14772.390	14780.966	14789.542	14798.119	14806.697
1960	14858.172	14866.752	14875.334	14883.916	14892.498
1970	14944.004	14952.590	14961.177	14969.764	14978.351
1980	15029.887	15038.478	15047.070	15055.662	15064.255
1990	15115.821	15124.417	15133.014	15141.611	15150.209

f	5	6	7	8	9
1500	11011.405	11019.722	11028.039	11036.358	11044.677
1510	11094.604	11102.927	11111.251	11119.576	11127.902
1520	11177.868	11186.198	11194.529	11202.860	11211.193
1530	11261.198	11269.535	11277.872	11286.210	11294.549
1540	11344.594	11352.937	11361.281	11369.625	11377.970
1550	11428.054	11436.403	11444.754	11453.105	11461.456
1560	11511.578	11519.934	11528.291	11536.648	11545.006
1570	11595.167	11603.529	11611.892	11620.256	11628.620
1580	11678.818	11687.187	11695.556	11703.926	11712.297
1590	11762.533	11770.908	11779.284	11787.660	11796.037
1600	11846.311	11854.692	11863.074	11871.456	11879.839
1610	11930.151	11938.538	11946.926	11955.315	11963.704
1620	12014.053	12022.446	12030.840	12039.235	12047.631
1630	12098.016	12106.416	12114.816	12123.217	12131.618
1640	12182.040	12190.446	12198.853	12207.260	12215.667
1650	12266.126	12274.538	12282.950	12291.363	12299.777
1660	12350.271	12358.689	12367.108	12375.527	12383.946
1670	12434.477	12442.901	12451.325	12459.751	12468.176
1680	12518.743	12527.172	12535.603	12544.034	12552.465
1690	12603.067	12611.503	12619.939	12628.376	12636.814
1700	12687.451	12695.893	12704.335	12712.778	12721.221
1710	12771.894	12780.341	12788.789	12797.238	12805.687
1720	12856.395	12864.848	12873.302	12881.756	12890.211
1730	12940.953	12949.412	12957.872	12966.332	12974.793
1740	13025.570	13034.034	13042.500	13050.966	13059.432
1750	13110.243	13118.714	13127.185	13135.657	13144.129
1760	13194.974	13203.450	13211.927	13220.404	13228.882
1770	13279.761	13288.243	13296.726	13305.209	13313.692
1780	13364.605	13373.092	13381.581	13390.069	13398.558
1790	13449.505	13457.998	13466.491	13474.986	13483.480
1800	13534.460	13542.959	13551.458	13559.958	13568.458
1810	13619.471	13627.975	13636.480	13644.985	13653.491
1820	13704.537	13713.046	13721.557	13730.067	13738.579
1830	13789.658	13798.173	13806.688	13815.204	13823.721
1840	13874.833	13883.353	13891.874	13900.396	13908.918
1850	13960.062	13968.588	13977.115	13985.641	13994.169
1860	14045.345	14053.877	14062.409	14070.941	14079.474
1870	14130.682	14139.219	14147.756	14156.294	14164.832
1880	14216.073	14224.615	14233.157	14241.700	14250.244
1890	14301.516	14310.063	14318.611	14327.159	14335.708
1900	14387.012	14395.565	14404.118	14412.671	14421.225
1910	14472.561	14481.118	14489.677	14498.235	14506.795
1920	14558.161	14566.724	14575.288	14583.852	14592.416
1930	14643.814	14652.382	14660.951	14669.520	14678.090
1940	14729.519	14738.092	14746.666	14755.240	14763.815
1950	14815.274	14823.853	14832.432	14841.011	14849.591
1960	14901.081	14909.665	14918.249	14926.833	14935.418
1970	14986.939	14995.528	15004.117	15012.707	15021.297
1980	15072.848	15081.441	15090.036	15098.630	15107.225
1990	15158.807	15167.405	15176.004	15184.604	15193.204

TABLE 2 $f \ln f$ as a function of f

f	0	1	2	3	4
2000	15201.805	15210.406	15219.008	15227.610	15236.213
2010	15287.839	15296.445	15305.052	15313.659	15322.266
2020	15373.923	15382.534	15391.145	15399.757	15408.370
2030	15460.056	15468.672	15477.288	15485.905	15494.523
2040	15546.238	15554.859	15563.481	15572.103	15580.725
2050	15632.470	15641.096	15649.722	15658.349	15666.976
2060	15718.750	15727.381	15736.012	15744.644	15753.276
2070	15805.079	15813.715	15822.351	15830.987	15839.624
2080	15891.456	15900.097	15908.737	15917.379	15926.021
2090	15977.881	15986.527	15995.172	16003.818	16012.465
2100	16064.355	16073.004	16081.655	16090.306	16098.957
2110	16150.875	16159.530	16168.185	16176.841	16185.497
2120	16237.443	16246.103	16254.763	16263.423	16272.084
2130	16324.059	16332.723	16341.387	16350.052	16358.718
2140	16410.721	16419.390	16428.059	16436.729	16445.399
2150	16497.430	16506.103	16514.777	16523.451	16532.126
2160	16584.185	16592.863	16601.542	16610.221	16618.900
2170	16670.987	16679.670	16688.353	16697.036	16705.721
2180	16757.835	16766.522	16775.210	16783.898	16792.587
2190	16844.728	16853.420	16862.113	16870.805	16879.499
2200	16931.668	16940.364	16949.061	16957.758	16966.456
2210	17018.653	17027.354	17036.055	17044.757	17053.459
2220	17105.683	17114.388	17123.094	17131.801	17140.507
2230	17192.758	17201.468	17210.178	17218.889	17227.600
2240	17279.878	17288.592	17297.307	17306.022	17314.738
2250	17367.042	17375.761	17384.481	17393.200	17401.921
2260	17454.251	17462.975	17471.699	17480.423	17489.147
2270	17541.505	17550.232	17558.961	17567.689	17576.418
2280	17628.802	17637.534	17646.267	17655.000	17663.733
2290	17716.143	17724.880	17733.617	17742.354	17751.092
2300	17803.528	17812.269	17821.010	17829.752	17838.494
2310	17890.956	17899.702	17908.447	17917.193	17925.940
2320	17978.428	17987.178	17995.928	18004.678	18013.429
2330	18065.943	18074.697	18083.451	18092.206	18100.961
2340	18153.501	18162.259	18171.017	18179.776	18188.536
2350	18241.101	18249.863	18258.626	18267.389	18276.153
2360	18328.744	18337.511	18346.278	18355.045	18363.813
2370	18416.429	18425.200	18433.971	18442.743	18451.515
2380	18504.157	18512.932	18521.707	18530.483	18539.260
2390	18591.926	18600.706	18609.485	18618.265	18627.046
2400	18679.738	18688.521	18697.305	18706.089	18714.874
2410	18767.591	18776.378	18785.166	18793.955	18802.744
2420	18855.485	18864.277	18873.069	18881.862	18890.655
2430	18943.421	18952.217	18961.013	18969.810	18978.607
2440	19031.398	19040.198	19048.998	19057.799	19066.600
2450	19119.416	19128.220	19137.025	19145.829	19154.635
2460	19207.475	19216.283	19225.092	19233.900	19242.710
2470	19295.574	19304.387	19313.199	19322.012	19330.826
2480	19383.714	19392.531	19401.347	19410.164	19418.982
2490	19471.895	19480.715	19489.535	19498.357	19507.178

5	6	7	8	9	f
15244.816	15253.419	15262.023	15270.628	15279.233	2000
15330.875	15339.483	15348.092	15356.702	15365.312	2010
15416.983	15425.597	15434.211	15442.825	15451.440	2020
15503.141	15511.759	15520.378	15528.998	15537.618	2030
15589.348	15597.971	15606.595	15615.220	15623.845	2040
15675.604	15684.232	15692.861	15701.490	15710.120	2050
15761.909	15770.542	15779.175	15787.809	15796.444	2060
15848.262	15856.900	15865.538	15874.177	15882.816	2070
15934.663	15943.306	15951.949	15960.593	15969.237	2080
16021.112	16029.760	16038.408	16047.056	16055.705	2090
16107.609	16116.261	16124.914	16133.567	16142.221	2100
16194.153	16202.810	16211.468	16220.126	16228.784	2110
16280.745	16289.407	16298.069	16306.732	16315.395	2120
16367.384	16376.050	16384.717	16393.385	16402.052	2130
16454.069	16462.741	16471.412	16480.084	16488.757	2140
16540.802	16549.477	16558.154	16566.830	16575.508	2150
16627.580	16636.261	16644.942	16653.623	16662.305	2160
16714.405	16723.090	16731.776	16740.461	16749.148	2170
16801.276	16809.965	16818.656	16827.346	16836.037	2180
16888.192	16896.887	16905.581	16914.276	16922.972	2190
16975.155	16983.853	16992.552	17001.252	17009.952	2200
17062.162	17070.865	17079.569	17088.273	17096.978	2210
17149.215	17157.922	17166.631	17175.339	17184.048	2220
17236.312	17245.024	17253.737	17262.450	17271.164	2230
17323.454	17332.171	17340.888	17349.606	17358.324	2240
17410.641	17419.362	17428.084	17436.806	17445.529	2250
17497.873	17506.598	17515.324	17524.051	17532.777	2260
17585.148	17593.878	17602.608	17611.339	17620.070	2270
17672.467	17681.202	17689.936	17698.672	17707.407	2280
17759.830	17768.569	17777.308	17786.048	17794.788	2290
17847.237	17855.980	17864.723	17873.467	17882.212	2300
17934.687	17943.434	17952.182	17960.930	17969.679	2310
18022.180	18030.932	18039.684	18048.436	18057.189	2320
18109.716	18118.472	18127.229	18135.986	18144.743	2330
18197.295	18206.056	18214.816	18223.577	18232.339	2340
18284.917	18293.682	18302.447	18311.212	18319.978	2350
18372.581	18381.350	18390.119	18398.889	18407.659	2360
18460.288	18469.061	18477.834	18486.608	18495.382	2370
18548.036	18556.813	18565.591	18574.369	18583.147	2380
18635.827	18644.608	18653.390	18662.172	18670.955	2390
18723.659	18732.444	18741.230	18750.017	18758.804	2400
18811.533	18820.322	18829.113	18837.903	18846.694	2410
18899.448	18908.242	18917.036	18925.831	18934.626	2420
18987.404	18996.202	19005.001	19013.799	19022.599	2430
19075.402	19084.204	19093.006	19101.809	19110.612	2440
19163.440	19172.246	19181.053	19189.860	19198.667	2450
19251.520	19260.330	19269.140	19277.951	19286.763	2460
19339.639	19348.453	19357.268	19366.083	19374.899	2470
19427.799	19436.618	19445.436	19454.255	19463.075	2480
19515.000	19524.822	19533.645	19542.468	19551.291	2490

TABLE 2 $f \ln f$ as a function of f

f	0	1	2	3	4
2500	19560.115	19568.939	19577.764	19586.589	19595.414
2510	19648.375	19657.204	19666.032	19674.861	19683.691
2520	19736.676	19745.508	19754.341	19763.174	19772.007
2530	19825.016	19833.852	19842.688	19851.525	19860.363
2540	19913.395	19922.235	19931.076	19939.917	19948.758
2550	20001.814	20010.658	20019.503	20028.347	20037.193
2560	20090.272	20099.120	20107.968	20116.817	20125.666
2570	20178.769	20187.621	20196.473	20205.326	20214.179
2580	20267.305	20276.161	20285.017	20293.874	20302.731
2590	20355.880	20364.740	20373.600	20382.460	20391.321
2600	20444.493	20453.357	20462.221	20471.085	20479.950
2610	20533.145	20542.013	20550.880	20559.748	20568.617
2620	20621.836	20630.707	20639.578	20648.450	20657.322
2630	20710.564	20719.439	20728.314	20737.190	20746.066
2640	20799.330	20808.209	20817.088	20825.968	20834.847
2650	20888.135	20897.017	20905.900	20914.783	20923.667
2660	20976.977	20985.863	20994.749	21003.636	21012.524
2670	21065.856	21074.746	21083.637	21092.527	21101.418
2680	21154.773	21163.667	21172.561	21181.456	21190.350
2690	21243.728	21252.625	21261.523	21270.421	21279.320
2700	21332.719	21341.620	21350.522	21359.424	21368.326
2710	21421.748	21430.652	21439.558	21448.463	21457.369
2720	21510.813	21519.722	21528.631	21537.540	21546.450
2730	21599.915	21608.828	21617.740	21626.653	21635.566
2740	21689.054	21697.970	21706.886	21715.803	21724.720
2750	21778.230	21787.149	21796.069	21804.989	21813.910
2760	21867.441	21876.364	21885.288	21894.212	21903.136
2770	21956.689	21965.616	21974.543	21983.471	21992.398
2780	22045.973	22054.904	22063.834	22072.765	22081.697
2790	22135.293	22144.227	22153.162	22162.096	22171.031
2800	22224.649	22233.587	22242.525	22251.463	22260.402
2810	22314.041	22322.982	22331.923	22340.865	22349.807
2820	22403.468	22412.413	22421.358	22430.303	22439.249
2830	22492.931	22501.879	22510.827	22519.776	22528.725
2840	22582.429	22591.380	22600.332	22609.285	22618.238
2850	22671.962	22680.917	22689.873	22698.828	22707.785
2860	22761.530	22770.489	22779.448	22788.407	22797.367
2870	22851.133	22860.095	22869.058	22878.021	22886.984
2880	22940.771	22949.737	22958.703	22967.669	22976.636
2890	23030.444	23039.413	23048.383	23057.353	23066.323
2900	23120.151	23129.124	23138.097	23147.070	23156.044
2910	23209.893	23218.869	23227.846	23236.823	23245.800
2920	23299.670	23308.649	23317.629	23326.609	23335.590
2930	23389.480	23398.463	23407.446	23416.430	23425.414
2940	23479.325	23488.311	23497.298	23506.285	23515.272
2950	23569.203	23578.193	23587.183	23596.174	23605.164
2960	23659.116	23668.109	23677.102	23686.096	23695.090
2970	23749.062	23758.059	23767.055	23776.053	23785.050
2980	23839.042	23848.042	23857.042	23866.043	23875.044
2990	23929.056	23938.059	23947.062	23956.066	23965.071

5	6	7	8	9	f
19604.240	19613.066	19621.893	19630.720	19639.548	2500
19692.521	19701.351	19710.181	19719.013	19727.844	2510
19780.841	19789.675	19798.510	19807.345	19816.180	2520
19869.201	19878.039	19886.877	19895.716	19904.555	2530
19957.600	19966.442	19975.284	19984.127	19992.970	2540
20046.038	20054.884	20063.731	20072.577	20081.425	2550
20134.516	20143.366	20152.216	20161.067	20169.918	2560
20223.032	20231.886	20240.740	20249.595	20258.450	2570
20311.588	20320.446	20329.304	20338.162	20347.021	2580
20400.182	20409.043	20417.905	20426.768	20435.630	2590
20488.815	20497.680	20506.546	20515.412	20524.278	2600
20577.486	20586.355	20595.224	20604.094	20612.965	2610
20666.195	20675.068	20683.941	20692.815	20701.689	2620
20754.942	20763.819	20772.696	20781.574	20790.452	2630
20843.728	20852.608	20861.489	20870.371	20879.252	2640
20932.551	20941.435	20950.320	20959.205	20968.091	2650
21021.412	21030.300	21039.188	21048.077	21056.966	2660
21110.310	21119.202	21128.094	21136.987	21145.880	2670
21199.246	21208.141	21217.037	21225.934	21234.830	2680
21288.219	21297.118	21306.018	21314.918	21323.818	2690
21377.229	21386.132	21395.035	21403.939	21412.843	2700
21466.276	21475.182	21484.090	21492.997	21501.905	2710
21555.360	21564.270	21573.181	21582.092	21591.003	2720
21644.480	21653.394	21662.309	21671.223	21680.139	2730
21733.637	21742.555	21751.473	21760.392	21769.310	2740
21822.831	21831.752	21840.674	21849.596	21858.518	2750
21912.061	21920.986	21929.911	21938.837	21947.763	2760
22001.327	22010.255	22019.184	22028.114	22037.043	2770
22090.629	22099.561	22108.494	22117.426	22126.360	2780
22179.967	22188.903	22197.839	22206.775	22215.712	2790
22269.340	22278.280	22287.220	22296.160	22305.100	2800
22358.750	22367.693	22376.636	22385.580	22394.524	2810
22448.195	22457.141	22466.088	22475.035	22483.983	2820
22537.675	22546.625	22555.575	22564.526	22573.477	2830
22627.191	22636.144	22645.098	22654.052	22663.007	2840
22716.741	22725.698	22734.656	22743.613	22752.572	2850
22806.327	22815.288	22824.249	22833.210	22842.171	2860
22895.948	22904.912	22913.876	22922.841	22931.806	2870
22985.603	22994.571	23003.539	23012.507	23021.475	2880
23075.293	23084.264	23093.236	23102.207	23111.179	2890
23165.018	23173.992	23182.967	23191.942	23200.918	2900
23254.777	23263.755	23272.733	23281.712	23290.690	2910
23344.571	23353.552	23362.533	23371.515	23380.497	2920
23434.398	23443.383	23452.368	23461.353	23470.339	2930
23524.260	23533.248	23542.236	23551.225	23560.214	2940
23614.155	23623.147	23632.139	23641.131	23650.123	2950
23704.085	23713.080	23722.075	23731.070	23740.066	2960
23794.048	23803.046	23812.045	23821.043	23830.043	2970
23884.045	23893.046	23902.048	23911.050	23920.053	2980
23974.075	23983.080	23992.085	24001.091	24010.097	2990

TABLE 2 $f \ln f$ as a function of f

f	0	1	2	3	4
3000	24019.103	24028.109	24037.116	24046.123	24055.131
3010	24109.183	24118.193	24127.203	24136.214	24145.224
3020	24199.297	24208.310	24217.323	24226.337	24235.351
3030	24289.443	24298.460	24307.477	24316.494	24325.511
3040	24379.623	24388.643	24397.663	24406.683	24415.704
3050	24469.835	24478.859	24487.882	24496.906	24505.930
3060	24560.081	24569.107	24578.134	24587.161	24596.188
3070	24650.359	24659.388	24668.418	24677.449	24686.479
3080	24740.669	24749.702	24758.735	24767.769	24776.803
3090	24831.012	24840.049	24849.085	24858.122	24867.159
3100	24921.388	24930.427	24939.467	24948.507	24957.547
3110	25011.796	25020.838	25029.881	25038.924	25047.968
3120	25102.235	25111.281	25120.327	25129.374	25138.420
3130	25192.707	25201.756	25210.806	25219.855	25228.905
3140	25283.211	25292.263	25301.316	25310.369	25319.422
3150	25373.747	25382.802	25391.858	25400.914	25409.970
3160	25464.314	25473.373	25482.432	25491.491	25500.550
3170	25554.913	25563.975	25573.037	25582.099	25591.162
3180	25645.544	25654.609	25663.674	25672.739	25681.805
3190	25736.206	25745.274	25754.342	25763.411	25772.480
3200	25826.899	25835.971	25845.042	25854.114	25863.186
3210	25917.624	25926.698	25935.773	25944.848	25953.923
3220	26008.380	26017.457	26026.535	26035.613	26044.691
3230	26099.167	26108.247	26117.328	26126.409	26135.490
3240	26189.985	26199.068	26208.152	26217.236	26226.320
3250	26280.833	26289.920	26299.007	26308.094	26317.181
3260	26371.713	26380.803	26389.892	26398.983	26408.073
3270	26462.623	26471.716	26480.809	26489.902	26498.996
3280	26553.564	26562.659	26571.756	26580.852	26589.949
3290	26644.535	26653.634	26662.733	26671.832	26680.932
3300	26735.537	26744.638	26753.741	26762.843	26771.946
3310	26826.568	26835.673	26844.778	26853.884	26862.990
3320	26917.631	26926.738	26935.847	26944.955	26954.064
3330	27008.723	27017.834	27026.945	27036.056	27045.168
3340	27099.845	27108.959	27118.073	27127.188	27136.302
3350	27190.997	27200.114	27209.231	27218.349	27227.467
3360	27282.179	27291.299	27300.419	27309.540	27318.661
3370	27373.391	27382.514	27391.637	27400.761	27409.884
3380	27464.633	27473.759	27482.885	27492.011	27501.138
3390	27555.904	27565.033	27574.162	27583.291	27592.421
3400	27647.204	27656.336	27665.468	27674.600	27683.733
3410	27738.534	27747.669	27756.804	27765.939	27775.075
3420	27829.894	27839.031	27848.169	27857.307	27866.446
3430	27921.282	27930.423	27939.564	27948.705	27957.846
3440	28012.700	28021.843	28030.987	28040.131	28049.275
3450	28104.147	28113.293	28122.440	28131.587	28140.734
3460	28195.623	28204.772	28213.921	28223.071	28232.221
3470	28287.127	28296.279	28305.432	28314.584	28323.737
3480	28378.661	28387.816	28396.971	28406.126	28415.282
3490	28470.223	28479.381	28488.539	28497.697	28506.856

5	6	7	8	9	f
24064.139	24073.147	24082.155	24091.164	24100.173	3000
24154.236	24163.247	24172.259	24181.271	24190.284	3010
24244.366	24253.381	24262.396	24271.411	24280.427	3020
24334.529	24343.547	24352.566	24361.584	24370.603	3030
24424.725	24433.746	24442.768	24451.790	24460.813	3040
24514.954	24523.979	24533.004	24542.029	24551.055	3050
24605.216	24614.244	24623.272	24632.301	24641.330	3060
24695.510	24704.541	24713.573	24722.605	24731.637	3070
24785.837	24794.871	24803.906	24812.941	24821.977	3080
24876.196	24885.234	24894.272	24903.310	24912.349	3090
24966.588	24975.629	24984.670	24993.711	25002.753	3100
25057.012	25066.056	25075.100	25084.145	25093.190	3110
25147.467	25156.515	25165.562	25174.610	25183.659	3120
25237.955	25247.006	25256.057	25265.108	25274.159	3130
25328.475	25337.529	25346.583	25355.637	25364.692	3140
25419.027	25428.084	25437.141	25446.198	25455.256	3150
25509.610	25518.670	25527.730	25536.791	25545.852	3160
25600.225	25609.288	25618.351	25627.415	25636.480	3170
25690.871	25699.937	25709.004	25718.071	25727.138	3180
25781.549	25790.618	25799.688	25808.758	25817.829	3190
25872.258	25881.331	25890.403	25899.477	25908.550	3200
25962.998	25972.074	25981.150	25990.226	25999.303	3210
26053.770	26062.848	26071.928	26081.007	26090.087	3220
26144.572	26153.654	26162.736	26171.819	26180.902	3230
26235.405	26244.490	26253.576	26262.661	26271.747	3240
26326.269	26335.357	26344.446	26353.535	26362.624	3250
26417.164	26426.255	26435.347	26444.439	26453.531	3260
26508.090	26517.184	26526.278	26535.373	26544.468	3270
26599.046	26608.143	26617.240	26626.338	26635.436	3280
26690.032	26699.132	26708.233	26717.334	26726.435	3290
26781.049	26790.152	26799.256	26808.360	26817.464	3300
26872.096	26881.202	26890.309	26899.416	26908.523	3310
26963.173	26972.282	26981.392	26990.502	26999.612	3320
27054.280	27063.393	27072.505	27081.618	27090.732	3330
27145.417	27154.533	27163.649	27172.765	27181.881	3340
27236.585	27245.703	27254.822	27263.941	27273.060	3350
27327.782	27336.903	27346.025	27355.146	27364.269	3360
27419.008	27428.133	27437.257	27446.382	27455.507	3370
27510.265	27519.392	27528.519	27537.647	27546.775	3380
27601.550	27610.681	27619.811	27628.942	27638.073	3390
27692.866	27701.999	27711.132	27720.266	27729.400	3400
27784.210	27793.346	27802.483	27811.620	27820.756	3410
27875.584	27884.723	27893.863	27903.002	27912.142	3420
27966.988	27976.129	27985.272	27994.414	28003.557	3430
28058.420	28067.565	28076.710	28085.855	28095.001	3440
28149.881	28159.029	28168.177	28177.325	28186.474	3450
28241.371	28250.522	28259.673	28268.824	28277.975	3460
28332.890	28342.044	28351.198	28360.352	28369.506	3470
28424.438	28433.595	28442.751	28451.908	28461.065	3480
28516.015	28525.174	28534.334	28543.493	28552.653	3490

TABLE 2 $f \ln f$ as a function of f

f	0	1	2	3	4
3500	28561.814	28570.975	28580.135	28589.297	28598.458
3510	28653.433	28662.597	28671.761	28680.925	28690.089
3520	28745.081	28754.248	28763.414	28772.581	28781.748
3530	28836.758	28845.927	28855.096	28864.266	28873.436
3540	28928.462	28937.634	28946.807	28955.979	28965.152
3550	29020.195	29029.370	29038.545	29047.721	29056.896
3560	29111.956	29121.134	29130.312	29139.490	29148.669
3570	29203.746	29212.926	29222.107	29231.288	29240.469
3580	29295.563	29304.746	29313.930	29323.113	29332.297
3590	29387.408	29396.594	29405.780	29414.967	29424.154
3600	29479.281	29488.470	29497.659	29506.848	29516.038
3610	29571.182	29580.373	29589.565	29598.757	29607.950
3620	29663.110	29672.304	29681.499	29690.694	29699.889
3630	29755.066	29764.263	29773.461	29782.658	29791.856
3640	29847.050	29856.250	29865.450	29874.650	29883.851
3650	29939.061	29948.264	29957.466	29966.670	29975.873
3660	30031.099	30040.305	30049.510	30058.716	30067.922
3670	30123.165	30132.373	30141.582	30150.790	30159.999
3680	30215.258	30224.469	30233.680	30242.892	30252.103
3690	30307.379	30316.592	30325.806	30335.020	30344.234
3700	30399.526	30408.742	30417.959	30427.175	30436.392
3710	30491.700	30500.919	30510.138	30519.358	30528.578
3720	30583.902	30593.123	30602.345	30611.567	30620.790
3730	30676.130	30685.354	30694.579	30703.804	30713.029
3740	30768.385	30777.612	30786.839	30796.067	30805.294
3750	30860.667	30869.896	30879.126	30888.356	30897.587
3760	30952.975	30962.207	30971.440	30980.673	30989.906
3770	31045.310	31054.545	31063.780	31073.016	31082.252
3780	31137.672	31146.909	31156.147	31165.385	31174.624
3790	31230.060	31239.300	31248.540	31257.781	31267.022
3800	31322.474	31331.717	31340.960	31350.204	31359.447
3810	31414.915	31424.160	31433.406	31442.652	31451.898
3820	31507.382	31516.630	31525.878	31535.127	31544.376
3830	31599.875	31609.126	31618.377	31627.628	31636.879
3840	31692.394	31701.648	31710.901	31720.155	31729.409
3850	31784.939	31794.195	31803.452	31812.708	31821.965
3860	31877.511	31886.769	31896.028	31905.287	31914.546
3870	31970.108	31979.369	31988.630	31997.892	32007.154
3880	32062.731	32071.995	32081.259	32090.523	32099.787
3890	32155.380	32164.646	32173.913	32183.179	32192.446
3900	32248.054	32257.323	32266.592	32275.861	32285.131
3910	32340.754	32350.026	32359.297	32368.569	32377.841
3920	32433.480	32442.754	32452.028	32461.303	32470.577
3930	32526.231	32535.508	32544.784	32554.062	32563.339
3940	32619.008	32628.287	32637.566	32646.846	32656.126
3950	32711.810	32721.091	32730.373	32739.655	32748.938
3960	32804.637	32813.921	32823.206	32832.490	32841.775
3970	32897.490	32906.777	32916.063	32925.351	32934.638
3980	32990.368	32999.657	33008.946	33018.236	33027.526
3990	33083.271	33092.562	33101.854	33111.146	33120.439

5	6	7	8	9	f
28607.620	28616.782	28625.944	28635.107	28644.270	3500
28699.254	28708.419	28717.584	28726.749	28735.915	3510
28790.916	28800.084	28809.252	28818.420	28827.589	3520
28882.606	28891.777	28900.948	28910.119	28919.291	3530
28974.325	28983.499	28992.672	29001.846	29011.021	3540
29066.072	29075.249	29084.425	29093.602	29102.779	3550
29157.847	29167.026	29176.206	29185.385	29194.565	3560
29249.651	29258.832	29268.015	29277.197	29286.380	3570
29341.482	29350.666	29359.851	29369.037	29378.222	3580
29433.341	29442.528	29451.716	29460.904	29470.092	3590
29525.228	29534.418	29543.608	29552.799	29561.990	3600
29617.142	29626.335	29635.529	29644.722	29653.916	3610
29709.085	29718.280	29727.476	29736.673	29745.869	3620
29801.055	29810.253	29819.452	29828.651	29837.850	3630
29893.052	29902.253	29911.455	29920.657	29929.859	3640
29985.077	29994.281	30003.485	30012.690	30021.894	3650
30077.129	30086.336	30095.543	30104.750	30113.957	3660
30169.208	30178.418	30187.628	30196.838	30206.048	3670
30261.315	30270.527	30279.740	30288.952	30298.165	3680
30353.449	30362.664	30371.879	30381.094	30390.310	3690
30445.610	30454.827	30464.045	30473.263	30482.482	3700
30537.798	30547.018	30556.238	30565.459	30574.680	3710
30630.012	30639.235	30648.459	30657.682	30666.906	3720
30722.254	30731.480	30740.706	30749.932	30759.158	3730
30814.522	30823.751	30832.979	30842.208	30851.437	3740
30906.818	30916.049	30925.280	30934.511	30943.743	3750
30999.139	31008.373	31017.607	31026.841	31036.075	3760
31091.488	31100.724	31109.960	31119.197	31128.434	3770
31183.862	31193.101	31202.341	31211.580	31220.820	3780
31276.264	31285.505	31294.747	31303.989	31313.231	3790
31368.691	31377.935	31387.180	31396.425	31405.670	3800
31461.145	31470.392	31479.639	31488.886	31498.134	3810
31553.625	31562.875	31572.124	31581.374	31590.624	3820
31646.131	31655.383	31664.636	31673.888	31683.141	3830
31738.664	31747.918	31757.173	31766.428	31775.684	3840
31831.222	31840.479	31849.737	31858.994	31868.252	3850
31923.806	31933.066	31942.326	31951.586	31960.847	3860
32016.416	32025.679	32034.941	32044.204	32053.467	3870
32109.052	32118.317	32127.582	32136.848	32146.114	3880
32201.714	32210.981	32220.249	32229.517	32238.786	3890
32294.401	32303.671	32312.942	32322.212	32331.483	3900
32387.114	32396.387	32405.660	32414.933	32424.206	3910
32479.852	32489.128	32498.403	32507.679	32516.955	3920
32572.616	32581.894	32591.172	32600.450	32609.729	3930
32665.406	32674.686	32683.967	32693.247	32702.529	3940
32758.220	32767.503	32776.786	32786.070	32795.353	3950
32851.060	32860.346	32869.631	32878.917	32888.203	3960
32943.926	32953.214	32962.502	32971.790	32981.079	3970
33036.816	33046.106	33055.397	33064.688	33073.979	3980
33129.731	33139.024	33148.318	33157.611	33166.905	3990

TABLE 2 $f \ln f$ as a function of f

f	0	1	2	3	4
4000	33176.199	33185.493	33194.787	33204.082	33213.377
4010	33269.152	33278.448	33287.745	33297.042	33306.340
4020	33362.129	33371.429	33380.728	33390.028	33399.328
4030	33455.132	33464.434	33473.736	33483.038	33492.340
4040	33548.160	33557.464	33566.768	33576.073	33585.378
4050	33641.212	33650.519	33659.826	33669.133	33678.440
4060	33734.289	33743.598	33752.908	33762.217	33771.527
4070	33827.391	33836.703	33846.014	33855.326	33864.639
4080	33920.517	33929.831	33939.145	33948.460	33957.775
4090	34013.668	34022.984	34032.301	34041.618	34050.935
4100	34106.843	34116.162	34125.481	34134.801	34144.120
4110	34200.043	34209.364	34218.686	34228.007	34237.330
4120	34293.267	34302.591	34311.914	34321.239	34330.563
4130	34386.515	34395.841	34405.168	34414.494	34423.821
4140	34479.787	34489.116	34498.445	34507.774	34517.103
4150	34573.084	34582.415	34591.746	34601.078	34610.409
4160	34666.405	34675.738	34685.072	34694.406	34703.740
4170	34759.749	34769.085	34778.421	34787.757	34797.094
4180	34853.118	34862.456	34871.795	34881.133	34890.472
4190	34946.511	34955.851	34965.192	34974.533	34983.874
4200	35039.927	35049.270	35058.613	35067.957	35077.300
4210	35133.367	35142.713	35152.058	35161.404	35170.750
4220	35226.832	35236.179	35245.527	35254.875	35264.224
4230	35320.319	35329.669	35339.020	35348.370	35357.721
4240	35413.831	35423.183	35432.536	35441.889	35451.242
4250	35507.366	35516.720	35526.075	35535.431	35544.786
4260	35600.924	35610.281	35619.639	35628.996	35638.354
4270	35694.506	35703.866	35713.225	35722.585	35731.945
4280	35788.111	35797.473	35806.835	35816.198	35825.560
4290	35881.740	35891.104	35900.469	35909.833	35919.198
4300	35975.392	35984.759	35994.126	36003.492	36012.860
4310	36069.068	36078.436	36087.805	36097.175	36106.544
4320	36162.766	36172.137	36181.509	36190.880	36200.252
4330	36256.488	36265.861	36275.235	36284.609	36293.983
4340	36350.233	36359.608	36368.984	36378.361	36387.737
4350	36444.000	36453.378	36462.757	36472.135	36481.514
4360	36537.791	36547.172	36556.552	36565.933	36575.314
4370	36631.605	36640.988	36650.370	36659.754	36669.137
4380	36725.442	36734.826	36744.212	36753.597	36762.983
4390	36819.301	36828.688	36838.076	36847.463	36856.851
4400	36913.183	36922.573	36931.962	36941.352	36950.742
4410	37007.088	37016.480	37025.872	37035.264	37044.656
4420	37101.016	37110.410	37119.804	37129.198	37138.593
4430	37194.966	37204.362	37213.759	37223.156	37232.552
4440	37288.939	37298.337	37307.736	37317.135	37326.534
4450	37382.934	37392.335	37401.736	37411.137	37420.539
4460	37476.952	37486.355	37495.758	37505.162	37514.565
4470	37570.992	37580.398	37589.803	37599.209	37608.615
4480	37665.055	37674.462	37683.870	37693.278	37702.686
4490	37759.140	37768.550	37777.959	37787.370	37796.780

5	6	7	8	9	*f*
33222.672	33231.967	33241.263	33250.559	33259.855	4000
33315.637	33324.935	33334.233	33343.532	33352.831	4010
33408.628	33417.928	33427.229	33436.530	33445.831	4020
33501.643	33510.946	33520.249	33529.552	33538.856	4030
33594.683	33603.988	33613.294	33622.600	33631.906	4040
33687.748	33697.056	33706.364	33715.672	33724.980	4050
33780.837	33790.147	33799.458	33808.769	33818.080	4060
33873.951	33883.264	33892.577	33901.890	33911.204	4070
33967.090	33976.405	33985.720	33995.036	34004.352	4080
34060.253	34069.570	34078.888	34088.206	34097.525	4090
34153.440	34162.760	34172.080	34181.401	34190.722	4100
34246.652	34255.974	34265.297	34274.620	34283.943	4110
34339.888	34349.213	34358.538	34367.863	34377.189	4120
34433.148	34442.476	34451.803	34461.131	34470.459	4130
34526.433	34535.762	34545.092	34554.423	34563.753	4140
34619.741	34629.074	34638.406	34647.739	34657.072	4150
34713.074	34722.409	34731.743	34741.079	34750.414	4160
34806.431	34815.768	34825.105	34834.442	34843.780	4170
34899.811	34909.151	34918.490	34927.830	34937.170	4180
34993.216	35002.558	35011.900	35021.242	35030.584	4190
35086.644	35095.989	35105.333	35114.678	35124.022	4200
35180.097	35189.443	35198.790	35208.137	35217.484	4210
35273.572	35282.921	35292.270	35301.620	35310.969	4220
35367.072	35376.423	35385.775	35395.126	35404.478	4230
35460.595	35469.949	35479.303	35488.657	35498.011	4240
35554.142	35563.498	35572.854	35582.211	35591.567	4250
35647.712	35657.070	35666.429	35675.788	35685.147	4260
35741.306	35750.667	35760.027	35769.389	35778.750	4270
35834.923	35844.286	35853.649	35863.013	35872.376	4280
35928.563	35937.929	35947.294	35956.660	35966.026	4290
36022.227	36031.595	36040.963	36050.331	36059.699	4300
36115.914	36125.284	36134.654	36144.025	36153.395	4310
36209.624	36218.996	36228.369	36237.742	36247.115	4320
36303.357	36312.732	36322.107	36331.482	36340.857	4330
36397.114	36406.491	36415.868	36425.245	36434.623	4340
36490.893	36500.272	36509.652	36519.031	36528.411	4350
36584.695	36594.077	36603.458	36612.840	36622.223	4360
36678.520	36687.904	36697.288	36706.672	36716.057	4370
36772.368	36781.754	36791.141	36800.527	36809.914	4380
36866.239	36875.628	36885.016	36894.405	36903.794	4390
36960.133	36969.523	36978.914	36988.305	36997.697	4400
37054.049	37063.442	37072.835	37082.228	37091.622	4410
37147.988	37157.383	37166.779	37176.174	37185.570	4420
37241.950	37251.347	37260.745	37270.143	37279.541	4430
37335.934	37345.333	37354.733	37364.133	37373.534	4440
37429.940	37439.342	37448.744	37458.147	37467.549	4450
37523.969	37533.373	37542.778	37552.182	37561.587	4460
37618.021	37627.427	37636.834	37646.241	37655.648	4470
37712.095	37721.503	37730.912	37740.321	37749.730	4480
37806.191	37815.601	37825.013	37834.424	37843.835	4490

TABLE 2 $f \ln f$ as a function of f

f	0	1	2	3	4
4500	37853.247	37862.659	37872.071	37881.484	37890.896
4510	37947.376	37956.791	37966.205	37975.620	37985.034
4520	38041.528	38050.944	38060.361	38069.778	38079.195
4530	38135.702	38145.120	38154.539	38163.958	38173.377
4540	38229.898	38239.318	38248.739	38258.161	38267.582
4550	38324.115	38333.538	38342.962	38352.385	38361.809
4560	38418.355	38427.780	38437.206	38446.631	38456.057
4570	38512.617	38522.044	38531.472	38540.900	38550.328
4580	38606.901	38616.330	38625.760	38635.190	38644.620
4590	38701.206	38710.638	38720.070	38729.502	38738.934
4600	38795.533	38804.967	38814.401	38823.836	38833.270
4610	38889.882	38899.318	38908.755	38918.191	38927.628
4620	38984.253	38993.691	39003.130	39012.568	39022.007
4630	39078.645	39088.086	39097.526	39106.967	39116.408
4640	39173.059	39182.502	39191.945	39201.388	39210.831
4650	39267.495	39276.939	39286.384	39295.829	39305.275
4660	39361.952	39371.398	39380.846	39390.293	39399.740
4670	39456.430	39465.879	39475.328	39484.778	39494.227
4680	39550.930	39560.381	39569.832	39579.284	39588.736
4690	39645.451	39654.904	39664.358	39673.812	39683.266
4700	39739.994	39749.449	39758.905	39768.361	39777.817
4710	39834.557	39844.015	39853.473	39862.931	39872.389
4720	39929.142	39938.602	39948.062	39957.522	39966.982
4730	40023.749	40033.210	40042.672	40052.135	40061.597
4740	40118.376	40127.840	40137.304	40146.768	40156.233
4750	40213.025	40222.491	40231.957	40241.423	40250.890
4760	40307.694	40317.162	40326.630	40336.099	40345.568
4770	40402.385	40411.855	40421.325	40430.796	40440.267
4780	40497.096	40506.568	40516.041	40525.514	40534.987
4790	40591.828	40601.303	40610.777	40620.252	40629.727
4800	40686.582	40696.058	40705.535	40715.012	40724.489
4810	40781.356	40790.834	40800.313	40809.792	40819.271
4820	40876.151	40885.631	40895.112	40904.593	40914.075
4830	40970.966	40980.449	40989.932	40999.415	41008.898
4840	41065.803	41075.288	41084.773	41094.258	41103.743
4850	41160.660	41170.147	41179.634	41189.121	41198.608
4860	41255.537	41265.026	41274.515	41284.005	41293.494
4870	41350.436	41359.927	41369.418	41378.909	41388.401
4880	41445.354	41454.847	41464.341	41473.834	41483.328
4890	41540.294	41549.789	41559.284	41568.779	41578.275
4900	41635.253	41644.750	41654.248	41663.745	41673.243
4910	41730.233	41739.733	41749.232	41758.731	41768.231
4920	41825.234	41834.735	41844.236	41853.738	41863.240
4930	41920.255	41929.758	41939.261	41948.765	41958.269
4940	42015.296	42024.801	42034.306	42043.812	42053.318
4950	42110.357	42119.864	42129.372	42138.879	42148.387
4960	42205.439	42214.948	42224.457	42233.967	42243.477
4970	42300.540	42310.052	42319.563	42329.075	42338.587
4980	42395.662	42405.175	42414.689	42424.203	42433.716
4990	42490.804	42500.319	42509.835	42519.351	42528.866

5	6	7	8	9	f
37900.309	37909.722	37919.135	37928.549	37937.963	4500
37994.450	38003.865	38013.280	38022.696	38032.112	4510
38088.612	38098.030	38107.447	38116.865	38126.283	4520
38182.797	38192.217	38201.637	38211.057	38220.477	4530
38277.004	38286.426	38295.848	38305.270	38314.693	4540
38371.233	38380.657	38390.081	38399.506	38408.930	4550
38465.483	38474.910	38484.336	38493.763	38503.190	4560
38559.756	38569.185	38578.613	38588.042	38597.471	4570
38654.051	38663.481	38672.912	38682.343	38691.775	4580
38748.367	38757.800	38767.233	38776.666	38786.100	4590
38842.705	38852.140	38861.575	38871.011	38880.446	4600
38937.065	38946.502	38955.939	38965.377	38974.815	4610
39031.446	39040.886	39050.325	39059.765	39069.205	4620
39125.850	39135.291	39144.733	39154.175	39163.617	4630
39220.274	39229.718	39239.162	39248.606	39258.050	4640
39314.720	39324.166	39333.612	39343.058	39352.505	4650
39409.188	39418.636	39428.084	39437.533	39446.981	4660
39503.677	39513.127	39522.578	39532.028	39541.479	4670
39598.188	39607.640	39617.092	39626.545	39635.998	4680
39692.720	39702.174	39711.629	39721.083	39730.538	4690
39787.273	39796.729	39806.186	39815.643	39825.100	4700
39881.847	39891.306	39900.765	39910.224	39919.683	4710
39976.443	39985.904	39995.365	40004.826	40014.287	4720
40071.060	40080.523	40089.986	40099.449	40108.912	4730
40165.698	40175.163	40184.628	40194.093	40203.559	4740
40260.357	40269.824	40279.291	40288.758	40298.226	4750
40355.037	40364.506	40373.975	40383.445	40392.915	4760
40449.738	40459.209	40468.680	40478.152	40487.624	4770
40544.460	40553.933	40563.407	40572.880	40582.354	4780
40639.202	40648.678	40658.154	40667.629	40677.105	4790
40733.966	40743.444	40752.921	40762.399	40771.878	4800
40828.751	40838.230	40847.710	40857.190	40866.670	4810
40923.556	40933.038	40942.520	40952.002	40961.484	4820
41018.382	41027.866	41037.350	41046.834	41056.318	4830
41113.229	41122.715	41132.201	41141.687	41151.173	4840
41208.096	41217.584	41227.072	41236.560	41246.049	4850
41302.984	41312.474	41321.964	41331.454	41340.945	4860
41397.892	41407.384	41416.877	41426.369	41435.862	4870
41492.821	41502.316	41511.810	41521.304	41530.799	4880
41587.771	41597.267	41606.763	41616.260	41625.756	4890
41682.741	41692.239	41701.737	41711.236	41720.735	4900
41777.731	41787.231	41796.732	41806.232	41815.733	4910
41872.742	41882.244	41891.746	41901.249	41910.752	4920
41967.773	41977.277	41986.781	41996.286	42005.791	4930
42062.824	42072.330	42081.837	42091.343	42100.850	4940
42157.895	42167.404	42176.912	42186.421	42195.930	4950
42252.987	42262.497	42272.008	42281.518	42291.029	4960
42348.099	42357.611	42367.123	42376.636	42386.149	4970
42443.231	42452.745	42462.259	42471.774	42481.289	4980
42538.382	42547.899	42557.415	42566.932	42576.449	4990

TABLE 2 $f \ln f$ as a function of f

f	0	1	2	3	4
5000	42585.966	42595.483	42605.001	42614.518	42624.036
5010	42681.148	42690.667	42700.187	42709.706	42719.226
5020	42776.350	42785.871	42795.393	42804.914	42814.436
5030	42871.572	42881.095	42890.618	42900.142	42909.666
5040	42966.813	42976.339	42985.864	42995.390	43004.915
5050	43062.075	43071.602	43081.129	43090.657	43100.185
5060	43157.356	43166.885	43176.415	43185.944	43195.474
5070	43252.657	43262.188	43271.720	43281.251	43290.783
5080	43347.978	43357.511	43367.045	43376.578	43386.112
5090	43443.319	43452.854	43462.389	43471.925	43481.460
5100	43538.679	43548.216	43557.753	43567.291	43576.828
5110	43634.058	43643.597	43653.137	43662.676	43672.216
5120	43729.458	43738.999	43748.540	43758.081	43767.623
5130	43824.877	43834.420	43843.963	43853.506	43863.050
5140	43920.315	43929.860	43939.405	43948.950	43958.496
5150	44015.773	44025.320	44034.867	44044.414	44053.961
5160	44111.250	44120.799	44130.348	44139.897	44149.446
5170	44206.747	44216.297	44225.848	44235.399	44244.951
5180	44302.263	44311.815	44321.368	44330.921	44340.474
5190	44397.798	44407.352	44416.907	44426.462	44436.017
5200	44493.352	44502.909	44512.466	44522.022	44531.579
5210	44588.926	44598.484	44608.043	44617.602	44627.161
5220	44684.519	44694.079	44703.640	44713.201	44722.762
5230	44780.131	44789.693	44799.256	44808.818	44818.381
5240	44875.762	44885.326	44894.891	44904.455	44914.020
5250	44971.413	44980.979	44990.545	45000.111	45009.678
5260	45067.082	45076.650	45086.218	45095.786	45105.355
5270	45162.770	45172.340	45181.910	45191.481	45201.051
5280	45258.478	45268.049	45277.621	45287.194	45296.766
5290	45354.204	45363.778	45373.351	45382.926	45392.500
5300	45449.949	45459.525	45469.100	45478.676	45488.252
5310	45545.713	45555.291	45564.868	45574.446	45584.024
5320	45641.496	45651.075	45660.655	45670.235	45679.814
5330	45737.298	45746.879	45756.460	45766.042	45775.624
5340	45833.118	45842.701	45852.285	45861.868	45871.452
5350	45928.957	45938.542	45948.127	45957.713	45967.298
5360	46024.815	46034.402	46043.989	46053.576	46063.164
5370	46120.692	46130.280	46139.869	46149.458	46159.048
5380	46216.587	46226.177	46235.768	46245.359	46254.950
5390	46312.501	46322.093	46331.686	46341.278	46350.871
5400	46408.433	46418.027	46427.622	46437.216	46446.811
5410	46504.384	46513.980	46523.576	46533.172	46542.769
5420	46600.353	46609.951	46619.549	46629.147	46638.746
5430	46696.341	46705.940	46715.540	46725.141	46734.741
5440	46792.347	46801.948	46811.550	46821.152	46830.754
5450	46888.371	46897.975	46907.578	46917.182	46926.786
5460	46984.414	46994.020	47003.625	47013.231	47022.836
5470	47080.475	47090.083	47099.690	47109.297	47118.905
5480	47176.555	47186.164	47195.773	47205.382	47214.992
5490	47272.653	47282.263	47291.874	47301.485	47311.097

5	6	7	8	9	f
42633.554	42643.073	42652.591	42662.110	42671.629	5000
42728.746	42738.267	42747.787	42757.308	42766.829	5010
42823.958	42833.480	42843.003	42852.526	42862.048	5020
42919.190	42928.714	42938.239	42947.763	42957.288	5030
43014.442	43023.968	43033.494	43043.021	43052.548	5040
43109.713	43119.241	43128.770	43138.298	43147.827	5050
43205.004	43214.534	43224.065	43233.595	43243.126	5060
43300.315	43309.847	43319.380	43328.912	43338.445	5070
43395.646	43405.180	43414.714	43424.249	43433.784	5080
43490.996	43500.532	43510.069	43519.605	43529.142	5090
43586.366	43595.904	43605.442	43614.981	43624.520	5100
43681.756	43691.296	43700.836	43710.376	43719.917	5110
43777.165	43786.707	43796.249	43805.791	43815.334	5120
43872.593	43882.137	43891.681	43901.226	43910.770	5130
43968.041	43977.587	43987.133	43996.680	44006.226	5140
44063.509	44073.057	44082.605	44092.153	44101.701	5150
44158.996	44168.546	44178.096	44187.646	44197.196	5160
44254.502	44264.054	44273.606	44283.158	44292.710	5170
44350.028	44359.581	44369.135	44378.689	44388.243	5180
44445.573	44455.128	44464.684	44474.240	44483.796	5190
44541.137	44550.694	44560.252	44569.810	44579.368	5200
44636.720	44646.280	44655.839	44665.399	44674.959	5210
44732.323	44741.884	44751.445	44761.007	44770.569	5220
44827.944	44837.508	44847.071	44856.635	44866.198	5230
44923.585	44933.150	44942.716	44952.281	44961.847	5240
45019.245	45028.812	45038.379	45047.947	45057.514	5250
45114.924	45124.493	45134.062	45143.631	45153.201	5260
45210.622	45220.192	45229.763	45239.335	45248.906	5270
45306.338	45315.911	45325.484	45335.057	45344.630	5280
45402.074	45411.649	45421.224	45430.799	45440.374	5290
45497.829	45507.405	45516.982	45526.559	45536.136	5300
45593.602	45603.181	45612.759	45622.338	45631.917	5310
45689.395	45698.975	45708.555	45718.136	45727.717	5320
45785.206	45794.788	45804.370	45813.953	45823.535	5330
45881.035	45890.619	45900.204	45909.788	45919.373	5340
45976.884	45986.470	45996.056	46005.642	46015.229	5350
46072.751	46082.339	46091.927	46101.515	46111.103	5360
46168.637	46178.227	46187.816	46197.406	46206.997	5370
46264.541	46274.133	46283.725	46293.316	46302.908	5380
46360.464	46370.058	46379.651	46389.245	46398.839	5390
46456.406	46466.001	46475.596	46485.192	46494.788	5400
46552.366	46561.963	46571.560	46581.158	46590.755	5410
46648.344	46657.943	46667.542	46677.142	46686.741	5420
46744.341	46753.942	46763.543	46773.144	46782.745	5430
46840.357	46849.959	46859.562	46869.165	46878.768	5440
46936.390	46945.995	46955.599	46965.204	46974.809	5450
47032.443	47042.049	47051.655	47061.262	47070.868	5460
47128.513	47138.121	47147.729	47157.338	47166.946	5470
47224.601	47234.211	47243.821	47253.432	47263.042	5480
47320.708	47330.320	47339.932	47349.544	47359.156	5490

TABLE 2 $f \ln f$ as a function of f

f	0	1	2	3	4
5500	47368.769	47378.381	47387.994	47397.607	47407.220
5510	47464.903	47474.517	47484.132	47493.746	47503.361
5520	47561.055	47570.671	47580.288	47589.904	47599.521
5530	47657.225	47666.843	47676.462	47686.080	47695.699
5540	47753.414	47763.034	47772.654	47782.274	47791.894
5550	47849.620	47859.242	47868.864	47878.486	47888.108
5560	47945.845	47955.468	47965.092	47974.716	47984.340
5570	48042.087	48051.713	48061.338	48070.964	48080.589
5580	48138.348	48147.975	48157.602	48167.229	48176.857
5590	48234.626	48244.255	48253.884	48263.513	48273.143
5600	48330.923	48340.553	48350.184	48359.815	48369.446
5610	48427.237	48436.869	48446.502	48456.134	48465.767
5620	48523.569	48533.203	48542.837	48552.472	48562.106
5630	48619.918	48629.554	48639.190	48648.827	48658.463
5640	48716.286	48725.924	48735.562	48745.200	48754.838
5650	48812.671	48822.311	48831.950	48841.590	48851.230
5660	48909.074	48918.715	48928.357	48937.998	48947.640
5670	49005.495	49015.138	49024.781	49034.424	49044.068
5680	49101.933	49111.578	49121.223	49130.868	49140.513
5690	49198.389	49208.035	49217.682	49227.329	49236.976
5700	49294.862	49304.511	49314.159	49323.808	49333.457
5710	49391.353	49401.003	49410.654	49420.304	49429.955
5720	49487.862	49497.514	49507.166	49516.818	49526.470
5730	49584.388	49594.041	49603.695	49613.349	49623.003
5740	49680.931	49690.586	49700.242	49709.898	49719.553
5750	49777.492	49787.149	49796.806	49806.464	49816.121
5760	49874.070	49883.729	49893.388	49903.047	49912.706
5770	49970.666	49980.326	49989.987	49999.648	50009.309
5780	50067.279	50076.941	50086.603	50096.266	50105.929
5790	50163.909	50173.573	50183.237	50192.901	50202.566
5800	50260.557	50270.222	50279.888	50289.554	50299.220
5810	50357.221	50366.889	50376.556	50386.224	50395.892
5820	50453.903	50463.572	50473.242	50482.911	50492.581
5830	50550.602	50560.273	50569.944	50579.615	50589.287
5840	50647.319	50656.991	50666.664	50676.337	50686.010
5850	50744.052	50753.726	50763.401	50773.075	50782.750
5860	50840.803	50850.479	50860.155	50869.831	50879.508
5870	50937.570	50947.248	50956.926	50966.604	50976.282
5880	51034.355	51044.034	51053.714	51063.394	51073.073
5890	51131.156	51140.838	51150.519	51160.200	51169.882
5900	51227.975	51237.658	51247.341	51257.024	51266.707
5910	51324.811	51334.495	51344.180	51353.865	51363.550
5920	51421.663	51431.349	51441.036	51450.722	51460.409
5930	51518.532	51528.220	51537.908	51547.596	51557.285
5940	51615.419	51625.108	51634.798	51644.488	51654.178
5950	51712.322	51722.013	51731.704	51741.396	51751.088
5960	51809.242	51818.934	51828.628	51838.321	51848.014
5970	51906.178	51915.873	51925.568	51935.262	51944.958
5980	52003.132	52012.828	52022.524	52032.221	52041.918
5990	52100.102	52109.800	52119.498	52129.196	52138.894

5	6	7	8	9	f
47416.833	47426.447	47436.061	47445.674	47455.288	5500
47512.977	47522.592	47532.207	47541.823	47551.439	5510
47609.138	47618.755	47628.372	47637.990	47647.607	5520
47705.317	47714.936	47724.555	47734.175	47743.794	5530
47801.515	47811.136	47820.756	47830.378	47839.999	5540
47897.730	47907.353	47916.976	47926.598	47936.222	5550
47993.964	48003.588	48013.213	48022.837	48032.462	5560
48090.215	48099.841	48109.468	48119.094	48128.721	5570
48186.485	48196.113	48205.741	48215.369	48224.998	5580
48282.772	48292.402	48302.032	48311.662	48321.292	5590
48379.077	48388.709	48398.341	48407.972	48417.604	5600
48475.400	48485.034	48494.667	48504.301	48513.935	5610
48571.741	48581.376	48591.012	48600.647	48610.283	5620
48668.100	48677.737	48687.374	48697.011	48706.648	5630
48764.476	48774.115	48783.754	48793.393	48803.032	5640
48860.870	48870.511	48880.151	48889.792	48899.433	5650
48957.282	48966.924	48976.567	48986.209	48995.852	5660
49053.712	49063.356	49072.000	49082.644	49092.288	5670
49150.159	49159.804	49169.450	49179.096	49188.742	5680
49246.623	49256.271	49265.918	49275.566	49285.214	5690
49343.106	49352.755	49362.404	49372.054	49381.703	5700
49439.605	49449.256	49458.907	49468.559	49478.210	5710
49536.123	49545.775	49555.428	49565.081	49574.734	5720
49632.657	49642.312	49651.966	49661.621	49671.276	5730
49729.209	49738.866	49748.522	49758.178	49767.835	5740
49825.779	49835.437	49845.095	49854.753	49864.412	5750
49922.366	49932.026	49941.685	49951.345	49961.006	5760
50018.970	50028.632	50038.293	50047.955	50057.617	5770
50115.592	50125.255	50134.918	50144.582	50154.245	5780
50212.231	50221.895	50231.560	50241.226	50250.891	5790
50308.887	50318.553	50328.220	50337.887	50347.554	5800
50405.560	50415.228	50424.897	50434.565	50444.234	5810
50502.251	50511.921	50521.591	50531.261	50540.932	5820
50598.958	50608.630	50618.302	50627.974	50637.646	5830
50695.683	50705.357	50715.030	50724.704	50734.378	5840
50792.425	50802.100	50811.776	50821.451	50831.127	5850
50889.184	50898.861	50908.538	50918.215	50927.893	5860
50985.960	50995.639	51005.318	51014.997	51024.676	5870
51082.753	51092.434	51102.114	51111.795	51121.475	5880
51179.564	51189.246	51198.928	51208.610	51218.292	5890
51276.391	51286.074	51295.758	51305.442	51315.126	5900
51373.235	51382.920	51392.606	51402.291	51411.977	5910
51470.096	51479.783	51489.470	51499.157	51508.845	5920
51566.973	51576.662	51586.351	51596.040	51605.729	5930
51663.868	51673.558	51683.249	51692.940	51702.631	5940
51760.779	51770.472	51780.164	51789.856	51799.549	5950
51857.708	51867.402	51877.095	51886.790	51896.484	5960
51954.653	51964.348	51974.044	51983.740	51993.435	5970
52051.615	52061.312	52071.009	52080.706	52090.404	5980
52148.593	52158.292	52167.991	52177.690	52187.389	5990

TABLE 2 $f \ln f$ as a function of f

f	0	1	2	3	4
6000	52197.088	52206.788	52216.488	52226.188	52235.888
6010	52294.092	52303.793	52313.495	52323.196	52332.898
6020	52391.112	52400.815	52410.518	52420.221	52429.925
6030	52488.149	52497.853	52507.558	52517.263	52526.968
6040	52585.202	52594.908	52604.615	52614.321	52624.028
6050	52682.272	52691.980	52701.688	52711.396	52721.105
6060	52779.358	52789.068	52798.778	52808.488	52818.198
6070	52876.461	52886.172	52895.884	52905.595	52915.307
6080	52973.581	52983.293	52993.006	53002.720	53012.433
6090	53070.716	53080.431	53090.146	53099.860	53109.575
6100	53167.869	53177.585	53187.301	53197.018	53206.734
6110	53265.037	53274.755	53284.473	53294.191	53303.909
6120	53362.222	53371.942	53381.661	53391.381	53401.101
6130	53459.424	53469.145	53478.866	53488.587	53498.309
6140	53556.641	53566.364	53576.087	53585.810	53595.533
6150	53653.875	53663.600	53673.324	53683.049	53692.773
6160	53751.125	53760.851	53770.577	53780.304	53790.030
6170	53848.392	53858.119	53867.847	53877.575	53887.303
6180	53945.675	53955.404	53965.133	53974.862	53984.592
6190	54042.973	54052.704	54062.435	54072.166	54081.897
6200	54140.288	54150.021	54159.753	54169.486	54179.219
6210	54237.619	54247.353	54257.088	54266.822	54276.556
6220	54334.967	54344.702	54354.438	54364.174	54373.910
6230	54432.330	54442.067	54451.805	54461.542	54471.280
6240	54529.709	54539.448	54549.187	54558.926	54568.666
6250	54627.105	54636.845	54646.586	54656.326	54666.067
6260	54724.516	54734.258	54744.000	54753.743	54763.485
6270	54821.943	54831.687	54841.431	54851.175	54860.919
6280	54919.387	54929.132	54938.877	54948.623	54958.368
6290	55016.846	55026.593	55036.340	55046.087	55055.834
6300	55114.321	55124.069	55133.818	55143.567	55153.315
6310	55211.812	55221.562	55231.312	55241.062	55250.813
6320	55309.319	55319.070	55328.822	55338.574	55348.326
6330	55406.841	55416.595	55426.348	55436.101	55445.855
6340	55504.380	55514.135	55523.889	55533.644	55543.400
6350	55601.934	55611.690	55621.447	55631.203	55640.960
6360	55699.504	55709.262	55719.020	55728.778	55738.536
6370	55797.090	55806.849	55816.609	55826.369	55836.128
6380	55894.691	55904.452	55914.213	55923.975	55933.736
6390	55992.308	56002.071	56011.833	56021.596	56031.359
6400	56089.941	56099.705	56109.469	56119.234	56128.998
6410	56187.589	56197.355	56207.121	56216.887	56226.653
6420	56285.253	56295.020	56304.788	56314.555	56324.323
6430	56382.933	56392.702	56402.470	56412.240	56422.009
6440	56480.628	56490.398	56500.169	56509.939	56519.710
6450	56578.338	56588.110	56597.882	56607.655	56617.427
6460	56676.064	56685.838	56695.612	56705.385	56715.159
6470	56773.806	56783.581	56793.356	56803.132	56812.907
6480	56871.563	56881.340	56891.116	56900.893	56910.670
6490	56969.336	56979.114	56988.892	56998.670	57008.449

5	6	7	8	9	*f*
52245.588	52255.289	52264.989	52274.690	52284.391	6000
52342.600	52352.302	52362.004	52371.707	52381.409	6010
52439.628	52449.332	52459.036	52468.740	52478.444	6020
52536.673	52546.379	52556.084	52565.790	52575.496	6030
52633.735	52643.442	52653.149	52662.857	52672.564	6040
52730.813	52740.522	52750.231	52759.940	52769.649	6050
52827.908	52837.618	52847.329	52857.039	52866.750	6060
52925.019	52934.731	52944.443	52954.155	52963.868	6070
53022.147	53031.860	53041.574	53051.288	53061.002	6080
53119.291	53129.006	53138.721	53148.437	53158.153	6090
53216.451	53226.168	53235.885	53245.602	53255.320	6100
53313.628	53323.346	53333.065	53342.784	53352.503	6110
53410.821	53420.541	53430.262	53439.982	53449.703	6120
53508.030	53517.752	53527.474	53537.196	53546.919	6130
53605.256	53614.980	53624.703	53634.427	53644.151	6140
53702.498	53712.223	53721.949	53731.674	53741.400	6150
53799.757	53809.483	53819.210	53828.937	53838.665	6160
53897.031	53906.760	53916.488	53926.217	53935.946	6170
53994.322	54004.052	54013.782	54023.512	54033.243	6180
54091.629	54101.360	54111.092	54120.824	54130.556	6190
54188.952	54198.685	54208.418	54218.152	54227.886	6200
54286.291	54296.026	54305.761	54315.496	54325.231	6210
54383.646	54393.383	54403.119	54412.856	54422.593	6220
54481.018	54490.756	54500.494	54510.232	54519.971	6230
54578.405	54588.145	54597.884	54607.624	54617.364	6240
54675.808	54685.550	54695.291	54705.032	54714.774	6250
54773.228	54782.970	54792.713	54802.457	54812.200	6260
54870.663	54880.407	54890.152	54899.897	54909.642	6270
54968.114	54977.860	54987.606	54997.353	55007.099	6280
55065.581	55075.329	55085.077	55094.825	55104.573	6290
55163.064	55172.814	55182.563	55192.312	55202.062	6300
55260.563	55270.314	55280.065	55289.816	55299.567	6310
55358.078	55367.830	55377.583	55387.336	55397.088	6320
55455.609	55465.363	55475.117	55484.871	55494.625	6330
55553.155	55562.911	55572.666	55582.422	55592.178	6340
55650.717	55660.474	55670.231	55679.989	55689.746	6350
55748.295	55758.054	55767.812	55777.571	55787.330	6360
55845.888	55855.649	55865.409	55875.170	55884.930	6370
55943.498	55953.260	55963.021	55972.784	55982.546	6380
56041.123	56050.886	56060.649	56070.413	56080.177	6390
56138.763	56148.528	56158.293	56168.058	56177.824	6400
56236.419	56246.186	56255.952	56265.719	56275.486	6410
56334.091	56343.859	56353.627	56363.396	56373.164	6420
56431.778	56441.548	56451.318	56461.088	56470.858	6430
56529.481	56539.252	56549.024	56558.795	56568.567	6440
56627.200	56636.972	56646.745	56656.518	56666.291	6450
56724.933	56734.708	56744.482	56754.257	56764.031	6460
56822.683	56832.458	56842.234	56852.010	56861.787	6470
56920.447	56930.225	56940.002	56949.780	56959.558	6480
57018.228	57028.006	57037.785	57047.565	57057.344	6490

6500–6990

TABLE 2 $f \ln f$ as a function of f

f	0	1	2	3	4
6500	57067.123	57076.903	57086.683	57096.463	57106.243
6510	57164.927	57174.708	57184.489	57194.271	57204.052
6520	57262.745	57272.528	57282.311	57292.094	57301.877
6530	57360.579	57370.364	57380.148	57389.932	57399.717
6540	57458.429	57468.214	57478.000	57487.786	57497.573
6550	57556.293	57566.080	57575.868	57585.656	57595.443
6560	57654.173	57663.962	57673.751	57683.540	57693.329
6570	57752.068	57761.858	57771.649	57781.440	57791.230
6580	57849.978	57859.770	57869.562	57879.354	57889.147
6590	57947.904	57957.697	57967.491	57977.284	57987.078
6600	58045.845	58055.639	58065.434	58075.230	58085.025
6610	58143.800	58153.597	58163.393	58173.190	58182.987
6620	58241.771	58251.569	58261.367	58271.166	58280.964
6630	58339.757	58349.557	58359.356	58369.156	58378.956
6640	58437.758	58447.559	58457.361	58467.162	58476.963
6650	58535.775	58545.577	58555.380	58565.182	58574.985
6660	58633.806	58643.610	58653.414	58663.218	58673.023
6670	58731.852	58741.658	58751.463	58761.269	58771.075
6680	58829.913	58839.720	58849.527	58859.335	58869.142
6690	58927.990	58937.798	58947.607	58957.415	58967.224
6700	59026.081	59035.891	59045.701	59055.511	59065.321
6710	59124.187	59133.998	59143.810	59153.622	59163.433
6720	59222.308	59232.121	59241.934	59251.747	59261.560
6730	59320.444	59330.258	59340.073	59349.887	59359.702
6740	59418.594	59428.410	59438.226	59448.043	59457.859
6750	59516.760	59526.577	59536.395	59546.213	59556.030
6760	59614.940	59624.759	59634.578	59644.397	59654.217
6770	59713.136	59722.956	59732.776	59742.597	59752.418
6780	59811.346	59821.167	59830.989	59840.811	59850.634
6790	59909.570	59919.394	59929.217	59939.041	59948.864
6800	60007.810	60017.634	60027.459	60037.284	60047.110
6810	60106.064	60115.890	60125.716	60135.543	60145.370
6820	60204.333	60214.160	60223.988	60233.816	60243.644
6830	60302.616	60312.445	60322.275	60332.104	60341.934
6840	60400.914	60410.745	60420.576	60430.406	60440.238
6850	60499.227	60509.059	60518.891	60528.724	60538.556
6860	60597.554	60607.388	60617.221	60627.055	60636.889
6870	60695.896	60705.731	60715.566	60725.402	60735.237
6880	60794.253	60804.089	60813.926	60823.762	60833.599
6890	60892.624	60902.462	60912.300	60922.138	60931.976
6900	60991.009	61000.849	61010.688	61020.528	61030.367
6910	61089.409	61099.250	61109.091	61118.932	61128.773
6920	61187.824	61197.666	61207.508	61217.351	61227.193
6930	61286.253	61296.096	61305.940	61315.784	61325.628
6940	61384.696	61394.541	61404.386	61414.232	61424.077
6950	61483.154	61493.000	61502.847	61512.694	61522.541
6960	61581.626	61591.474	61601.322	61611.170	61621.019
6970	61680.112	61689.962	61699.811	61709.661	61719.511
6980	61778.613	61788.464	61798.315	61808.166	61818.018
6990	61877.128	61886.981	61896.833	61906.686	61916.539

5	6	7	8	9	f
57116.023	57125.804	57135.584	57145.365	57155.146	6500
57213.834	57223.616	57233.398	57243.180	57252.963	6510
57311.660	57321.444	57331.228	57341.011	57350.795	6520
57409.502	57419.287	57429.072	57438.858	57448.643	6530
57507.359	57517.145	57526.932	57536.719	57546.506	6540
57605.231	57615.019	57624.807	57634.596	57644.384	6550
57703.119	57712.908	57722.698	57732.488	57742.278	6560
57801.021	57810.812	57820.604	57830.395	57840.187	6570
57898.939	57908.732	57918.525	57928.318	57938.111	6580
57996.872	58006.666	58016.461	58026.255	58036.050	6590
58094.821	58104.616	58114.412	58124.208	58134.004	6600
58192.784	58202.581	58212.378	58222.176	58231.974	6610
58290.762	58300.561	58310.360	58320.159	58329.958	6620
58388.756	58398.556	58408.357	58418.157	58427.958	6630
58486.765	58496.566	58506.368	58516.170	58525.972	6640
58584.788	58594.592	58604.395	58614.198	58624.002	6650
58682.827	58692.632	58702.437	58712.242	58722.047	6660
58780.881	58790.687	58800.493	58810.300	58820.107	6670
58878.950	58888.757	58898.565	58908.373	58918.181	6680
58977.033	58986.843	58996.652	59006.461	59016.271	6690
59075.132	59084.943	59094.753	59104.564	59114.376	6700
59173.246	59183.058	59192.870	59202.682	59212.495	6710
59271.374	59281.188	59291.001	59300.815	59310.629	6720
59369.517	59379.332	59389.148	59398.963	59408.779	6730
59467.675	59477.492	59487.309	59497.126	59506.943	6740
59565.848	59575.666	59585.485	59595.303	59605.122	6750
59664.036	59673.856	59683.675	59693.495	59703.315	6760
59762.239	59772.060	59781.881	59791.702	59801.524	6770
59860.456	59870.279	59880.101	59889.924	59899.747	6780
59958.688	59968.512	59978.336	59988.161	59997.985	6790
60056.935	60066.760	60076.586	60086.412	60096.238	6800
60155.196	60165.023	60174.850	60184.678	60194.505	6810
60253.473	60263.301	60273.129	60282.958	60292.787	6820
60351.763	60361.593	60371.423	60381.253	60391.084	6830
60450.069	60459.900	60469.732	60479.563	60489.395	6840
60548.389	60558.222	60568.055	60577.888	60587.721	6850
60646.723	60656.558	60666.392	60676.227	60686.061	6860
60745.073	60754.908	60764.744	60774.580	60784.416	6870
60843.436	60853.274	60863.111	60872.948	60882.786	6880
60941.815	60951.653	60961.492	60971.331	60981.170	6890
61040.207	61050.047	61059.888	61069.728	61079.569	6900
61138.615	61148.456	61158.298	61168.140	61177.982	6910
61237.036	61246.879	61256.722	61266.566	61276.409	6920
61335.472	61345.317	61355.161	61365.006	61374.851	6930
61433.923	61443.769	61453.615	61463.461	61473.307	6940
61532.388	61542.235	61552.083	61561.930	61571.778	6950
61630.867	61640.716	61650.565	61660.414	61670.263	6960
61729.361	61739.211	61749.062	61758.912	61768.763	6970
61827.869	61837.721	61847.572	61857.424	61867.276	6980
61926.391	61936.244	61946.098	61955.951	61965.804	6990

TABLE 2 $f \ln f$ as a function of f

f	0	1	2	3	4
7000	61975.658	61985.512	61995.366	62005.220	62015.074
7010	62074.202	62084.057	62093.912	62103.768	62113.623
7020	62172.760	62182.616	62192.473	62202.330	62212.187
7030	62271.332	62281.190	62291.048	62300.907	62310.765
7040	62369.919	62379.778	62389.638	62399.497	62409.357
7050	62468.519	62478.380	62488.241	62498.102	62507.964
7060	62567.134	62576.997	62586.859	62596.722	62606.584
7070	62665.763	62675.627	62685.491	62695.355	62705.219
7080	62764.407	62774.272	62784.137	62794.002	62803.868
7090	62863.064	62872.931	62882.797	62892.664	62902.531
7100	62961.735	62971.603	62981.471	62991.340	63001.208
7110	63060.421	63070.290	63080.160	63090.029	63099.899
7120	63159.121	63168.991	63178.862	63188.733	63198.604
7130	63257.834	63267.706	63277.579	63287.451	63297.324
7140	63356.562	63366.435	63376.309	63386.183	63396.057
7150	63455.304	63465.179	63475.054	63484.929	63494.804
7160	63554.059	63563.936	63573.812	63583.689	63593.565
7170	63652.829	63662.707	63672.584	63682.463	63692.341
7180	63751.612	63761.492	63771.371	63781.250	63791.130
7190	63850.410	63860.290	63870.171	63880.052	63889.933
7200	63949.221	63959.103	63968.985	63978.868	63988.750
7210	64048.047	64057.930	64067.813	64077.697	64087.581
7220	64146.886	64156.771	64166.655	64176.540	64186.425
7230	64245.739	64255.625	64265.511	64275.398	64285.284
7240	64344.606	64354.493	64364.381	64374.269	64384.156
7250	64443.486	64453.375	64463.264	64473.153	64483.043
7260	64542.381	64552.271	64562.161	64572.052	64581.943
7270	64641.289	64651.181	64661.072	64670.964	64680.856
7280	64740.211	64750.104	64759.997	64769.890	64779.784
7290	64839.147	64849.041	64858.936	64868.830	64878.725
7300	64938.096	64947.992	64957.888	64967.784	64977.680
7310	65037.059	65046.956	65056.854	65066.751	65076.649
7320	65136.036	65145.935	65155.833	65165.732	65175.631
7330	65235.027	65244.927	65254.826	65264.727	65274.627
7340	65334.031	65343.932	65353.833	65363.735	65373.636
7350	65433.049	65442.951	65452.854	65462.757	65472.660
7360	65532.080	65541.984	65551.888	65561.792	65571.696
7370	65631.125	65641.030	65650.936	65660.841	65670.747
7380	65730.183	65740.090	65749.997	65759.904	65769.811
7390	65829.255	65839.163	65849.072	65858.980	65868.888
7400	65928.341	65938.250	65948.160	65958.069	65967.979
7410	66027.440	66037.351	66047.262	66057.173	66067.084
7420	66126.553	66136.465	66146.377	66156.289	66166.202
7430	66225.679	66235.592	66245.506	66255.419	66265.333
7440	66324.818	66334.733	66344.648	66354.563	66364.478
7450	66423.971	66433.887	66443.804	66453.720	66463.636
7460	66523.138	66533.055	66542.973	66552.890	66562.808
7470	66622.318	66632.236	66642.155	66652.074	66661.993
7480	66721.511	66731.431	66741.351	66751.271	66761.192
7490	66820.717	66830.639	66840.560	66850.482	66860.404

5	6	7	8	9	*f*
62024.928	62034.783	62044.637	62054.492	62064.347	7000
62123.479	62133.335	62143.191	62153.047	62162.903	7010
62222.044	62231.902	62241.759	62251.617	62261.474	7020
62320.624	62330.482	62340.341	62350.200	62360.059	7030
62419.217	62429.077	62438.938	62448.798	62458.659	7040
62517.825	62527.687	62537.548	62547.410	62557.272	7050
62616.447	62626.310	62636.173	62646.036	62655.900	7060
62715.083	62724.948	62734.812	62744.677	62754.542	7070
62813.734	62823.599	62833.465	62843.331	62853.198	7080
62912.398	62922.265	62932.133	62942.000	62951.868	7090
63011.076	63020.945	63030.814	63040.683	63050.552	7100
63109.769	63119.639	63129.509	63139.380	63149.250	7110
63208.476	63218.347	63228.219	63238.090	63247.962	7120
63307.196	63317.069	63326.942	63336.815	63346.689	7130
63405.931	63415.805	63425.680	63435.554	63445.429	7140
63504.680	63514.555	63524.431	63534.307	63544.183	7150
63603.442	63613.319	63623.197	63633.074	63642.951	7160
63702.219	63712.097	63721.976	63731.855	63741.733	7170
63801.009	63810.889	63820.769	63830.649	63840.530	7180
63899.814	63909.695	63919.577	63929.458	63939.340	7190
63998.632	64008.515	64018.398	64028.281	64038.164	7200
64097.465	64107.349	64117.233	64127.117	64137.001	7210
64196.311	64206.196	64216.082	64225.967	64235.853	7220
64295.171	64305.057	64314.944	64324.831	64334.718	7230
64394.044	64403.932	64413.821	64423.709	64433.598	7240
64492.932	64502.821	64512.711	64522.601	64532.491	7250
64591.833	64601.724	64611.615	64621.506	64631.398	7260
64690.748	64700.641	64710.533	64720.426	64730.318	7270
64789.677	64799.571	64809.465	64819.359	64829.253	7280
64888.620	64898.515	64908.410	64918.305	64928.201	7290
64987.576	64997.473	65007.369	65017.266	65027.162	7300
65086.546	65096.444	65106.342	65116.240	65126.138	7310
65185.530	65195.429	65205.328	65215.228	65225.127	7320
65284.527	65294.428	65304.328	65314.229	65324.130	7330
65383.538	65393.440	65403.342	65413.244	65423.146	7340
65482.563	65492.466	65502.369	65512.273	65522.176	7350
65581.601	65591.505	65601.410	65611.315	65621.220	7360
65680.652	65690.558	65700.464	65710.371	65720.277	7370
65779.718	65789.625	65799.532	65809.440	65819.348	7380
65878.797	65888.705	65898.614	65908.523	65918.432	7390
65977.889	65987.799	65997.709	66007.619	66017.530	7400
66076.995	66086.906	66096.818	66106.729	66116.641	7410
66176.114	66186.027	66195.940	66205.853	66215.766	7420
66275.247	66285.161	66295.075	66304.989	66314.904	7430
66374.393	66384.309	66394.224	66404.140	66414.055	7440
66473.553	66483.470	66493.386	66503.303	66513.221	7450
66572.726	66582.644	66592.562	66602.481	66612.399	7460
66671.913	66681.832	66691.751	66701.671	66711.591	7470
66771.112	66781.033	66790.954	66800.875	66810.796	7480
66870.326	66880.248	66890.170	66900.092	66910.015	7490

TABLE 2 $f \ln f$ as a function of f

f	0	1	2	3	4
7500	66919.937	66929.860	66939.783	66949.706	66959.629
7510	67019.170	67029.095	67039.019	67048.943	67058.868
7520	67118.417	67128.342	67138.268	67148.194	67158.119
7530	67217.677	67227.604	67237.530	67247.457	67257.385
7540	67316.950	67326.878	67336.806	67346.735	67356.663
7550	67416.236	67426.166	67436.095	67446.025	67455.955
7560	67515.536	67525.467	67535.398	67545.329	67555.260
7570	67614.849	67624.781	67634.713	67644.645	67654.578
7580	67714.175	67724.108	67734.042	67743.975	67753.909
7590	67813.514	67823.449	67833.384	67843.319	67853.254
7600	67912.867	67922.803	67932.739	67942.675	67952.611
7610	68012.232	68022.170	68032.107	68042.045	68051.982
7620	68111.611	68121.550	68131.488	68141.427	68151.366
7630	68211.003	68220.943	68230.883	68240.823	68250.763
7640	68310.408	68320.349	68330.291	68340.232	68350.174
7650	68409.826	68419.769	68429.711	68439.654	68449.597
7660	68509.257	68519.201	68529.145	68539.089	68549.033
7670	68608.701	68618.647	68628.592	68638.537	68648.483
7680	68708.159	68718.105	68728.052	68737.998	68747.945
7690	68807.629	68817.577	68827.525	68837.473	68847.421
7700	68907.112	68917.061	68927.010	68936.960	68946.909
7710	69006.608	69016.559	69026.509	69036.460	69046.411
7720	69106.118	69116.069	69126.021	69135.973	69145.925
7730	69205.640	69215.593	69225.546	69235.499	69245.452
7740	69305.175	69315.129	69325.083	69335.038	69344.993
7750	69404.723	69414.678	69424.634	69434.590	69444.546
7760	69504.284	69514.241	69524.198	69534.155	69544.112
7770	69603.858	69613.816	69623.774	69633.732	69643.691
7780	69703.444	69713.404	69723.363	69733.323	69743.283
7790	69803.044	69813.005	69822.965	69832.926	69842.887
7800	69902.656	69912.618	69922.580	69932.543	69942.505
7810	70002.281	70012.245	70022.208	70032.172	70042.135
7820	70101.919	70111.884	70121.849	70131.813	70141.778
7830	70201.570	70211.536	70221.502	70231.468	70241.434
7840	70301.234	70311.201	70321.168	70331.135	70341.103
7850	70400.910	70410.878	70420.847	70430.816	70440.784
7860	70500.599	70510.569	70520.539	70530.508	70540.478
7870	70600.301	70610.272	70620.243	70630.214	70640.185
7880	70700.015	70709.988	70719.960	70729.932	70739.905
7890	70799.743	70809.716	70819.690	70829.663	70839.637
7900	70899.483	70909.457	70919.432	70929.407	70939.382
7910	70999.235	71009.211	71019.187	71029.163	71039.140
7920	71099.000	71108.977	71118.955	71128.932	71138.910
7930	71198.778	71208.756	71218.735	71228.714	71238.693
7940	71298.568	71308.548	71318.528	71328.508	71338.488
7950	71398.371	71408.352	71418.333	71428.315	71438.296
7960	71498.187	71508.169	71518.151	71528.134	71538.117
7970	71598.015	71607.998	71617.982	71627.966	71637.950
7980	71697.856	71707.840	71717.825	71727.810	71737.795
7990	71797.709	71807.695	71817.681	71827.667	71837.654

5	6	7	8	9	*f*
66969.552	66979.476	66989.399	66999.323	67009.247	7500
67068.792	67078.717	67088.642	67098.567	67108.492	7510
67168.045	67177.971	67187.898	67197.824	67207.750	7520
67267.312	67277.239	67287.167	67297.094	67307.022	7530
67366.592	67376.520	67386.449	67396.378	67406.307	7540
67465.885	67475.815	67485.745	67495.675	67505.606	7550
67565.191	67575.122	67585.054	67594.985	67604.917	7560
67664.510	67674.443	67684.376	67694.309	67704.242	7570
67763.843	67773.777	67783.711	67793.645	67803.580	7580
67863.189	67873.124	67883.060	67892.995	67902.931	7590
67962.548	67972.485	67982.421	67992.358	68002.295	7600
68061.920	68071.858	68081.796	68091.734	68101.673	7610
68161.305	68171.245	68181.184	68191.124	68201.063	7620
68260.704	68270.644	68280.585	68290.526	68300.467	7630
68360.115	68370.057	68379.999	68389.941	68399.884	7640
68459.540	68469.483	68479.427	68489.370	68499.314	7650
68558.978	68568.922	68578.867	68588.812	68598.756	7660
68658.428	68668.374	68678.320	68688.266	68698.212	7670
68757.892	68767.839	68777.786	68787.734	68797.681	7680
68857.369	68867.317	68877.266	68887.214	68897.163	7690
68956.859	68966.808	68976.758	68986.708	68996.658	7700
69056.361	69066.312	69076.264	69086.215	69096.166	7710
69155.877	69165.829	69175.782	69185.734	69195.687	7720
69255.406	69265.359	69275.313	69285.267	69295.221	7730
69354.947	69364.902	69374.857	69384.812	69394.768	7740
69454.502	69464.458	69474.414	69484.371	69494.327	7750
69554.069	69564.027	69573.984	69583.942	69593.900	7760
69653.649	69663.608	69673.567	69683.526	69693.485	7770
69753.243	69763.203	69773.163	69783.123	69793.083	7780
69852.849	69862.810	69872.771	69882.733	69892.694	7790
69952.467	69962.430	69972.393	69982.355	69992.318	7800
70052.099	70062.063	70072.027	70081.991	70091.955	7810
70151.743	70161.708	70171.674	70181.639	70191.605	7820
70251.400	70261.367	70271.333	70281.300	70291.267	7830
70351.070	70361.038	70371.006	70380.974	70390.942	7840
70450.753	70460.722	70470.691	70480.660	70490.630	7850
70550.449	70560.419	70570.389	70580.360	70590.330	7860
70650.157	70660.128	70670.100	70680.072	70690.043	7870
70749.877	70759.850	70769.823	70779.796	70789.769	7880
70849.611	70859.585	70869.559	70879.534	70889.508	7890
70949.357	70959.332	70969.308	70979.283	70989.259	7900
71049.116	71059.093	71069.069	71079.046	71089.023	7910
71148.887	71158.865	71168.843	71178.821	71188.800	7920
71248.672	71258.651	71268.630	71278.609	71288.589	7930
71348.468	71358.449	71368.429	71378.410	71388.390	7940
71448.278	71458.259	71468.241	71478.223	71488.205	7950
71548.099	71558.082	71568.065	71578.048	71588.032	7960
71647.934	71657.918	71667.902	71677.887	71687.871	7970
71747.781	71757.766	71767.752	71777.737	71787.723	7980
71847.640	71857.627	71867.614	71877.600	71887.587	7990

TABLE 2 $f \ln f$ as a function of f

f	0	1	2	3	4
8000	71897.575	71907.562	71917.549	71927.537	71937.524
8010	71997.453	72007.441	72017.430	72027.419	72037.408
8020	72097.343	72107.333	72117.323	72127.313	72137.303
8030	72197.247	72207.238	72217.229	72227.220	72237.211
8040	72297.162	72307.155	72317.147	72327.139	72337.132
8050	72397.090	72407.084	72417.077	72427.071	72437.065
8060	72497.031	72507.026	72517.020	72527.015	72537.010
8070	72596.984	72606.980	72616.976	72626.972	72636.968
8080	72696.949	72706.946	72716.944	72726.941	72736.939
8090	72796.927	72806.925	72816.924	72826.922	72836.921
8100	72896.917	72906.916	72916.916	72926.916	72936.916
8110	72996.919	73006.920	73016.921	73026.922	73036.923
8120	73096.934	73106.936	73116.938	73126.941	73136.943
8130	73196.961	73206.964	73216.968	73226.971	73236.975
8140	73297.000	73307.005	73317.009	73327.014	73337.019
8150	73397.052	73407.057	73417.063	73427.070	73437.076
8160	73497.115	73507.123	73517.130	73527.137	73537.144
8170	73597.192	73607.200	73617.208	73627.217	73637.225
8180	73697.280	73707.289	73717.299	73727.309	73737.319
8190	73797.381	73807.391	73817.402	73827.413	73837.424
8200	73897.493	73907.505	73917.517	73927.530	73937.542
8210	73997.618	74007.632	74017.645	74027.658	74037.672
8220	74097.756	74107.770	74117.784	74127.799	74137.814
8230	74197.905	74207.920	74217.936	74227.952	74237.968
8240	74298.066	74308.083	74318.100	74328.117	74338.134
8250	74398.240	74408.258	74418.276	74428.294	74438.313
8260	74498.426	74508.445	74518.464	74528.484	74538.503
8270	74598.624	74608.644	74618.665	74628.685	74638.706
8280	74698.833	74708.855	74718.877	74728.899	74738.921
8290	74799.056	74809.078	74819.101	74829.124	74839.148
8300	74899.290	74909.314	74919.338	74929.362	74939.387
8310	74999.536	75009.561	75019.586	75029.612	75039.638
8320	75099.794	75109.820	75119.847	75129.874	75139.901
8330	75200.064	75210.092	75220.120	75230.147	75240.175
8340	75300.346	75310.375	75320.404	75330.433	75340.462
8350	75400.640	75410.671	75420.701	75430.731	75440.761
8360	75500.947	75510.978	75521.009	75531.041	75541.072
8370	75601.265	75611.297	75621.330	75631.362	75641.395
8380	75701.595	75711.628	75721.662	75731.696	75741.730
8390	75801.937	75811.972	75822.007	75832.042	75842.077
8400	75902.291	75912.327	75922.363	75932.399	75942.436
8410	76002.656	76012.694	76022.731	76032.769	76042.806
8420	76103.034	76113.073	76123.111	76133.150	76143.189
8430	76203.424	76213.463	76223.503	76233.543	76243.583
8440	76303.825	76313.866	76323.907	76333.948	76343.989
8450	76404.239	76414.281	76424.323	76434.365	76444.407
8460	76504.664	76514.707	76524.750	76534.794	76544.837
8470	76605.101	76615.145	76625.189	76635.234	76645.279
8480	76705.549	76715.595	76725.641	76735.686	76745.732
8490	76806.010	76816.057	76826.103	76836.150	76846.197

5	6	7	8	9	f
71947.512	71957.500	71967.488	71977.476	71987.464	8000
72047.397	72057.386	72067.375	72077.364	72087.354	8010
72147.294	72157.284	72167.274	72177.265	72187.256	8020
72247.203	72257.195	72267.186	72277.178	72287.170	8030
72347.125	72357.118	72367.111	72377.104	72387.097	8040
72447.059	72457.053	72467.047	72477.042	72487.036	8050
72547.006	72557.001	72566.997	72576.992	72586.988	8060
72646.965	72656.961	72666.958	72676.955	72686.952	8070
72746.936	72756.934	72766.932	72776.930	72786.928	8080
72846.920	72856.919	72866.918	72876.918	72886.917	8090
72946.916	72956.917	72966.917	72976.918	72986.918	8100
73046.925	73056.926	73066.928	73076.930	73086.932	8110
73146.946	73156.948	73166.951	73176.954	73186.957	8120
73246.979	73256.983	73266.987	73276.991	73286.996	8130
73347.024	73357.030	73367.035	73377.040	73387.046	8140
73447.082	73457.088	73467.095	73477.102	73487.109	8150
73547.152	73557.160	73567.167	73577.175	73587.183	8160
73647.234	73657.243	73667.252	73677.261	73687.271	8170
73747.329	73757.339	73767.349	73777.359	73787.370	8180
73847.435	73857.447	73867.458	73877.470	73887.482	8190
73947.554	73957.567	73967.580	73977.592	73987.605	8200
74047.685	74057.699	74067.713	74077.727	74087.741	8210
74147.829	74157.844	74167.859	74177.874	74187.889	8220
74247.984	74258.000	74268.017	74278.033	74288.050	8230
74348.152	74358.169	74368.187	74378.204	74388.222	8240
74448.331	74458.350	74468.369	74478.388	74488.407	8250
74548.523	74558.543	74568.563	74578.583	74588.603	8260
74648.727	74658.748	74668.769	74678.791	74688.812	8270
74748.943	74758.965	74768.988	74779.010	74789.033	8280
74849.171	74859.195	74869.218	74879.242	74889.266	8290
74949.411	74959.436	74969.461	74979.486	74989.511	8300
75049.663	75059.689	75069.715	75079.741	75089.768	8310
75149.927	75159.955	75169.982	75180.009	75190.037	8320
75250.204	75260.232	75270.260	75280.289	75290.317	8330
75350.492	75360.521	75370.551	75380.581	75390.610	8340
75450.792	75460.823	75470.853	75480.884	75490.915	8350
75551.104	75561.136	75571.168	75581.200	75591.232	8360
75651.428	75661.461	75671.494	75681.528	75691.561	8370
75751.764	75761.799	75771.833	75781.867	75791.902	8380
75852.112	75862.148	75872.183	75882.219	75892.255	8390
75952.472	75962.509	75972.545	75982.582	75992.619	8400
76052.844	76062.882	76072.920	76082.958	76092.996	8410
76153.228	76163.267	76173.306	76183.345	76193.384	8420
76253.623	76263.663	76273.704	76283.744	76293.785	8430
76354.030	76364.072	76374.113	76384.155	76394.197	8440
76454.450	76464.492	76474.535	76484.578	76494.621	8450
76554.881	76564.924	76574.968	76585.012	76595.056	8460
76655.324	76665.368	76675.414	76685.459	76695.504	8470
76755.778	76765.824	76775.871	76785.917	76795.963	8480
76856.245	76866.292	76876.339	76886.387	76896.434	8490

8500–8990

TABLE 2 $f \ln f$ as a function of f

f	0	1	2	3	4
8500	76906.482	76916.530	76926.578	76936.626	76946.674
8510	77006.966	77017.015	77027.065	77037.114	77047.163
8520	77107.462	77117.512	77127.563	77137.613	77147.664
8530	77207.970	77218.021	77228.073	77238.124	77248.176
8540	77308.489	77318.542	77328.594	77338.647	77348.700
8550	77409.020	77419.074	77429.128	77439.182	77449.236
8560	77509.563	77519.618	77529.673	77539.728	77549.783
8570	77610.117	77620.173	77630.229	77640.286	77650.342
8580	77710.683	77720.741	77730.798	77740.855	77750.913
8590	77811.261	77821.319	77831.378	77841.437	77851.495
8600	77911.850	77921.910	77931.970	77942.029	77952.089
8610	78012.451	78022.512	78032.573	78042.634	78052.695
8620	78113.064	78123.126	78133.188	78143.250	78153.312
8630	78213.688	78223.751	78233.814	78243.878	78253.941
8640	78314.324	78324.388	78334.452	78344.517	78354.581
8650	78414.971	78425.037	78435.102	78445.168	78455.233
8660	78515.630	78525.697	78535.763	78545.830	78555.897
8670	78616.301	78626.368	78636.436	78646.504	78656.572
8680	78716.983	78727.052	78737.120	78747.190	78757.259
8690	78817.676	78827.746	78837.816	78847.887	78857.957
8700	78918.381	78928.452	78938.524	78948.595	78958.666
8710	79019.098	79029.170	79039.242	79049.315	79059.388
8720	79119.826	79129.899	79139.973	79150.046	79160.120
8730	79220.565	79230.640	79240.715	79250.789	79260.864
8740	79321.316	79331.392	79341.468	79351.544	79361.620
8750	79422.079	79432.155	79442.232	79452.310	79462.387
8760	79522.852	79532.930	79543.009	79553.087	79563.165
8770	79623.638	79633.717	79643.796	79653.875	79663.955
8780	79724.434	79734.514	79744.595	79754.675	79764.756
8790	79825.242	79835.324	79845.405	79855.487	79865.569
8800	79926.062	79936.144	79946.227	79956.310	79966.393
8810	80026.892	80036.976	80047.060	80057.144	80067.228
8820	80127.734	80137.819	80147.904	80157.989	80168.074
8830	80228.588	80238.674	80248.760	80258.846	80268.932
8840	80329.453	80339.540	80349.627	80359.714	80369.802
8850	80430.329	80440.417	80450.505	80460.594	80470.682
8860	80531.216	80541.305	80551.395	80561.485	80571.574
8870	80632.115	80642.205	80652.296	80662.387	80672.477
8880	80733.025	80743.116	80753.208	80763.300	80773.392
8890	80833.946	80844.039	80854.131	80864.224	80874.318
8900	80934.878	80944.972	80955.066	80965.160	80975.254
8910	81035.822	81045.917	81056.012	81066.107	81076.203
8920	81136.777	81146.873	81156.969	81167.066	81177.162
8930	81237.743	81247.840	81257.938	81268.035	81278.133
8940	81338.720	81348.819	81358.917	81369.016	81379.114
8950	81439.709	81449.808	81459.908	81470.008	81480.107
8960	81540.709	81550.809	81560.910	81571.011	81581.112
8970	81641.719	81651.821	81661.923	81672.025	81682.127
8980	81742.741	81752.844	81762.947	81773.050	81783.153
8990	81843.774	81853.878	81863.982	81874.087	81884.191

5	6	7	8	9	f
76956.723	76966.771	76976.820	76986.869	76996.917	8500
77057.213	77067.262	77077.312	77087.362	77097.412	8510
77157.715	77167.765	77177.816	77187.867	77197.918	8520
77258.228	77268.280	77278.332	77288.384	77298.437	8530
77358.753	77368.806	77378.860	77388.913	77398.966	8540
77459.290	77469.344	77479.399	77489.453	77499.508	8550
77559.839	77569.894	77579.950	77590.005	77600.061	8560
77660.399	77670.455	77680.512	77690.569	77700.626	8570
77760.971	77771.029	77781.086	77791.145	77801.203	8580
77861.554	77871.613	77881.672	77891.732	77901.791	8590
77962.149	77972.210	77982.270	77992.330	78002.391	8600
78062.756	78072.818	78082.879	78092.940	78103.002	8610
78163.375	78173.437	78183.500	78193.562	78203.625	8620
78264.005	78274.068	78284.132	78294.196	78304.260	8630
78364.646	78374.711	78384.776	78394.841	78404.906	8640
78465.299	78475.365	78485.431	78495.498	78505.564	8650
78565.964	78576.031	78586.098	78596.166	78606.233	8660
78666.640	78676.709	78686.777	78696.845	78706.914	8670
78767.328	78777.397	78787.467	78797.537	78807.606	8680
78868.027	78878.098	78888.169	78898.239	78908.310	8690
78968.738	78978.810	78988.882	78998.954	79009.026	8700
79069.460	79079.533	79089.606	79099.679	79109.752	8710
79170.194	79180.268	79190.342	79200.416	79210.491	8720
79270.939	79281.014	79291.090	79301.165	79311.241	8730
79371.696	79381.772	79391.849	79401.925	79412.002	8740
79472.464	79482.541	79492.619	79502.697	79512.774	8750
79573.244	79583.322	79593.401	79603.480	79613.559	8760
79674.034	79684.114	79694.194	79704.274	79714.354	8770
79774.837	79784.918	79794.999	79805.080	79815.161	8780
79875.650	79885.732	79895.815	79905.897	79915.979	8790
79976.476	79986.559	79996.642	80006.725	80016.809	8800
80077.312	80087.396	80097.481	80107.565	80117.650	8810
80178.160	80188.245	80198.331	80208.416	80218.502	8820
80279.019	80289.105	80299.192	80309.279	80319.366	8830
80379.889	80389.977	80400.065	80410.153	80420.241	8840
80480.771	80490.860	80500.949	80511.038	80521.127	8850
80581.664	80591.754	80601.844	80611.934	80622.024	8860
80682.568	80692.659	80702.751	80712.842	80722.933	8870
80783.484	80793.576	80803.668	80813.761	80823.853	8880
80884.411	80894.504	80904.597	80914.691	80924.785	8890
80985.349	80995.443	81005.538	81015.632	81025.727	8900
81086.298	81096.394	81106.489	81116.585	81126.681	8910
81187.259	81197.355	81207.452	81217.549	81227.646	8920
81288.230	81298.328	81308.426	81318.524	81328.622	8930
81389.213	81399.312	81409.411	81419.510	81429.610	8940
81490.207	81500.307	81510.407	81520.508	81530.608	8950
81591.213	81601.314	81611.415	81621.516	81631.618	8960
81692.229	81702.331	81712.434	81722.536	81732.639	8970
81793.257	81803.360	81813.463	81823.567	81833.671	8980
81894.295	81904.400	81914.504	81924.609	81934.714	8990

TABLE 2 $f \ln f$ as a function of f

f	0	1	2	3	4
9000	81944.819	81954.924	81965.029	81975.134	81985.240
9010	82045.874	82055.980	82066.086	82076.193	82086.299
9020	82146.941	82157.048	82167.155	82177.263	82187.370
9030	82248.018	82258.126	82268.235	82278.343	82288.452
9040	82349.107	82359.216	82369.326	82379.435	82389.545
9050	82450.206	82460.317	82470.428	82480.538	82490.649
9060	82551.317	82561.429	82571.541	82581.652	82591.764
9070	82652.439	82662.552	82672.664	82682.777	82692.891
9080	82753.572	82763.685	82773.799	82783.914	82794.028
9090	82854.715	82864.830	82874.945	82885.061	82895.176
9100	82955.870	82965.986	82976.102	82986.219	82996.335
9110	83057.036	83067.153	83077.270	83087.388	83097.505
9120	83158.213	83168.331	83178.449	83188.568	83198.687
9130	83259.400	83269.520	83279.639	83289.759	83299.879
9140	83360.599	83370.720	83380.840	83390.961	83401.082
9150	83461.809	83471.930	83482.052	83492.174	83502.296
9160	83563.029	83573.152	83583.275	83593.398	83603.521
9170	83664.261	83674.385	83684.508	83694.632	83704.756
9180	83765.503	83775.628	83785.753	83795.878	83806.003
9190	83866.756	83876.882	83887.008	83897.135	83907.261
9200	83968.021	83978.148	83988.275	83998.402	84008.529
9210	84069.296	84079.424	84089.552	84099.680	84109.809
9220	84170.582	84180.711	84190.840	84200.969	84211.099
9230	84271.878	84282.009	84292.139	84302.269	84312.400
9240	84373.186	84383.317	84393.449	84403.580	84413.712
9250	84474.504	84484.637	84494.769	84504.902	84515.035
9260	84575.833	84585.967	84596.101	84606.234	84616.368
9270	84677.173	84687.308	84697.443	84707.577	84717.712
9280	84778.524	84788.660	84798.796	84808.931	84819.067
9290	84879.886	84890.022	84900.159	84910.296	84920.433
9300	84981.258	84991.396	85001.534	85011.672	85021.810
9310	85082.641	85092.780	85102.919	85113.058	85123.197
9320	85184.035	85194.175	85204.315	85214.455	85224.595
9330	85285.439	85295.580	85305.722	85315.863	85326.004
9340	85386.855	85396.997	85407.139	85417.281	85427.424
9350	85488.281	85498.424	85508.567	85518.711	85528.854
9360	85589.717	85599.862	85610.006	85620.150	85630.295
9370	85691.165	85701.310	85711.455	85721.601	85731.747
9380	85792.623	85802.769	85812.916	85823.062	85833.209
9390	85894.091	85904.239	85914.386	85924.534	85934.682
9400	85995.571	86005.719	86015.868	86026.017	86036.165
9410	86097.061	86107.210	86117.360	86127.510	86137.660
9420	86198.561	86208.712	86218.863	86229.014	86239.164
9430	86300.072	86310.224	86320.376	86330.528	86340.680
9440	86401.594	86411.747	86421.900	86432.053	86442.206
9450	86503.127	86513.281	86523.434	86533.588	86543.743
9460	86604.670	86614.825	86624.980	86635.135	86645.290
9470	86706.223	86716.379	86726.535	86736.691	86746.848
9480	86807.787	86817.944	86828.101	86838.259	86848.416
9490	86909.362	86919.520	86929.678	86939.836	86949.995

5	6	7	8	9	f
81995.345	82005.451	82015.556	82025.662	82035.768	9000
82096.406	82106.513	82116.619	82126.726	82136.833	9010
82197.478	82207.586	82217.694	82227.802	82237.910	9020
82298.561	82308.670	82318.779	82328.888	82338.997	9030
82399.655	82409.765	82419.875	82429.986	82440.096	9040
82500.760	82510.871	82520.983	82531.094	82541.205	9050
82601.877	82611.989	82622.101	82632.214	82642.326	9060
82703.004	82713.117	82723.231	82733.344	82743.458	9070
82804.142	82814.257	82824.371	82834.486	82844.601	9080
82905.291	82915.407	82925.523	82935.638	82945.754	9090
83006.452	83016.568	83026.685	83036.802	83046.919	9100
83107.623	83117.741	83127.859	83137.977	83148.095	9110
83208.805	83218.924	83229.043	83239.162	83249.281	9120
83309.998	83320.118	83330.238	83340.359	83350.479	9130
83411.203	83421.324	83431.445	83441.566	83451.687	9140
83512.418	83522.540	83532.662	83542.784	83552.907	9150
83613.644	83623.767	83633.890	83644.014	83654.137	9160
83714.881	83725.005	83735.129	83745.254	83755.378	9170
83816.128	83826.254	83836.379	83846.505	83856.631	9180
83917.387	83927.514	83937.640	83947.767	83957.894	9190
84018.657	84028.784	84038.912	84049.040	84059.168	9200
84119.937	84130.066	84140.195	84150.323	84160.452	9210
84221.229	84231.358	84241.488	84251.618	84261.748	9220
84322.531	84332.661	84342.792	84352.923	84363.055	9230
84423.844	84433.976	84444.108	84454.240	84464.372	9240
84525.167	84535.300	84545.433	84555.567	84565.700	9250
84626.502	84636.636	84646.770	84656.905	84667.039	9260
84727.847	84737.983	84748.118	84758.253	84768.389	9270
84829.204	84839.340	84849.476	84859.613	84869.749	9280
84930.571	84940.708	84950.845	84960.983	84971.120	9290
85031.948	85042.087	85052.225	85062.364	85072.502	9300
85133.337	85143.476	85153.616	85163.755	85173.895	9310
85234.736	85244.876	85255.017	85265.158	85275.299	9320
85336.146	85346.287	85356.429	85366.571	85376.713	9330
85437.566	85447.709	85457.852	85467.995	85478.138	9340
85538.998	85549.141	85559.285	85569.429	85579.573	9350
85640.440	85650.584	85660.729	85670.874	85681.019	9360
85741.892	85752.038	85762.184	85772.330	85782.476	9370
85843.356	85853.503	85863.650	85873.797	85883.944	9380
85944.830	85954.978	85965.126	85975.274	85985.422	9390
86046.314	86056.463	86066.613	86076.762	86086.911	9400
86147.810	86157.960	86168.110	86178.260	86188.411	9410
86249.316	86259.467	86269.618	86279.769	86289.921	9420
86350.832	86360.984	86371.137	86381.289	86391.442	9430
86452.359	86462.512	86472.666	86482.819	86492.973	9440
86553.897	86564.051	86574.206	86584.360	86594.515	9450
86655.445	86665.601	86675.756	86685.912	86696.067	9460
86757.004	86767.160	86777.317	86787.474	86797.630	9470
86858.573	86868.731	86878.889	86889.046	86899.204	9480
86960.153	86970.312	86980.471	86990.629	87000.788	9490

TABLE 2 $f \ln f$ as a function of f

f	0	1	2	3	4
9500	87010.947	87021.106	87031.266	87041.425	87051.584
9510	87112.543	87122.703	87132.863	87143.024	87153.184
9520	87214.149	87224.310	87234.472	87244.633	87254.795
9530	87315.766	87325.928	87336.091	87346.253	87356.416
9540	87417.393	87427.557	87437.720	87447.883	87458.047
9550	87519.031	87529.195	87539.360	87549.524	87559.689
9560	87620.679	87630.845	87641.010	87651.176	87661.341
9570	87722.338	87732.504	87742.671	87752.837	87763.004
9580	87824.007	87834.174	87844.342	87854.510	87864.677
9590	87925.686	87935.855	87946.024	87956.192	87966.361
9600	88027.376	88037.546	88047.716	88057.885	88068.055
9610	88129.077	88139.247	88149.418	88159.589	88169.760
9620	88230.788	88240.959	88251.131	88261.303	88271.475
9630	88332.509	88342.681	88352.854	88363.027	88373.200
9640	88434.240	88444.414	88454.588	88464.762	88474.936
9650	88535.982	88546.157	88556.332	88566.507	88576.682
9660	88637.735	88647.910	88658.086	88668.262	88678.438
9670	88739.497	88749.674	88759.851	88770.028	88780.205
9680	88841.270	88851.448	88861.626	88871.804	88881.982
9690	88943.054	88953.233	88963.412	88973.591	88983.770
9700	89044.847	89055.027	89065.207	89075.387	89085.568
9710	89146.651	89156.832	89167.013	89177.194	89187.376
9720	89248.466	89258.648	89268.830	89279.012	89289.194
9730	89350.290	89360.473	89370.656	89380.839	89391.023
9740	89452.125	89462.309	89472.493	89482.677	89492.862
9750	89553.970	89564.155	89574.340	89584.526	89594.711
9760	89655.825	89666.011	89676.198	89686.384	89696.570
9770	89757.691	89767.878	89778.065	89788.253	89798.440
9780	89859.567	89869.755	89879.943	89890.132	89900.320
9790	89961.453	89971.642	89981.831	89992.021	90002.210
9800	90063.349	90073.539	90083.730	90093.920	90104.110
9810	90165.256	90175.447	90185.638	90195.830	90206.021
9820	90267.172	90277.364	90287.557	90297.749	90307.942
9830	90369.099	90379.292	90389.486	90399.679	90409.873
9840	90471.036	90481.230	90491.425	90501.619	90511.814
9850	90572.983	90583.179	90593.374	90603.569	90613.765
9860	90674.941	90685.137	90695.333	90705.530	90715.726
9870	90776.908	90787.105	90797.303	90807.500	90817.698
9880	90878.886	90889.084	90899.283	90909.481	90919.680
9890	90980.874	90991.073	91001.272	91011.472	91021.671
9900	91082.871	91093.072	91103.272	91113.473	91123.673
9910	91184.879	91195.081	91205.282	91215.484	91225.685
9920	91286.897	91297.100	91307.302	91317.505	91327.707
9930	91388.925	91399.129	91409.332	91419.536	91429.740
9940	91490.964	91501.168	91511.373	91521.577	91531.782
9950	91593.012	91603.217	91613.423	91623.628	91633.834
9960	91695.070	91705.277	91715.483	91725.690	91735.896
9970	91797.139	91807.346	91817.553	91827.761	91837.969
9980	91899.217	91909.425	91919.634	91929.842	91940.051
9990	92001.305	92011.515	92021.724	92031.934	92042.143

5	6	7	8	9	f
87061.744	87071.903	87082.063	87092.223	87102.383	9500
87163.345	87173.505	87183.666	87193.827	87203.988	9510
87264.956	87275.118	87285.280	87295.442	87305.604	9520
87366.578	87376.741	87386.904	87397.067	87407.230	9530
87468.211	87478.375	87488.539	87498.703	87508.867	9540
87569.854	87580.019	87590.184	87600.349	87610.514	9550
87671.507	87681.673	87691.839	87702.005	87712.171	9560
87773.171	87783.338	87793.505	87803.672	87813.840	9570
87874.845	87885.013	87895.181	87905.350	87915.518	9580
87976.530	87986.699	87996.868	88007.038	88017.207	9590
88078.225	88088.395	88098.566	88108.736	88118.906	9600
88179.931	88190.102	88200.273	88210.445	88220.616	9610
88281.647	88291.819	88301.991	88312.164	88322.336	9620
88383.373	88393.547	88403.720	88413.893	88424.067	9630
88485.110	88495.284	88505.459	88515.633	88525.808	9640
88586.857	88597.032	88607.208	88617.383	88627.559	9650
88688.615	88698.791	88708.967	88719.144	88729.321	9660
88790.383	88800.560	88810.737	88820.915	88831.093	9670
88892.161	88902.339	88912.518	88922.696	88932.875	9680
88993.949	89004.129	89014.308	89024.488	89034.667	9690
89095.748	89105.928	89116.109	89126.290	89136.470	9700
89197.557	89207.739	89217.920	89228.102	89238.284	9710
89299.377	89309.559	89319.742	89329.924	89340.107	9720
89401.206	89411.390	89421.573	89431.757	89441.941	9730
89503.046	89513.231	89523.415	89533.600	89543.785	9740
89604.896	89615.082	89625.268	89635.453	89645.639	9750
89706.757	89716.943	89727.130	89737.317	89747.504	9760
89808.628	89818.815	89829.003	89839.191	89849.379	9770
89910.509	89920.697	89930.886	89941.075	89951.264	9780
90012.400	90022.589	90032.779	90042.969	90053.159	9790
90114.301	90124.492	90134.683	90144.873	90155.064	9800
90216.213	90226.404	90236.596	90246.788	90256.980	9810
90318.134	90328.327	90338.520	90348.713	90358.906	9820
90420.066	90430.260	90440.454	90450.648	90460.842	9830
90522.008	90532.203	90542.398	90552.593	90562.788	9840
90623.961	90634.157	90644.352	90654.548	90664.744	9850
90725.923	90736.120	90746.317	90756.514	90766.711	9860
90827.896	90838.094	90848.291	90858.489	90868.688	9870
90929.878	90940.077	90950.276	90960.475	90970.674	9880
91031.871	91042.071	91052.271	91062.471	91072.671	9890
91133.874	91144.075	91154.276	91164.477	91174.678	9900
91235.887	91246.089	91256.291	91266.493	91276.695	9910
91337.910	91348.113	91358.316	91368.519	91378.722	9920
91439.943	91450.147	91460.351	91470.555	91480.759	9930
91541.987	91552.191	91562.396	91572.601	91582.807	9940
91644.040	91654.246	91664.452	91674.658	91684.864	9950
91746.103	91756.310	91766.517	91776.724	91786.931	9960
91848.176	91858.384	91868.592	91878.800	91889.009	9970
91950.260	91960.469	91970.678	91980.887	91991.096	9980
92052.353	92062.563	92072.773	92082.983	92093.193	9990

TABLE **3** $(f + \frac{1}{2}) \ln (f + \frac{1}{2})$ as a function of f

Table of $(f + \frac{1}{2}) \ln (f + \frac{1}{2})$ for integers f ranging from 0 to 499. Three-decimal-place accuracy is furnished.

This table is similar to Table **2** and is looked up in a similar manner. As an example, find the value of 284.5 ln 284.5. The entry for argument 284 is 1607.634, which is the desired function.

This table is used when applying the correction for continuity to the G-test for goodness of fit (Section 17.2) or Yates' correction to a 2×2 test of independence (Section 17.4).

Entries for this table were computed from the FORTRAN IV library function for natural logarithms, using double precision arithmetic.

TABLE 3 $(f + \frac{1}{2}) \ln (f + \frac{1}{2})$ as a function of f

f	0	1	2	3	4
0	-0.347	0.608	2.291	4.385	6.768
10	24.689	28.087	31.572	35.136	38.775
20	61.919	65.963	70.054	74.190	78.367
30	104.241	108.675	113.140	117.637	122.163
40	149.903	154.616	159.354	164.115	168.899
50	198.060	202.991	207.943	212.913	217.902
60	248.210	253.321	258.448	263.591	268.750
70	300.021	305.283	310.560	315.850	321.155
80	353.255	358.649	364.056	369.475	374.906
90	407.734	413.245	418.767	424.299	429.843
100	463.321	468.936	474.561	480.196	485.840
110	519.904	525.614	531.332	537.060	542.796
120	577.394	583.190	588.994	594.806	600.626
130	635.714	641.589	647.472	653.363	659.260
140	694.802	700.750	706.706	712.669	718.639
150	754.601	760.619	766.643	772.673	778.710
160	815.066	821.148	827.235	833.329	839.429
170	876.154	882.296	888.444	894.597	900.756
180	937.829	944.028	950.232	956.441	962.656
190	1000.059	1006.311	1012.569	1018.831	1025.099
200	1062.813	1069.117	1075.425	1081.738	1088.056
210	1126.067	1132.419	1138.775	1145.136	1151.502
220	1189.795	1196.194	1202.596	1209.003	1215.415
230	1253.978	1260.420	1266.867	1273.318	1279.773
240	1318.594	1325.079	1331.568	1338.061	1344.558
250	1383.626	1390.152	1396.681	1403.215	1409.752
260	1449.058	1455.623	1462.191	1468.763	1475.339
270	1514.874	1521.476	1528.082	1534.691	1541.304
280	1581.059	1587.697	1594.339	1600.985	1607.634
290	1647.601	1654.274	1660.951	1667.631	1674.315
300	1714.487	1721.194	1727.905	1734.618	1741.335
310	1781.706	1788.446	1795.189	1801.935	1808.684
320	1849.247	1856.019	1862.793	1869.571	1876.352
330	1917.101	1923.903	1930.708	1937.516	1944.327
340	1985.256	1992.088	1998.923	2005.761	2012.601
350	2053.706	2060.567	2067.430	2074.297	2081.166
360	2122.441	2129.330	2136.221	2143.116	2150.013
370	2191.453	2198.369	2205.288	2212.210	2219.134
380	2260.735	2267.678	2274.624	2281.572	2288.522
390	2330.281	2337.249	2344.221	2351.194	2358.171
400	2400.082	2407.076	2414.072	2421.071	2428.073
410	2470.133	2477.151	2484.172	2491.196	2498.222
420	2540.427	2547.470	2554.515	2561.562	2568.612
430	2610.960	2618.026	2625.094	2632.165	2639.238
440	2681.725	2688.814	2695.905	2702.998	2710.094
450	2752.716	2759.828	2766.941	2774.057	2781.175
460	2823.930	2831.063	2838.199	2845.337	2852.477
470	2895.361	2902.516	2909.673	2916.832	2923.993
480	2967.004	2974.180	2981.358	2988.538	2995.720
490	3038.856	3046.053	3053.251	3060.452	3067.654

5	6	7	8	9	*f*
9.376	12.167	15.112	18.191	21.387	0
42.483	46.255	50.089	53.979	57.923	10
82.586	86.844	91.140	95.472	99.840	20
126.718	131.302	135.913	140.550	145.214	30
173.706	178.535	183.385	188.256	193.148	40
222.909	227.935	232.978	238.038	243.116	50
273.924	279.114	284.319	289.538	294.772	60
326.472	331.803	337.147	342.503	347.873	70
380.348	385.802	391.268	396.746	402.234	80
435.397	440.961	446.536	452.121	457.716	90
491.494	497.157	502.830	508.512	514.204	100
548.541	554.294	560.057	565.827	571.606	110
606.454	612.291	618.135	623.987	629.847	120
665.166	671.078	676.998	682.926	688.860	130
724.616	730.599	736.590	742.587	748.591	140
784.753	790.803	796.860	802.922	808.991	150
845.535	851.647	857.765	863.889	870.019	160
906.921	913.091	919.267	925.449	931.636	170
968.877	975.102	981.334	987.570	993.812	180
1031.372	1037.650	1043.933	1050.222	1056.515	190
1094.379	1100.707	1107.040	1113.377	1119.720	200
1157.873	1164.248	1170.628	1177.013	1183.402	210
1221.831	1228.252	1234.677	1241.106	1247.540	220
1286.233	1292.697	1299.165	1305.637	1312.114	230
1351.059	1357.565	1364.074	1370.588	1377.105	240
1416.293	1422.839	1429.388	1435.941	1442.497	250
1481.919	1488.502	1495.089	1501.680	1508.275	260
1547.921	1554.541	1561.165	1567.793	1574.424	270
1614.286	1620.942	1627.602	1634.265	1640.931	280
1681.002	1687.692	1694.386	1701.083	1707.783	290
1748.056	1754.779	1761.506	1768.236	1774.970	300
1815.437	1822.193	1828.952	1835.714	1842.479	310
1883.135	1889.922	1896.712	1903.505	1910.301	320
1951.141	1957.958	1964.778	1971.601	1978.427	330
2019.445	2026.291	2033.141	2039.993	2046.848	340
2088.038	2094.913	2101.791	2108.671	2115.555	350
2156.913	2163.815	2170.721	2177.629	2184.540	360
2226.061	2232.991	2239.923	2246.858	2253.795	370
2295.476	2302.431	2309.390	2316.351	2323.314	380
2365.150	2372.131	2379.115	2386.101	2393.090	390
2435.077	2442.083	2449.092	2456.103	2463.117	400
2505.250	2512.281	2519.314	2526.349	2533.387	410
2575.664	2582.719	2589.775	2596.835	2603.896	420
2646.313	2653.391	2631.395	2738.500	2745.607	440
2717.192	2724.293	2731.395	2738.500	2745.607	440
2788.296	2795.418	2802.543	2809.670	2816.799	450
2859.619	2866.763	2873.909	2881.058	2888.208	460
2931.156	2938.322	2945.489	2952.659	2959.831	470
3002.905	3010.091	3017.279	3024.469	3031.662	480
3074.859	3082.065	3089.274	3096.484	3103.697	490

TABLE 4 Orthogonal polynomials

This table furnishes coefficients of first-, second-, and third-order orthogonal polynomials for sample sizes from $n = 3$ to $n = 12$. The orthogonal polynomials given here are numerical values of the following polynomial expressions $\xi_1' = \lambda_1 x$; $\xi_2' = \lambda_2[x^2 - \frac{1}{12}(n^2 - 1)]$; and $\xi_3' = \lambda_3[x^3 - \frac{1}{20}(3n^2 - 7)x]$, where the x's are deviations $x = X - \bar{X}$ for the integral values of $X = 1, 2, 3, \ldots, n$ and the λ_i's are arbitrary coefficients multiplying the polynomials to make them into single integers. The table gives the appropriate numerical values for the n coefficients of the polynomials ξ_1', ξ_2', and ξ_3' for sample sizes from $n = 3$ to $n = 12$. Below each column of coefficients is the quantity $\Sigma\xi_i'^2$, and underneath it the numerical value of coefficient λ_i.

To look up a series of polynomials in this table, select the columns corresponding to the number of (equally spaced) X-points in the sample. Thus for $n = 8$ equally spaced points, the coefficients of the linear (first-order) orthogonal polynomials will be $-7, -5, -3, -1, +1, +3, +5, +7$ for the eight X-values arrayed by increasing order of magnitude. Similarly, the coefficients of the second-degree polynomials will be $+7, +1, -3, -5, -5, -3, +1, +7$, while the third-order polynomials will be $-7, +5, +7, +3, -3, -7, -5, +7$. The corresponding quantities $\Sigma\xi_i'^2$ are 168, 168, and 264, and the coefficients λ_i are 2, 1, and $\frac{2}{3}$.

The orthogonal polynomials tabled here can be employed for making linear, quadratic, and cubic contrasts among treatment effects in an analysis of variance (Section 14.10), for estimating quadratic and cubic components of regression (see Section 16.6), and for fitting quadratic and cubic regression equations by the method of orthogonal polynomials.

This table was generated by means of Tschebyscheff polynomials for discrete variables for the case of equal weights. This yields a least squares fit for the n points, where n equals the sample size. We obtain the following polynomial expressions for $X = 0, 1, 2, \ldots, n - 1$:

$$F_0(X) = 1 \qquad F_1(X) = 1 - \frac{2X}{n - 1}$$

$$F_{k+1}(X) = \frac{[(2k + 1)(n - 2X - 1)F_k(X)] - [k(n + k)F_{k-1}(X)]}{(k + 1)(n - k - 1)}$$

This leads to a recurrence relationship in which the fractions are simplified in the interest of presenting a simpler table. More extensive tables of orthogonal polynomials furnishing polynomials up to the sixth degree and for sample sizes up to $n = 45$ can be found in table XXIII of R. A. Fisher and F. Yates, *Statistical Tables for Biological, Agricultural and Medical Research*, 5th ed. (Oliver & Boyd, Edinburgh, 1958).

TABLE 4 Orthogonal polynomials

	$n = 3$		$n = 4$			$n = 5$			$n = 6$		
	ξ_1'	ξ_2'	ξ_1'	ξ_2'	ξ_3'	ξ_1'	ξ_2'	ξ_3'	ξ_1'	ξ_2'	ξ_3'
	−1	+1	−3	+1	−1	−2	+2	−1	−5	+5	−5
	0	−2	−1	−1	+3	−1	−1	+2	−3	−1	+7
	+1	+1	+1	−1	−3	0	−2	0	−1	−4	+4
			+3	+1	+1	+1	−1	−2	+1	−4	−4
						+2	+2	+1	+3	−1	−7
									+5	+5	+5
$\Sigma\xi_i'^2$	2	6	20	4	20	10	14	10	70	84	180
λ_i	1	3	2	1	10/3	1	1	5/6	2	3/2	5/3

	$n = 7$			$n = 8$			$n = 9$			$n = 10$		
	ξ_1'	ξ_2'	ξ_3'	ξ_1'	ξ_2'	ξ_3'	ξ_1'	ξ_2'	ξ_3'	ξ_1'	ξ_2'	ξ_3'
	−3	+5	−1	−7	+7	−7	−4	+28	−14	−9	+6	−42
	−2	0	+1	−5	+1	+5	−3	+ 7	+ 7	−7	+2	+14
	−1	−3	+1	−3	−3	+7	−2	− 8	+13	−5	−1	+35
	0	−4	0	−1	−5	+3	−1	−17	+ 9	−3	−3	+31
	+1	−3	−1	+1	−5	−3	0	−20	0	−1	−4	+12
	+2	0	−1	+3	−3	−7	+1	−17	− 9	+1	−4	−12
	+3	+5	+1	+5	+1	−5	+2	− 8	−13	+3	−3	−31
				+7	+7	+7	+3	+ 7	− 7	+5	−1	−35
							+4	+28	+14	+7	+2	−14
										+9	+6	+42
$\Sigma\xi_i'^2$	28	84	6	168	168	264	60	2772	990	330	132	8580
λ_i	1	1	1/6	2	1	2/3	1	3	5/6	2	1/2	5/3

	$n = 11$			$n = 12$		
	ξ_1'	ξ_2'	ξ_3'	ξ_1'	ξ_2'	ξ_3'
	−5	+15	−30	−11	+55	−33
	−4	+ 6	+ 6	− 9	+25	+ 3
	−3	− 1	+22	− 7	+ 1	+21
	−2	− 6	+23	− 5	−17	+25
	−1	− 9	+14	− 3	−29	+19
	0	−10	0	− 1	−35	+ 7
	+1	− 9	−14	+ 1	−35	− 7
	+2	− 6	−23	+ 3	−29	−19
	+3	− 1	−22	+ 5	−17	−25
	+4	+ 6	− 6	+ 7	+ 1	−21
	+5	+15	+30	+ 9	+25	− 3
				+11	+55	+35
$\Sigma\xi_i'^2$	110	858	4290	572	12012	5140
λ_i	1	1	5/6	2	3	2/3

TABLE 5 The angular transformation

This is a table of $\theta = \arcsin \sqrt{p}$, where p is a proportion from 0 to 1. The arcsine is the angle, in degrees, whose sine corresponds to the value given. The argument is given in increments of 0.0001 between $p = 0$ and $p = 0.01$ and again between $p = 0.99$ and $p = 1.0$. All other values of p are given in increments of 0.001.

To find $\arcsin \sqrt{p}$ for the proportion $p = 0.735$, we obtain a function of 59.02°. Rarely will it be necessary to interpolate for a fourth decimal place in the argument. If so, linear interpolation is sufficient. At either tail of the transformation scale, four decimal places are given for the argument. Thus, when $p = 0.9962$, $\arcsin \sqrt{p} = 86.47°$. Often, the proportions are given as percentages; however, the function is the same. Thus, for 52%, look up $p = 0.52$ and record $\arcsin \sqrt{p} = 46.15°$.

This table is employed for transforming percentages and proportions into a variable meeting the assumptions of the analysis of variance (see Section 13.10).

This table was generated by use of the arcsine and square root library functions of the FORTRAN IV compiler by means of the equation $\theta = \sin^{-1}(\sqrt{p})$.

TABLE 5 The angular transformation

p	0	1	2	3	4	5	6	7	8	9	p
.000	0.00	0.57	0.81	0.99	1.15	1.28	1.40	1.52	1.62	1.72	.000
.001	1.81	1.90	1.99	2.07	2.14	2.22	2.29	2.36	2.43	2.50	.001
.002	2.56	2.63	2.69	2.75	2.81	2.87	2.92	2.98	3.03	3.09	.002
.003	3.14	3.19	3.24	3.29	3.34	3.39	3.44	3.49	3.53	3.58	.003
.004	3.63	3.67	3.72	3.76	3.80	3.85	3.89	3.93	3.97	4.01	.004
.005	4.05	4.10	4.14	4.17	4.21	4.25	4.29	4.33	4.37	4.41	.005
.006	4.44	4.48	4.52	4.55	4.59	4.62	4.66	4.70	4.73	4.76	.006
.007	4.80	4.83	4.87	4.90	4.93	4.97	5.00	5.03	5.07	5.10	.007
.008	5.13	5.16	5.20	5.23	5.26	5.29	5.32	5.35	5.38	5.41	.008
.009	5.44	5.47	5.50	5.53	5.56	5.59	5.62	5.65	5.68	5.71	.009
.01	5.74	6.02	6.29	6.55	6.80	7.03	7.27	7.49	7.71	7.92	.01
.02	8.13	8.33	8.53	8.72	8.91	9.10	9.28	9.46	9.63	9.80	.02
.03	9.97	10.14	10.30	10.47	10.63	10.78	10.94	11.09	11.24	11.39	.03
.04	11.54	11.68	11.83	11.97	12.11	12.25	12.38	12.52	12.66	12.79	.04
.05	12.92	13.05	13.18	13.31	13.44	13.56	13.69	13.81	13.94	14.06	.05
.06	14.18	14.30	14.42	14.54	14.65	14.77	14.89	15.00	15.12	15.23	.06
.07	15.34	15.45	15.56	15.68	15.79	15.89	16.00	16.11	16.22	16.32	.07
.08	16.43	16.54	16.64	16.74	16.85	16.95	17.05	17.15	17.26	17.36	.08
.09	17.46	17.56	17.66	17.76	17.85	17.95	18.05	18.15	18.24	18.34	.09
.10	18.43	18.53	18.63	18.72	18.81	18.91	19.00	19.09	19.19	19.28	.10
.11	19.37	19.46	19.55	19.64	19.73	19.82	19.91	20.00	20.09	20.18	.11
.12	20.27	20.36	20.44	20.53	20.62	20.70	20.79	20.88	20.96	21.05	.12
.13	21.13	21.22	21.30	21.39	21.47	21.56	21.64	21.72	21.81	21.89	.13
.14	21.97	22.06	22.14	22.22	22.30	22.38	22.46	22.54	22.63	22.71	.14
.15	22.79	22.87	22.95	23.03	23.11	23.18	23.26	23.34	23.42	23.50	.15
.16	23.58	23.66	23.73	23.81	23.89	23.97	24.04	24.12	24.20	24.27	.16
.17	24.35	24.43	24.50	24.58	24.65	24.73	24.80	24.88	24.95	25.03	.17
.18	25.10	25.18	25.25	25.33	25.40	25.47	25.55	25.62	25.70	25.77	.18
.19	25.84	25.91	25.99	26.06	26.13	26.21	26.28	26.35	26.42	26.49	.19
.20	26.57	26.64	26.71	26.78	26.85	26.92	26.99	27.06	27.13	27.20	.20
.21	27.27	27.35	27.42	27.49	27.56	27.62	27.69	27.76	27.83	27.90	.21
.22	27.97	28.04	28.11	28.18	28.25	28.32	28.39	28.45	28.52	28.59	.22
.23	28.66	28.73	28.79	28.86	28.93	29.00	29.06	29.13	29.20	29.27	.23
.24	29.33	29.40	29.47	29.53	29.60	29.67	29.73	29.80	29.87	29.93	.24
.25	30.00	30.07	30.13	30.20	30.26	30.33	30.40	30.46	30.53	30.59	.25
.26	30.66	30.72	30.79	30.85	30.92	30.98	31.05	31.11	31.18	31.24	.26
.27	31.31	31.37	31.44	31.50	31.56	31.63	31.69	31.76	31.82	31.88	.27
.28	31.95	32.01	32.08	32.14	32.20	32.27	32.33	32.39	32.46	32.52	.28
.29	32.58	32.65	32.71	32.77	32.83	32.90	32.96	33.02	33.09	33.15	.29

TABLE 5 The angular transformation

p	0	1	2	3	4	5	6	7	8	9	p
.30	33.21	33.27	33.34	33.40	33.46	33.52	33.58	33.65	33.71	33.77	.30
.31	33.83	33.90	33.96	34.02	34.08	34.14	34.20	34.27	34.33	34.39	.31
.32	34.45	34.51	34.57	34.63	34.70	34.76	34.82	34.88	34.94	35.00	.32
.33	35.06	35.12	35.18	35.24	35.30	35.37	35.43	35.49	35.55	35.61	.33
.34	35.67	35.73	35.79	35.85	35.91	35.97	36.03	36.09	36.15	36.21	.34
.35	36.27	36.33	36.39	36.45	36.51	36.57	36.63	36.69	36.75	36.81	.35
.36	36.87	36.93	36.99	37.05	37.11	37.17	37.23	37.29	37.35	37.41	.36
.37	37.46	37.52	37.58	37.64	37.70	37.76	37.82	37.88	37.94	38.00	.37
.38	38.06	38.12	38.17	38.23	38.29	38.35	38.41	38.47	38.53	38.59	.38
.39	38.65	38.70	38.76	38.82	38.88	38.94	39.00	39.06	39.11	39.17	.39
.40	39.23	39.29	39.35	39.41	39.47	39.52	39.58	39.64	39.70	39.76	.40
.41	39.82	39.87	39.93	39.99	40.05	40.11	40.16	40.22	40.28	40.34	.41
.42	40.40	40.45	40.51	40.57	40.63	40.69	40.74	40.80	40.86	40.92	.42
.43	40.98	41.03	41.09	41.15	41.21	41.27	41.32	41.38	41.44	41.50	.43
.44	41.55	41.61	41.67	41.73	41.78	41.84	41.90	41.96	42.02	42.07	.44
.45	42.13	42.19	42.25	42.30	42.36	42.42	42.48	42.53	42.59	42.65	.45
.46	42.71	42.76	42.82	42.88	42.94	42.99	43.05	43.11	43.17	43.22	.46
.47	43.28	43.34	43.39	43.45	43.51	43.57	43.62	43.68	43.74	43.80	.47
.48	43.85	43.91	43.97	44.03	44.08	44.14	44.20	44.26	44.31	44.37	.48
.49	44.43	44.48	44.54	44.60	44.66	44.71	44.77	44.83	44.89	44.94	.49
.50	45.00	45.06	45.11	45.17	45.23	45.29	45.34	45.40	45.46	45.52	.50
.51	45.57	45.63	45.69	45.74	45.80	45.86	45.92	45.97	46.03	46.09	.51
.52	46.15	46.20	46.26	46.32	46.38	46.43	46.49	46.55	46.61	46.66	.52
.53	46.72	46.78	46.83	46.89	46.95	47.01	47.06	47.12	47.18	47.24	.53
.54	47.29	47.35	47.41	47.47	47.52	47.58	47.64	47.70	47.75	47.81	.54
.55	47.87	47.93	47.98	48.04	48.10	48.16	48.22	48.27	48.33	48.39	.55
.56	48.45	48.50	48.56	48.62	48.68	48.73	48.79	48.85	48.91	48.97	.56
.57	49.02	49.08	49.14	49.20	49.26	49.31	49.37	49.43	49.49	49.55	.57
.58	49.60	49.66	49.72	49.78	49.84	49.89	49.95	50.01	50.07	50.13	.58
.59	50.18	50.24	50.30	50.36	50.42	50.48	50.53	50.59	50.65	50.71	.59
.60	50.77	50.83	50.89	50.94	51.00	51.06	51.12	51.18	51.24	51.30	.60
.61	51.35	51.41	51.47	51.53	51.59	51.65	51.71	51.77	51.83	51.88	.61
.62	51.94	52.00	52.06	52.12	52.18	52.24	52.30	52.36	52.42	52.48	.62
.63	52.54	52.59	52.65	52.71	52.77	52.83	52.89	52.95	53.01	53.07	.63
.64	53.13	53.19	53.25	53.31	53.37	53.43	53.49	53.55	53.61	53.67	.64
.65	53.73	53.79	53.85	53.91	53.97	54.03	54.09	54.15	54.21	54.27	.65
.66	54.33	54.39	54.45	54.51	54.57	54.63	54.70	54.76	54.82	54.88	.66
.67	54.94	55.00	55.06	55.12	55.18	55.24	55.30	55.37	55.43	55.49	.67
.68	55.55	55.61	55.67	55.73	55.80	55.86	55.92	55.98	56.04	56.10	.68
.69	56.17	56.23	56.29	56.35	56.42	56.48	56.54	56.60	56.66	56.73	.69

TABLE 5 The angular transformation

p	0	1	2	3	4	5	6	7	8	9	*p*
.70	56.79	56.85	56.91	56.98	57.04	57.10	57.17	57.23	57.29	57.35	.70
.71	57.42	57.48	57.54	57.61	57.67	57.73	57.80	57.86	57.92	57.99	.71
.72	58.05	58.12	58.18	58.24	58.31	58.37	58.44	58.50	58.56	58.63	.72
.73	58.69	58.76	58.82	58.89	58.95	59.02	59.08	59.15	59.21	59.28	.73
.74	59.34	59.41	59.47	59.54	59.60	59.67	59.74	59.80	59.87	59.93	.74
.75	60.00	60.07	60.13	60.20	60.27	60.33	60.40	60.47	60.53	60.60	.75
.76	60.67	60.73	60.80	60.87	60.94	61.00	61.07	61.14	61.21	61.27	.76
.77	61.34	61.41	61.48	61.55	61.61	61.68	61.75	61.82	61.89	61.96	.77
.78	62.03	62.10	62.17	62.24	62.31	62.38	62.44	62.51	62.58	62.65	.78
.79	62.73	62.80	62.87	62.94	63.01	63.08	63.15	63.22	63.29	63.36	.79
.80	63.43	63.51	63.58	63.65	63.72	63.79	63.87	63.94	64.01	64.09	.80
.81	64.16	64.23	64.30	64.38	64.45	64.53	64.60	64.67	64.75	64.82	.81
.82	64.90	64.97	65.05	65.12	65.20	65.27	65.35	65.42	65.50	65.57	.82
.83	65.65	65.73	65.80	65.88	65.96	66.03	66.11	66.19	66.27	66.34	.83
.84	66.42	66.50	66.58	66.66	66.74	66.82	66.89	66.97	67.05	67.13	.84
.85	67.21	67.29	67.37	67.46	67.54	67.62	67.70	67.78	67.86	67.94	.85
.86	68.03	68.11	68.19	68.28	68.36	68.44	68.53	68.61	68.70	68.78	.86
.87	68.87	68.95	69.04	69.12	69.21	69.30	69.38	69.47	69.56	69.64	.87
.88	69.73	69.82	69.91	70.00	70.09	70.18	70.27	70.36	70.45	70.54	.88
.89	70.63	70.72	70.81	70.91	71.00	71.09	71.19	71.28	71.37	71.47	.89
.90	71.57	71.66	71.76	71.85	71.95	72.05	72.15	72.24	72.34	72.44	.90
.91	72.54	72.64	72.74	72.85	72.95	73.05	73.15	73.26	73.36	73.46	.91
.92	73.57	73.68	73.78	73.89	74.00	74.11	74.21	74.32	74.44	74.55	.92
.93	74.66	74.77	74.88	75.00	75.11	75.23	75.35	75.46	75.58	75.70	.93
.94	75.82	75.94	76.06	76.19	76.31	76.44	76.56	76.69	76.82	76.95	.94
.95	77.08	77.21	77.34	77.48	77.62	77.75	77.89	78.03	78.17	78.32	.95
.96	78.46	78.61	78.76	78.91	79.06	79.22	79.37	79.53	79.70	79.86	.96
.97	80.03	80.20	80.37	80.54	80.72	80.90	81.09	81.28	81.47	81.67	.97
.98	81.87	82.08	82.29	82.51	82.73	82.97	83.20	83.45	83.71	83.98	.98
.990	84.26	84.29	84.32	84.35	84.38	84.41	84.44	84.47	84.50	84.53	.990
.991	84.56	84.59	84.62	84.65	84.68	84.71	84.74	84.77	84.80	84.84	.991
.992	84.87	84.90	84.93	84.97	85.00	85.03	85.07	85.10	85.13	85.17	.992
.993	85.20	85.24	85.27	85.30	85.34	85.38	85.41	85.45	85.48	85.52	.993
.994	85.56	85.59	85.63	85.67	85.71	85.75	85.79	85.83	85.86	85.90	.994
.995	85.95	85.99	86.03	86.07	86.11	86.15	86.20	86.24	86.28	86.33	.995
.996	86.37	86.42	86.47	86.51	86.56	86.61	86.66	86.71	86.76	86.81	.996
.997	86.86	86.91	86.97	87.02	87.08	87.13	87.19	87.25	87.31	87.37	.997
.998	87.44	87.50	87.57	87.64	87.71	87.78	87.86	87.93	88.01	88.10	.998
.999	88.19	88.28	88.38	88.48	88.60	88.72	88.85	89.01	89.19	89.43	.999
1.000	90.00										

TABLE **6** Proportions corresponding to angle θ

This table is the inverse of Table **5.** It furnishes a proportion corresponding to a given angle θ ranging from 0° to 90° and tabulated in increments of 0.1.

An arcsine value (angle) of 34.5° corresponds to a proportion of 0.3208 (or to 32.08%). When necessary, linear interpolation can be used for a second decimal place in the argument.

After the arcsine transformation of Table **5** has been applied to a set of data (see Section 13.10), statistics such as means are obtained in transformed scale. We look up the corresponding proportions or percentages for these transformed values in Table **6**.

The sine library function of the FORTRAN IV compiler was employed in preparation of this table, following the equation $p = (\sin\theta)^2$.

TABLE 6 Proportions corresponding to angle θ

θ	0	1	2	3	4	5	6	7	8	9	θ
0.0	.0000	.0000	.0000	.0000	.0000	.0001	.0001	.0001	.0002	.0002	0.0
1.0	.0003	.0004	.0004	.0005	.0006	.0007	.0008	.0009	.0010	.0011	1.0
2.0	.0012	.0013	.0015	.0016	.0018	.0019	.0021	.0022	.0024	.0026	2.0
3.0	.0027	.0029	.0031	.0033	.0035	.0037	.0039	.0042	.0044	.0046	3.0
4.0	.0049	.0051	.0054	.0056	.0059	.0062	.0064	.0067	.0070	.0073	4.0
5.0	.0076	.0079	.0082	.0085	.0089	.0092	.0095	.0099	.0102	.0106	5.0
6.0	.0109	.0113	.0117	.0120	.0124	.0128	.0132	.0136	.0140	.0144	6.0
7.0	.0149	.0153	.0157	.0161	.0166	.0170	.0175	.0180	.0184	.0189	7.0
8.0	.0194	.0199	.0203	.0208	.0213	.0218	.0224	.0229	.0234	.0239	8.0
9.0	.0245	.0250	.0256	.0261	.0267	.0272	.0278	.0284	.0290	.0296	9.0
10.0	.0302	.0308	.0314	.0320	.0326	.0332	.0338	.0345	.0351	.0358	10.0
11.0	.0364	.0371	.0377	.0384	.0391	.0397	.0404	.0411	.0418	.0425	11.0
12.0	.0432	.0439	.0447	.0454	.0461	.0468	.0476	.0483	.0491	.0498	12.0
13.0	.0506	.0514	.0521	.0529	.0537	.0545	.0553	.0561	.0569	.0577	13.0
14.0	.0585	.0593	.0602	.0610	.0618	.0627	.0635	.0644	.0653	.0661	14.0
15.0	.0670	.0679	.0687	.0696	.0705	.0714	.0723	.0732	.0741	.0751	15.0
16.0	.0760	.0769	.0778	.0788	.0797	.0807	.0816	.0826	.0835	.0845	16.0
17.0	.0855	.0865	.0874	.0884	.0894	.0904	.0914	.0924	.0934	.0945	17.0
18.0	.0955	.0965	.0976	.0986	.0996	.1007	.1017	.1028	.1039	.1049	18.0
19.0	.1060	.1071	.1082	.1092	.1103	.1114	.1125	.1136	.1147	.1159	19.0
20.0	.1170	.1181	.1192	.1204	.1215	.1226	.1238	.1249	.1261	.1273	20.0
21.0	.1284	.1296	.1308	.1320	.1331	.1343	.1355	.1367	.1379	.1391	21.0
22.0	.1403	.1415	.1428	.1440	.1452	.1464	.1477	.1489	.1502	.1514	22.0
23.0	.1527	.1539	.1552	.1565	.1577	.1590	.1603	.1616	.1628	.1641	23.0
24.0	.1654	.1667	.1680	.1693	.1707	.1720	.1733	.1746	.1759	.1773	24.0
25.0	.1786	.1799	.1813	.1826	.1840	.1853	.1867	.1881	.1894	.1908	25.0
26.0	.1922	.1935	.1949	.1963	.1977	.1991	.2005	.2019	.2033	.2047	26.0
27.0	.2061	.2075	.2089	.2104	.2118	.2132	.2146	.2161	.2175	.2190	27.0
28.0	.2204	.2219	.2233	.2248	.2262	.2277	.2291	.2306	.2321	.2336	28.0
29.0	.2350	.2365	.2380	.2395	.2410	.2425	.2440	.2455	.2470	.2485	29.0
30.0	.2500	.2515	.2530	.2545	.2561	.2576	.2591	.2607	.2622	.2637	30.0
31.0	.2653	.2668	.2684	.2699	.2715	.2730	.2746	.2761	.2777	.2792	31.0
32.0	.2808	.2824	.2840	.2855	.2871	.2887	.2903	.2919	.2934	.2950	32.0
33.0	.2966	.2982	.2998	.3014	.3030	.3046	.3062	.3079	.3095	.3111	33.0
34.0	.3127	.3143	.3159	.3176	.3192	.3208	.3224	.3241	.3257	.3274	34.0
35.0	.3290	.3306	.3323	.3339	.3356	.3372	.3389	.3405	.3422	.3438	35.0
36.0	.3455	.3472	.3488	.3505	.3521	.3538	.3555	.3572	.3588	.3605	36.0
37.0	.3622	.3639	.3655	.3672	.3689	.3706	.3723	.3740	.3757	.3773	37.0
38.0	.3790	.3807	.3824	.3841	.3858	.3875	.3892	.3909	.3926	.3943	38.0
39.0	.3960	.3978	.3995	.4012	.4029	.4046	.4063	.4080	.4097	.4115	39.0
40.0	.4132	.4149	.4166	.4183	.4201	.4218	.4235	.4252	.4270	.4287	40.0
41.0	.4304	.4321	.4339	.4356	.4373	.4391	.4408	.4425	.4443	.4460	41.0
42.0	.4477	.4495	.4512	.4529	.4547	.4564	.4582	.4599	.4616	.4634	42.0
43.0	.4651	.4669	.4686	.4703	.4721	.4738	.4756	.4773	.4791	.4808	43.0
44.0	.4826	.4843	.4860	.4878	.4895	.4913	.4930	.4948	.4965	.4983	44.0

TABLE 6 Proportions corresponding to angle θ

θ	0	1	2	3	4	5	6	7	8	9	θ
45.0	.5000	.5017	.5035	.5052	.5070	.5087	.5105	.5122	.5140	.5157	45.0
46.0	.5174	.5192	.5209	.5227	.5244	.5262	.5279	.5297	.5314	.5331	46.0
47.0	.5349	.5366	.5384	.5401	.5418	.5436	.5453	.5471	.5488	.5505	47.0
48.0	.5523	.5540	.5557	.5575	.5592	.5609	.5627	.5644	.5661	.5679	48.0
49.0	.5696	.5713	.5730	.5748	.5765	.5782	.5799	.5817	.5834	.5851	49.0
50.0	.5868	.5885	.5903	.5920	.5937	.5954	.5971	.5988	.6005	.6022	50.0
51.0	.6040	.6057	.6074	.6091	.6108	.6125	.6142	.6159	.6176	.6193	51.0
52.0	.6210	.6227	.6243	.6260	.6277	.6294	.6311	.6328	.6345	.6361	52.0
53.0	.6378	.6395	.6412	.6428	.6445	.6462	.6479	.6495	.6512	.6528	53.0
54.0	.6545	.6562	.6578	.6595	.6611	.6628	.6644	.6661	.6677	.6694	54.0
55.0	.6710	.6726	.6743	.6759	.6776	.6792	.6808	.6824	.6841	.6857	55.0
56.0	.6873	.6889	.6905	.6921	.6938	.6954	.6970	.6986	.7002	.7018	56.0
57.0	.7034	.7050	.7066	.7081	.7097	.7113	.7129	.7145	.7160	.7176	57.0
58.0	.7192	.7208	.7223	.7239	.7254	.7270	.7285	.7301	.7316	.7332	58.0
59.0	.7347	.7363	.7378	.7393	.7409	.7424	.7439	.7455	.7470	.7485	59.0
60.0	.7500	.7515	.7530	.7545	.7560	.7575	.7590	.7605	.7620	.7635	60.0
61.0	.7650	.7664	.7679	.7694	.7709	.7723	.7738	.7752	.7767	.7781	61.0
62.0	.7796	.7810	.7825	.7839	.7854	.7868	.7882	.7896	.7911	.7925	62.0
63.0	.7939	.7953	.7967	.7981	.7995	.8009	.8023	.8037	.8051	.8065	63.0
64.0	.8078	.8092	.8106	.8119	.8133	.8147	.8160	.8174	.8187	.8201	64.0
65.0	.8214	.8227	.8241	.8254	.8267	.8280	.8293	.8307	.8320	.8333	65.0
66.0	.8346	.8359	.8372	.8384	.8397	.8410	.8423	.8435	.8448	.8461	66.0
67.0	.8473	.8486	.8498	.8511	.8523	.8536	.8548	.8560	.8572	.8585	67.0
68.0	.8597	.8609	.8621	.8633	.8645	.8657	.8669	.8680	.8692	.8704	68.0
69.0	.8716	.8727	.8739	.8751	.8762	.8774	.8785	.8796	.8808	.8819	69.0
70.0	.8830	.8841	.8853	.8864	.8875	.8886	.8897	.8908	.8918	.8929	70.0
71.0	.8940	.8951	.8961	.8972	.8983	.8993	.9004	.9014	.9024	.9035	71.0
72.0	.9045	.9055	.9066	.9076	.9086	.9096	.9106	.9116	.9126	.9135	72.0
73.0	.9145	.9155	.9165	.9174	.9184	.9193	.9203	.9212	.9222	.9231	73.0
74.0	.9240	.9249	.9259	.9268	.9277	.9286	.9295	.9304	.9313	.9321	74.0
75.0	.9330	.9339	.9347	.9356	.9365	.9373	.9382	.9390	.9398	.9407	75.0
76.0	.9415	.9423	.9431	.9439	.9447	.9455	.9463	.9471	.9479	.9486	76.0
77.0	.9494	.9502	.9509	.9517	.9524	.9532	.9539	.9546	.9553	.9561	77.0
78.0	.9568	.9575	.9582	.9589	.9596	.9603	.9609	.9616	.9623	.9629	78.0
79.0	.9636	.9642	.9649	.9655	.9662	.9668	.9674	.9680	.9686	.9692	79.0
80.0	.9698	.9704	.9710	.9716	.9722	.9728	.9733	.9739	.9744	.9750	80.0
81.0	.9755	.9761	.9766	.9771	.9776	.9782	.9787	.9792	.9797	.9801	81.0
82.0	.9806	.9811	.9816	.9820	.9825	.9830	.9834	.9839	.9843	.9847	82.0
83.0	.9851	.9856	.9860	.9864	.9868	.9872	.9876	.9880	.9883	.9887	83.0
84.0	.9891	.9894	.9898	.9901	.9905	.9908	.9911	.9915	.9918	.9921	84.0
85.0	.9924	.9927	.9930	.9933	.9936	.9938	.9941	.9944	.9946	.9949	85.0
86.0	.9951	.9954	.9956	.9958	.9961	.9963	.9965	.9967	.9969	.9971	86.0
87.0	.9973	.9974	.9976	.9978	.9979	.9981	.9982	.9984	.9985	.9987	87.0
88.0	.9988	.9989	.9990	.9991	.9992	.9993	.9994	.9995	.9996	.9996	88.0
89.0	.9997	.9998	.9998	.9999	.9999	.9999	1.0000	1.0000	1.0000	1.0000	89.0
90.0	1.0000										

TABLE 7 The z-transformation of correlation coefficient r

This transformation is the inverse hyperbolic tangent of the correlation coefficient $z = \frac{1}{2} \ln [(1 + r)/(1 - r)]$. It is tabulated for the range of the correlation coefficient from 0 to 0.999 in increments of 0.001. The function is given to four significant decimal places.

The values of z for positive values of r can be looked up directly in the table. Thus, when $r = 0.772$, $z = 1.0253$. Since the transformation is symmetrical about the origin, to find z for a negative r, one simply looks up the z corresponding to $|r|$ and then makes it negative. For example, when $r = -0.511$, $z = -0.5641$. When $r = 0$, $z = 0$; when $r = \pm 1$, $z = \pm\infty$.

This transformation, due to Fisher, is employed to transform correlation coefficients for tests of significance (see Section 15.5). When the sample size, n, is greater than 25, z's will be approximately normally distributed with a standard deviation of $1/\sqrt{n - 3}$.

In the preparation of this table, the natural logarithm library function of the FORTRAN IV compiler was employed.

TABLE 7 The z-transformation of correlation coefficient r

r	0	1	2	3	4	5	6	7	8	9	r
.00	.0000	.0010	.0020	.0030	.0040	.0050	.0060	.0070	.0080	.0090	.00
.01	.0100	.0110	.0120	.0130	.0140	.0150	.0160	.0170	.0180	.0190	.01
.02	.0200	.0210	.0220	.0230	.0240	.0250	.0260	.0270	.0280	.0290	.02
.03	.0300	.0310	.0320	.0330	.0340	.0350	.0360	.0370	.0380	.0390	.03
.04	.0400	.0410	.0420	.0430	.0440	.0450	.0460	.0470	.0480	.0490	.04
.05	.0500	.0510	.0520	.0530	.0541	.0551	.0561	.0571	.0581	.0591	.05
.06	.0601	.0611	.0621	.0631	.0641	.0651	.0661	.0671	.0681	.0691	.06
.07	.0701	.0711	.0721	.0731	.0741	.0751	.0761	.0772	.0782	.0792	.07
.08	.0802	.0812	.0822	.0832	.0842	.0852	.0862	.0872	.0882	.0892	.08
.09	.0902	.0913	.0923	.0933	.0943	.0953	.0963	.0973	.0983	.0993	.09
.10	.1003	.1013	.1024	.1034	.1044	.1054	.1064	.1074	.1084	.1094	.10
.11	.1104	.1115	.1125	.1135	.1145	.1155	.1165	.1175	.1186	.1196	.11
.12	.1206	.1216	.1226	.1236	.1246	.1257	.1267	.1277	.1287	.1297	.12
.13	.1307	.1318	.1328	.1338	.1348	.1358	.1368	.1379	.1389	.1399	.13
.14	.1409	.1419	.1430	.1440	.1450	.1460	.1471	.1481	.1491	.1501	.14
.15	.1511	.1522	.1532	.1542	.1552	.1563	.1573	.1583	.1593	.1604	.15
.16	.1614	.1624	.1634	.1645	.1655	.1665	.1676	.1686	.1696	.1706	.16
.17	.1717	.1727	.1737	.1748	.1758	.1768	.1779	.1789	.1799	.1809	.17
.18	.1820	.1830	.1841	.1851	.1861	.1872	.1882	.1892	.1903	.1913	.18
.19	.1923	.1934	.1944	.1955	.1965	.1975	.1986	.1996	.2007	.2017	.19
.20	.2027	.2038	.2048	.2059	.2069	.2079	.2090	.2100	.2111	.2121	.20
.21	.2132	.2142	.2153	.2163	.2174	.2184	.2195	.2205	.2216	.2226	.21
.22	.2237	.2247	.2258	.2268	.2279	.2289	.2300	.2310	.2321	.2331	.22
.23	.2342	.2352	.2363	.2374	.2384	.2395	.2405	.2416	.2427	.2437	.23
.24	.2448	.2458	.2469	.2480	.2490	.2501	.2512	.2522	.2533	.2543	.24
.25	.2554	.2565	.2575	.2586	.2597	.2608	.2618	.2629	.2640	.2650	.25
.26	.2661	.2672	.2683	.2693	.2704	.2715	.2726	.2736	.2747	.2758	.26
.27	.2769	.2779	.2790	.2801	.2812	.2823	.2833	.2844	.2855	.2866	.27
.28	.2877	.2888	.2899	.2909	.2920	.2931	.2942	.2953	.2964	.2975	.28
.29	.2986	.2997	.3008	.3018	.3029	.3040	.3051	.3062	.3073	.3084	.29
.30	.3095	.3106	.3117	.3128	.3139	.3150	.3161	.3172	.3183	.3194	.30
.31	.3205	.3217	.3228	.3239	.3250	.3261	.3272	.3283	.3294	.3305	.31
.32	.3316	.3328	.3339	.3350	.3361	.3372	.3383	.3395	.3406	.3417	.32
.33	.3428	.3440	.3451	.3462	.3473	.3484	.3496	.3507	.3518	.3530	.33
.34	.3541	.3552	.3564	.3575	.3586	.3598	.3609	.3620	.3632	.3643	.34
.35	.3654	.3666	.3677	.3689	.3700	.3712	.3723	.3734	.3746	.3757	.35
.36	.3769	.3780	.3792	.3803	.3815	.3826	.3838	.3850	.3861	.3873	.36
.37	.3884	.3896	.3907	.3919	.3931	.3942	.3954	.3966	.3977	.3989	.37
.38	.4001	.4012	.4024	.4036	.4047	.4059	.4071	.4083	.4094	.4106	.38
.39	.4118	.4130	.4142	.4153	.4165	.4177	.4189	.4201	.4213	.4225	.39
.40	.4236	.4248	.4260	.4272	.4284	.4296	.4308	.4320	.4332	.4344	.40
.41	.4356	.4368	.4380	.4392	.4404	.4416	.4428	.4441	.4453	.4465	.41
.42	.4477	.4489	.4501	.4513	.4526	.4538	.4550	.4562	.4574	.4587	.42
.43	.4599	.4611	.4[illegible]4	.4636	.4648	.4660	.4673	.4685	.4698	.4710	.43
.44	.4722	.4735	.4747	.4760	.4772	.4784	.4797	.4809	.4822	.4834	.44
.45	.4847	.4860	.4872	.4885	.4897	.4910	.4922	.4935	.4948	.4960	.45
.46	.4973	.4986	.4999	.5011	.5024	.5037	.5049	.5062	.5075	.5088	.46
.47	.5101	.5114	.5126	.5139	.5152	.5165	.5178	.5191	.5204	.5217	.47
.48	.5230	.5243	.5256	.5269	.5282	.5295	.5308	.5321	.5334	.5347	.48
.49	.5361	.5374	.5387	.5400	.5413	.5427	.5440	.5453	.5466	.5480	.49

TABLE 7 The z-transformation of correlation coefficient r

r	0	1	2	3	4	5	6	7	8	9	r
.50	.5493	.5506	.5520	.5533	.5547	.5560	.5573	.5587	.5600	.5614	.50
.51	.5627	.5641	.5654	.5668	.5682	.5695	.5709	.5722	.5736	.5750	.51
.52	.5763	.5777	.5791	.5805	.5818	.5832	.5846	.5860	.5874	.5888	.52
.53	.5901	.5915	.5929	.5943	.5957	.5971	.5985	.5999	.6013	.6027	.53
.54	.6042	.6056	.6070	.6084	.6098	.6112	.6127	.6141	.6155	.6169	.54
.55	.6184	.6198	.6213	.6227	.6241	.6256	.6270	.6285	.6299	.6314	.55
.56	.6328	.6343	.6358	.6372	.6387	.6401	.6416	.6431	.6446	.6460	.56
.57	.6475	.6490	.6505	.6520	.6535	.6550	.6565	.6580	.6595	.6610	.57
.58	.6625	.6640	.6655	.6670	.6685	.6700	.6716	.6731	.6746	.6761	.58
.59	.6777	.6792	.6807	.6823	.6838	.6854	.6869	.6885	.6900	.6916	.59
.60	.6931	.6947	.6963	.6978	.6994	.7010	.7026	.7042	.7057	.7073	.60
.61	.7089	.7105	.7121	.7137	.7153	.7169	.7185	.7201	.7218	.7234	.61
.62	.7250	.7266	.7283	.7299	.7315	.7332	.7348	.7365	.7381	.7398	.62
.63	.7414	.7431	.7447	.7464	.7481	.7498	.7514	.7531	.7548	.7565	.63
.64	.7582	.7599	.7616	.7633	.7650	.7667	.7684	.7701	.7718	.7736	.64
.65	.7753	.7770	.7788	.7805	.7823	.7840	.7858	.7875	.7893	.7910	.65
.66	.7928	.7946	.7964	.7981	.7999	.8017	.8035	.8053	.8071	.8089	.66
.67	.8107	.8126	.8144	.8162	.8180	.8199	.8217	.8236	.8254	.8273	.67
.68	.8291	.8310	.8328	.8347	.8366	.8385	.8404	.8423	.8441	.8460	.68
.69	.8480	.8499	.8518	.8537	.8556	.8576	.8595	.8614	.8634	.8653	.69
.70	.8673	.8693	.8712	.8732	.8752	.8772	.8792	.8812	.8832	.8852	.70
.71	.8872	.8892	.8912	.8933	.8953	.8973	.8994	.9014	.9035	.9056	.71
.72	.9076	.9097	.9118	.9139	.9160	.9181	.9202	.9223	.9245	.9266	.72
.73	.9287	.9309	.9330	.9352	.9373	.9395	.9417	.9439	.9461	.9483	.73
.74	.9505	.9527	.9549	.9571	.9594	.9616	.9639	.9661	.9684	.9707	.74
.75	.9730	.9752	.9775	.9798	.9822	.9845	.9868	.9892	.9915	.9939	.75
.76	.9962	.9986	1.0010	1.0034	1.0058	1.0082	1.0106	1.0130	1.0154	1.0179	.76
.77	1.0203	1.0228	1.0253	1.0277	1.0302	1.0327	1.0352	1.0378	1.0403	1.0428	.77
.78	1.0454	1.0479	1.0505	1.0531	1.0557	1.0583	1.0609	1.0635	1.0661	1.0688	.78
.79	1.0714	1.0741	1.0768	1.0795	1.0822	1.0849	1.0876	1.0903	1.0931	1.0958	.79
.80	1.0986	1.1014	1.1042	1.1070	1.1098	1.1127	1.1155	1.1184	1.1212	1.1241	.80
.81	1.1270	1.1299	1.1329	1.1358	1.1388	1.1417	1.1447	1.1477	1.1507	1.1538	.81
.82	1.1568	1.1599	1.1630	1.1660	1.1692	1.1723	1.1754	1.1786	1.1817	1.1849	.82
.83	1.1881	1.1914	1.1946	1.1979	1.2011	1.2044	1.2077	1.2111	1.2144	1.2178	.83
.84	1.2212	1.2246	1.2280	1.2315	1.2349	1.2384	1.2419	1.2454	1.2490	1.2526	.84
.85	1.2562	1.2598	1.2634	1.2671	1.2707	1.2745	1.2782	1.2819	1.2857	1.2895	.85
.86	1.2933	1.2972	1.3011	1.3050	1.3089	1.3129	1.3169	1.3209	1.3249	1.3290	.86
.87	1.3331	1.3372	1.3414	1.3456	1.3498	1.3540	1.3583	1.3626	1.3670	1.3714	.87
.88	1.3758	1.3802	1.3847	1.3892	1.3938	1.3984	1.4030	1.4077	1.4124	1.4171	.88
.89	1.4219	1.4268	1.4316	1.4365	1.4415	1.4465	1.4516	1.4567	1.4618	1.4670	.89
.90	1.4722	1.4775	1.4828	1.4882	1.4937	1.4992	1.5047	1.5103	1.5160	1.5217	.90
.91	1.5275	1.5334	1.5393	1.5453	1.5513	1.5574	1.5636	1.5698	1.5762	1.5826	.91
.92	1.5890	1.5956	1.6022	1.6089	1.6157	1.6226	1.6296	1.6366	1.6438	1.6510	.92
.93	1.6584	1.6658	1.6734	1.6811	1.6888	1.6967	1.7047	1.7129	1.7211	1.7295	.93
.94	1.7380	1.7467	1.7555	1.7645	1.7736	1.7828	1.7923	1.8019	1.8117	1.8216	.94
.95	1.8318	1.8421	1.8527	1.8635	1.8745	1.8857	1.8972	1.9090	1.9210	1.9333	.95
.96	1.9459	1.9588	1.9721	1.9857	1.9996	2.0139	2.0287	2.0439	2.0595	2.0756	.96
.97	2.0923	2.1095	2.1273	2.1457	2.1649	2.1847	2.2054	2.2269	2.2494	2.2729	.97
.98	2.2976	2.3235	2.3507	2.3796	2.4101	2.4427	2.4774	2.5147	2.5550	2.5987	.98
.99	2.6467	2.6996	2.7587	2.8257	2.9031	2.9945	3.1063	3.2504	3.4534	3.8002	.99

TABLE **8** Correlation coefficient r as a function of transform z

This table is the inverse of Table **7**, and furnishes the hyperbolic tangents, $r = \tanh z$. Argument z is tabulated in increments of 0.001 from 0 to 1.0 and in increments of 0.01 from 1.0 to 3.49. The function r is given to four significant decimal places. A few values of r for z greater than 3.49 are also given.

The correlation coefficient corresponding to a positive value of z is looked up directly. For example, when $z = 1.82$, $r = 0.9488$. For negative z's, look up $|z|$ and make the function negative. When $z = -0.26$, $r = -0.2543$.

This table is used to convert the results of computations, using the z-transformation, back into correlation coefficient units (see Section 15.5).

The exponential library function of the FORTRAN IV compiler was used in computing this function by means of the equation $r = (e^z - e^{-z})/(e^z + e^{-z})$.

TABLE 8 Correlation coefficient r as a function of transform z

z	0	1	2	3	4	5	6	7	8	9	z
.00	.0000	.0010	.0020	.0030	.0040	.0050	.0060	.0070	.0080	.0090	.00
.01	.0100	.0110	.0120	.0130	.0140	.0150	.0160	.0170	.0180	.0190	.01
.02	.0200	.0210	.0220	.0230	.0240	.0250	.0260	.0270	.0280	.0290	.02
.03	.0300	.0310	.0320	.0330	.0340	.0350	.0360	.0370	.0380	.0390	.03
.04	.0400	.0410	.0420	.0430	.0440	.0450	.0460	.0470	.0480	.0490	.04
.05	.0500	.0510	.0520	.0530	.0539	.0549	.0559	.0569	.0579	.0589	.05
.06	.0599	.0609	.0619	.0629	.0639	.0649	.0659	.0669	.0679	.0689	.06
.07	.0699	.0709	.0719	.0729	.0739	.0749	.0759	.0768	.0778	.0788	.07
.08	.0798	.0808	.0818	.0828	.0838	.0848	.0858	.0868	.0878	.0888	.08
.09	.0898	.0907	.0917	.0927	.0937	.0947	.0957	.0967	.0977	.0987	.09
.10	.0997	.1007	.1016	.1026	.1036	.1046	.1056	.1066	.1076	.1086	.10
.11	.1096	.1105	.1115	.1125	.1135	.1145	.1155	.1165	.1175	.1184	.11
.12	.1194	.1204	.1214	.1224	.1234	.1244	.1253	.1263	.1273	.1283	.12
.13	.1293	.1303	.1312	.1322	.1332	.1342	.1352	.1361	.1371	.1381	.13
.14	.1391	.1401	.1411	.1420	.1430	.1440	.1450	.1460	.1469	.1479	.14
.15	.1489	.1499	.1508	.1518	.1528	.1538	.1547	.1557	.1567	.1577	.15
.16	.1586	.1596	.1606	.1616	.1625	.1635	.1645	.1655	.1664	.1674	.16
.17	.1684	.1694	.1703	.1713	.1723	.1732	.1742	.1752	.1761	.1771	.17
.18	.1781	.1790	.1800	.1810	.1820	.1829	.1839	.1849	.1858	.1868	.18
.19	.1877	.1887	.1897	.1906	.1916	.1926	.1935	.1945	.1955	.1964	.19
.20	.1974	.1983	.1993	.2003	.2012	.2022	.2031	.2041	.2051	.2060	.20
.21	.2070	.2079	.2089	.2098	.2108	.2117	.2127	.2137	.2146	.2156	.21
.22	.2165	.2175	.2184	.2194	.2203	.2213	.2222	.2232	.2241	.2251	.22
.23	.2260	.2270	.2279	.2289	.2298	.2308	.2317	.2327	.2336	.2346	.23
.24	.2355	.2364	.2374	.2383	.2393	.2402	.2412	.2421	.2430	.2440	.24
.25	.2449	.2459	.2468	.2477	.2487	.2496	.2506	.2515	.2524	.2534	.25
.26	.2543	.2552	.2562	.2571	.2580	.2590	.2599	.2608	.2618	.2627	.26
.27	.2636	.2646	.2655	.2664	.2673	.2683	.2692	.2701	.2711	.2720	.27
.28	.2729	.2738	.2748	.2757	.2766	.2775	.2784	.2794	.2803	.2812	.28
.29	.2821	.2831	.2840	.2849	.2858	.2867	.2876	.2886	.2895	.2904	.29
.30	.2913	.2922	.2931	.2941	.2950	.2959	.2968	.2977	.2986	.2995	.30
.31	.3004	.3013	.3023	.3032	.3041	.3050	.3059	.3068	.3077	.3086	.31
.32	.3095	.3104	.3113	.3122	.3131	.3140	.3149	.3158	.3167	.3176	.32
.33	.3185	.3194	.3203	.3212	.3221	.3230	.3239	.3248	.3257	.3266	.33
.34	.3275	.3284	.3293	.3302	.3310	.3319	.3328	.3337	.3346	.3355	.34
.35	.3364	.3373	.3381	.3390	.3399	.3408	.3417	.3426	.3435	.3443	.35
.36	.3452	.3461	.3470	.3479	.3487	.3496	.3505	.3514	.3522	.3531	.36
.37	.3540	.3549	.3557	.3566	.3575	.3584	.3592	.3601	.3610	.3618	.37
.38	.3627	.3636	.3644	.3653	.3662	.3670	.3679	.3688	.3696	.3705	.38
.39	.3714	.3722	.3731	.3739	.3748	.3757	.3765	.3774	.3782	.3791	.39
.40	.3799	.3808	.3817	.3825	.3834	.3842	.3851	.3859	.3868	.3876	.40
.41	.3885	.3893	.3902	.3910	.3919	.3927	.3936	.3944	.3952	.3961	.41
.42	.3969	.3978	.3986	.3995	.4003	.4011	.4020	.4028	.4036	.4045	.42
.43	.4053	.4062	.4070	.4078	.4087	.4095	.4103	.4112	.4120	.4128	.43
.44	.4136	.4145	.4153	.4161	.4170	.4178	.4186	.4194	.4203	.4211	.44
.45	.4219	.4227	.4235	.4244	.4252	.4260	.4268	.4276	.4285	.4293	.45
.46	.4301	.4309	.4317	.4325	.4333	.4342	.4350	.4358	.4366	.4374	.46
.47	.4382	.4390	.4398	.4406	.4414	.4422	.4430	.4438	.4446	.4454	.47
.48	.4462	.4470	.4478	.4486	.4494	.4502	.4510	.4518	.4526	.4534	.48
.49	.4542	.4550	.4558	.4566	.4574	.4582	.4590	.4598	.4605	.4613	.49

TABLE 8 Correlation coefficient r as a function of transform z

z	0	1	2	3	4	5	6	7	8	9	z
.50	.4621	.4629	.4637	.4645	.4653	.4660	.4668	.4676	.4684	.4692	.50
.51	.4699	.4707	.4715	.4723	.4731	.4738	.4746	.4754	.4762	.4769	.51
.52	.4777	.4785	.4792	.4800	.4808	.4815	.4823	.4831	.4839	.4846	.52
.53	.4854	.4861	.4869	.4877	.4884	.4892	.4900	.4907	.4915	.4922	.53
.54	.4930	.4937	.4945	.4953	.4960	.4968	.4975	.4983	.4990	.4998	.54
.55	.5005	.5013	.5020	.5028	.5035	.5043	.5050	.5057	.5065	.5072	.55
.56	.5080	.5087	.5095	.5102	.5109	.5117	.5124	.5132	.5139	.5146	.56
.57	.5154	.5161	.5168	.5176	.5183	.5190	.5198	.5205	.5212	.5219	.57
.58	.5227	.5234	.5241	.5248	.5256	.5263	.5270	.5277	.5285	.5292	.58
.59	.5299	.5306	.5313	.5320	.5328	.5335	.5342	.5349	.5356	.5363	.59
.60	.5370	.5378	.5385	.5392	.5399	.5406	.5413	.5420	.5427	.5434	.60
.61	.5441	.5448	.5455	.5462	.5469	.5476	.5483	.5490	.5497	.5504	.61
.62	.5511	.5518	.5525	.5532	.5539	.5546	.5553	.5560	.5567	.5574	.62
.63	.5581	.5587	.5594	.5601	.5608	.5615	.5622	.5629	.5635	.5642	.63
.64	.5649	.5656	.5663	.5669	.5676	.5683	.5690	.5696	.5703	.5710	.64
.65	.5717	.5723	.5730	.5737	.5744	.5750	.5757	.5764	.5770	.5777	.65
.66	.5784	.5790	.5797	.5804	.5810	.5817	.5823	.5830	.5837	.5843	.66
.67	.5850	.5856	.5863	.5869	.5876	.5883	.5889	.5896	.5902	.5909	.67
.68	.5915	.5922	.5928	.5935	.5941	.5948	.5954	.5961	.5967	.5973	.68
.69	.5980	.5986	.5993	.5999	.6005	.6012	.6018	.6025	.6031	.6037	.69
.70	.6044	.6050	.6056	.6063	.6069	.6075	.6082	.6088	.6094	.6100	.70
.71	.6107	.6113	.6119	.6126	.6132	.6138	.6144	.6150	.6157	.6163	.71
.72	.6169	.6175	.6181	.6188	.6194	.6200	.6206	.6212	.6218	.6225	.72
.73	.6231	.6237	.6243	.6249	.6255	.6261	.6267	.6273	.6279	.6285	.73
.74	.6291	.6297	.6304	.6310	.6316	.6322	.6328	.6334	.6340	.6346	.74
.75	.6351	.6357	.6363	.6369	.6375	.6381	.6387	.6393	.6399	.6405	.75
.76	.6411	.6417	.6423	.6428	.6434	.6440	.6446	.6452	.6458	.6463	.76
.77	.6469	.6475	.6481	.6487	.6492	.6498	.6504	.6510	.6516	.6521	.77
.78	.6527	.6533	.6539	.6544	.6550	.6556	.6561	.6567	.6573	.6578	.78
.79	.6584	.6590	.6595	.6601	.6607	.6612	.6618	.6624	.6629	.6635	.79
.80	.6640	.6646	.6652	.6657	.6663	.6668	.6674	.6679	.6685	.6690	.80
.81	.6696	.6701	.6707	.6712	.6718	.6723	.6729	.6734	.6740	.6745	.81
.82	.6751	.6756	.6762	.6767	.6772	.6778	.6783	.6789	.6794	.6799	.82
.83	.6805	.6810	.6815	.6821	.6826	.6832	.6837	.6842	.6847	.6853	.83
.84	.6858	.6863	.6869	.6874	.6879	.6884	.6890	.6895	.6900	.6905	.84
.85	.6911	.6916	.6921	.6926	.6932	.6937	.6942	.6947	.6952	.6957	.85
.86	.6963	.6968	.6973	.6978	.6983	.6988	.6993	.6998	.7004	.7009	.86
.87	.7014	.7019	.7024	.7029	.7034	.7039	.7044	.7049	.7054	.7059	.87
.88	.7064	.7069	.7074	.7079	.7084	.7089	.7094	.7099	.7104	.7109	.88
.89	.7114	.7119	.7124	.7129	.7134	.7139	.7143	.7148	.7153	.7158	.89
.90	.7163	.7168	.7173	.7178	.7182	.7187	.7192	.7197	.7202	.7207	.90
.91	.7211	.7216	.7221	.7226	.7230	.7235	.7240	.7245	.7249	.7254	.91
.92	.7259	.7264	.7268	.7273	.7278	.7283	.7287	.7292	.7297	.7301	.92
.93	.7306	.7311	.7315	.7320	.7325	.7329	.7334	.7338	.7343	.7348	.93
.94	.7352	.7357	.7361	.7366	.7371	.7375	.7380	.7384	.7389	.7393	.94
.95	.7398	.7402	.7407	.7411	.7416	.7420	.7425	.7429	.7434	.7438	.95
.96	.7443	.7447	.7452	.7456	.7461	.7465	.7469	.7474	.7478	.7483	.96
.97	.7487	.7491	.7496	.7500	.7505	.7509	.7513	.7518	.7522	.7526	.97
.98	.7531	.7535	.7539	.7544	.7548	.7552	.7557	.7561	.7565	.7569	.98
.99	.7574	.7578	.7582	.7586	.7591	.7595	.7599	.7603	.7608	.7612	.99

TABLE 8 Correlation coefficient r as a function of transform z

z	**0**	**1**	**2**	**3**	**4**	**5**	**6**	**7**	**8**	**9**	z
1.00	.7616	.7658	.7699	.7739	.7779	.7818	.7857	.7895	.7932	.7969	1.00
1.10	.8005	.8041	.8076	.8110	.8144	.8178	.8210	.8243	.8275	.8306	1.10
1.20	.8337	.8367	.8397	.8426	.8455	.8483	.8511	.8538	.8565	.8591	1.20
1.30	.8617	.8643	.8668	.8692	.8717	.8741	.8764	.8787	.8810	.8832	1.30
1.40	.8854	.8875	.8896	.8917	.8937	.8957	.8977	.8996	.9015	.9033	1.40
1.50	.9051	.9069	.9087	.9104	.9121	.9138	.9154	.9170	.9186	.9201	1.50
1.60	.9217	.9232	.9246	.9261	.9275	.9289	.9302	.9316	.9329	.9341	1.60
1.70	.9354	.9366	.9379	.9391	.9402	.9414	.9425	.9436	.9447	.9458	1.70
1.80	.9468	.9478	.9488	.9498	.9508	.9517	.9527	.9536	.9545	.9554	1.80
1.90	.9562	.9571	.9579	.9587	.9595	.9603	.9611	.9618	.9626	.9633	1.90
2.00	.9640	.9647	.9654	.9661	.9667	.9674	.9680	.9687	.9693	.9699	2.00
2.10	.9705	.9710	.9716	.9721	.9727	.9732	.9737	.9743	.9748	.9753	2.10
2.20	.9757	.9762	.9767	.9771	.9776	.9780	.9785	.9789	.9793	.9797	2.20
2.30	.9801	.9805	.9809	.9812	.9816	.9820	.9823	.9827	.9830	.9833	2.30
2.40	.9837	.9840	.9843	.9846	.9849	.9852	.9855	.9858	.9861	.9863	2.40
2.50	.9866	.9869	.9871	.9874	.9876	.9879	.9881	.9884	.9886	.9888	2.50
2.60	.9890	.9892	.9895	.9897	.9899	.9901	.9903	.9905	.9906	.9908	2.60
2.70	.9910	.9912	.9914	.9915	.9917	.9919	.9920	.9922	.9923	.9925	2.70
2.80	.9926	.9928	.9929	.9931	.9932	.9933	.9935	.9936	.9937	.9938	2.80
2.90	.9940	.9941	.9942	.9943	.9944	.9945	.9946	.9947	.9949	.9950	2.90
3.00	.9951	.9952	.9952	.9953	.9954	.9955	.9956	.9957	.9958	.9959	3.00
3.10	.9959	.9960	.9961	.9962	.9963	.9963	.9964	.9965	.9965	.9966	3.10
3.20	.9967	.9967	.9968	.9969	.9969	.9970	.9971	.9971	.9972	.9972	3.20
3.30	.9973	.9973	.9974	.9974	.9975	.9975	.9976	.9976	.9977	.9977	3.30
3.40	.9978	.9978	.9979	.9979	.9979	.9980	.9980	.9981	.9981	.9981	3.40
3.50	.9982										
4.00	.9993										
4.50	.9998										
5.00	.9999										

TABLE **9** C_n – Gurland and Tripathi's correction for the standard deviation

This table furnishes values of C_n, Gurland and Tripathi's correction for bias in estimates of the standard deviation σ. Values of C_n are given for all sample sizes n from 2 to 30.

For a sample size of $n = 10$, the correction factor C_n is 1.02811. When $n >$ 30, a satisfactory approximation is $C_n \approx 1 + 1/4(n - 1)$.

The correction for bias is applied by multiplying s, the estimate of the standard deviation, by the correction factor C_n (as further explained in Section 4.7).

The table was generated by computing the function

$$C_n = \left(\frac{n-1}{2}\right)^{\frac{1}{2}} \Gamma\left(\frac{n-1}{2}\right) \Big/ \Gamma \frac{n}{2}$$

in double precision arithmetic. The expression was developed by J. Gurland and R. C. Tripathi (*Am. Stat.*, **25**:30–32, 1971).

n	C_n
2	1.25331
3	1.12838
4	1.08540
5	1.06385
6	1.05094
7	1.04235
8	1.03624
9	1.03166
10	1.02811
11	1.02527
12	1.02296
13	1.02103
14	1.01940
15	1.01800
16	1.01679
17	1.01574
18	1.01481
19	1.01398
20	1.01324
21	1.01257
22	1.01197
23	1.01142
24	1.01093
25	1.01047
26	1.01005
27	1.00966
28	1.00930
29	1.00897
30	1.00866

TABLE **10** Ten thousand random digits

The 10,000 random digits are arranged in ten columns of 5 digits and blocked out in rows of 5, providing 25 digits per block.

The table of random digits should be entered at random. Choose the page by a random procedure, determine the row and column number by blindly pointing to it with a pencil, and proceed from there in some predetermined fashion, either horizontally or vertically.

A table of random numbers is generally useful for a variety of sampling operations. Extended use of the same set of random numbers in the same sampling experiment is ill advised. In such cases, new random numbers should be looked up in a different table or preferably generated on the computer.

A sequence of pseudorandom numbers between 0 and 1 was generated by the multiplicative congruential method. Since a computer with a 36-bit word length was used, the following recurrence equation was employed: $x_{i+1} = [x_i \times 5^{13}] \bmod (2^{35})$, where x_i is the ith pseudorandom number using integer arithmetic. Mod (2^{35}) means divide $x_i \times 5^{13}$ by 2^{35} and use the remainder, not the quotient, in subsequent computations. These numbers were then converted to floating point and divided by 2^{35} to scale them to between 0 and 1.0. Next, they were multiplied by 10 and truncated for the table. Since an analysis of the table revealed that some digits (3 and 7) were slightly too common and the digits 2 and 8 were a little too rare (although within the range of ordinary sampling error), a transformation was developed to adjust the sequence of numbers so that they would be more uniformly distributed.

TABLE 10 Ten thousand random digits

	1	2	3	4	5	6	7	8	9	10	
1	48461	14952	72619	73689	52059	37086	60050	86192	67049	64739	1
2	76534	38149	49692	31366	52093	15422	20498	33901	10319	43397	2
3	70437	25861	38504	14752	23757	59660	67844	78815	23758	86814	3
4	59584	03370	42806	11393	71722	93804	09095	07856	55589	46020	4
5	04285	58554	16085	51555	27501	73883	33427	33343	45507	50063	5
6	77340	10412	69189	85171	29082	44785	83638	02583	96483	76553	6
7	59183	62687	91778	80354	23512	97219	65921	02035	59847	91403	7
8	91800	04281	39979	03927	82564	28777	59049	97532	54540	79472	8
9	12066	24817	81099	48940	69554	55925	48379	12866	51232	21580	9
10	69907	91751	53512	23748	65906	91385	84983	27915	48491	91068	10
11	80467	04873	54053	25955	48518	13815	37707	68687	15570	08890	11
12	78057	67835	28302	45048	56761	97725	58438	91528	24645	18544	12
13	05648	39387	78191	88415	60269	94880	58812	42931	71898	61534	13
14	22304	39246	01350	99451	61862	78688	30339	60222	74052	25740	14
15	61346	50269	67005	40442	33100	16742	61640	21046	31909	72641	15
16	66793	37696	27965	30459	91011	51426	31006	77468	61029	57108	16
17	86411	48809	36698	42453	83061	43769	39948	87031	30767	13953	17
18	62098	12825	81744	28882	27369	88183	65846	92545	09065	22655	18
19	68775	06261	54265	16203	23340	84750	16317	88686	86842	00879	19
20	52679	19595	13687	74872	89181	01939	18447	10787	76246	80072	20
21	84096	87152	20719	25215	04349	54434	72344	93008	83282	31670	21
22	63964	55937	21417	49944	38356	98404	14850	17994	17161	98981	22
23	31191	75131	72386	11689	95727	05414	88727	45583	22568	77700	23
24	30545	68523	29850	67833	05622	89975	79042	27142	99257	32349	24
25	52573	91001	52315	26430	54175	30122	31796	98842	37600	26025	25
26	16586	81842	01076	99414	31574	94719	34656	80018	86988	79234	26
27	81841	88481	61191	25013	30272	23388	22463	65774	10029	58376	27
28	43563	66829	72838	08074	57080	15446	11034	98143	74989	26885	28
29	19945	84193	57581	77252	85604	45412	43556	27518	90572	00563	29
30	79374	23796	16919	99691	80276	32818	62953	78831	54395	30705	30
31	48503	26615	43980	09810	38289	66679	73799	48418	12647	40044	31
32	32049	65541	37937	41105	70106	89706	40829	40789	59547	00783	32
33	18547	71562	95493	34112	76895	46766	96395	31718	48302	45893	33
34	03180	96742	61486	43305	34183	99605	67803	13491	09243	29557	34
35	94822	24738	67749	83748	59799	25210	31093	62925	72061	69991	35
36	34330	60599	85828	19152	68499	27977	35611	96240	62747	89529	36
37	43770	81537	59527	95674	76692	86420	69930	10020	72881	12532	37
38	56908	77192	50623	41215	14311	42834	80651	93750	59957	31211	38
39	32787	07189	80539	75927	75475	73965	11796	72140	48944	74156	39
40	52441	78392	11733	57703	29133	71164	55355	31006	25526	55790	40
41	22377	54723	18227	28449	04570	18882	00023	67101	06895	08915	41
42	18376	73460	88841	39602	34049	20589	05701	08249	74213	25220	42
43	53201	28610	87957	21497	64729	64983	71551	99016	87903	63875	43
44	34919	78901	59710	27396	02593	05665	11964	44134	00273	76358	44
45	33617	92159	21971	16901	57383	34262	41744	60891	57624	06962	45
46	70010	40964	98780	72418	52571	18415	64362	90636	38034	04909	46
47	19282	68447	35665	31530	59832	49181	21914	65742	89815	39231	47
48	91429	73328	13266	54898	68795	40948	80808	63887	89939	47938	48
49	97637	78393	33021	05867	86520	45363	43066	00988	64040	09803	49
50	95150	07625	05255	83254	93943	52325	93230	62668	79529	65964	50

TABLE 10 Ten thousand random digits

	1	2	3	4	5	6	7	8	9	10	
51	58237	81333	12573	36181	84900	39614	61303	05086	97670	07961	51
52	54789	75554	36795	42649	02971	97584	38223	52643	25027	56849	52
53	55373	14272	62729	25659	84359	02654	08409	52703	88803	31919	53
54	74251	66100	10773	71393	80972	45092	07932	83065	06585	16454	54
55	86077	90904	14779	75116	50267	52217	08539	08345	52750	22815	55
56	13506	84170	08716	28894	20133	99489	87768	55582	96081	20774	56
57	13226	41411	16074	15438	68840	17064	96917	25404	47708	17861	57
58	01642	25456	69804	29277	99473	07912	90488	73325	88266	18082	58
59	23715	70933	37381	20388	34929	96585	14146	81617	36664	25060	59
60	98436	61100	45346	94664	30677	18677	99524	70767	39525	34023	60
61	96571	81879	50387	77316	18874	00763	99457	70858	79674	95618	61
62	70677	59632	22985	95166	54904	61995	63423	65335	13807	96638	62
63	33725	31717	04704	13669	91697	00107	33667	24770	60044	49107	63
64	40910	75631	56653	42858	85768	21254	01295	21507	33687	02404	64
65	67947	27522	14066	14943	19696	93933	52432	90569	14856	30580	65
66	41797	38840	68744	59348	05120	30184	35212	14348	37661	51451	66
67	41770	94218	52578	36238	40575	16793	77152	23382	11570	87276	67
68	34918	50080	97862	84932	57596	33749	78745	73377	72328	63074	68
69	47898	91359	10606	33735	46812	96239	23815	36757	17882	96143	69
70	45021	03882	94463	96369	56001	16348	89408	84563	66422	62636	70
71	92752	67479	72696	20645	78439	11224	17405	27884	15573	12490	71
72	13229	29631	01944	76916	30063	98507	11345	29576	20215	53972	72
73	39111	13161	85208	70273	24016	04960	46728	60292	24831	06403	73
74	50972	92325	72258	57577	32886	56062	62370	99461	64680	38080	74
75	08190	28700	87859	03684	83762	03810	12325	75445	41946	36420	75
76	98140	55201	89156	99277	78211	78692	96992	80163	67882	36674	76
77	14110	02500	54140	43371	06930	26853	56025	73530	97542	19287	77
78	35106	03726	35458	69204	47084	22676	62125	66443	73712	82879	78
79	04446	93672	24118	40460	79678	51259	37345	49666	08518	04251	79
80	42890	46488	42713	76138	82275	59529	98821	60243	51840	02294	80
81	64856	84896	77627	86920	59181	24162	34918	77203	55518	17174	81
82	29739	02885	14169	81125	59048	59396	35494	25220	91424	94750	82
83	70750	97663	18316	16741	75016	01404	78331	76908	68195	18714	83
84	28695	56066	21108	23021	31950	37840	33674	81877	92490	35488	84
85	58892	85844	04181	58470	13348	64277	43838	50362	08531	83388	85
86	79508	78596	96537	20553	41148	37805	14553	09919	11490	70231	86
87	13617	66975	68598	95450	06285	84134	62474	22329	82134	08283	87
88	15123	55351	28631	77941	90178	35876	27833	92494	88899	41558	88
89	34604	18686	05179	31756	47258	14945	98839	82051	86608	07022	89
90	51863	00432	27846	54577	84476	06652	88250	40187	29735	32621	90
91	65489	54535	60256	91285	81743	48426	58351	89166	52478	23935	91
92	86119	57900	04979	16358	06281	25469	02454	20658	08869	29293	92
93	01963	62421	86788	54260	61287	69893	56446	11608	04760	85532	93
94	52164	39397	60568	88382	01561	78861	02849	31033	76875	08260	94
95	91444	35684	80387	44827	71027	16850	04079	02394	49058	60509	95
96	09881	00351	14759	58624	50470	03348	23703	10959	39733	84954	96
97	09045	29309	17864	27687	21691	59354	56599	99735	57626	55645	97
98	13036	51186	32490	45564	53813	25529	66293	50352	41356	44697	98
99	96114	75971	70563	31203	85747	86469	78133	40310	48223	85933	99
100	47657	02006	33820	89370	60192	08248	10221	27754	31323	59019	100

TABLE 10 Ten thousand random digits

	1	2	3	4	5	6	7	8	9	10	
101	20992	99405	64724	68632	51100	05310	71468	25066	51115	11450	101
102	55109	29812	28047	77266	72131	31420	69336	35633	91527	58904	102
103	78521	74623	05802	80834	59282	21784	63995	71057	43191	39005	103
104	04663	28601	24573	98343	72061	11632	11364	50351	97030	03019	104
105	06008	67645	55979	21012	46444	88351	77264	83331	54233	72933	105
106	01971	82763	53029	54861	78099	93127	64114	08341	60164	50548	106
107	69487	17195	47802	76193	54244	75523	50502	23090	64761	32106	107
108	20480	62789	23756	26759	05657	40569	80333	10691	65575	75659	108
109	53736	32574	76781	99231	99766	36951	44813	21302	42681	20511	109
110	30683	57894	57235	19019	86941	22906	62998	31313	68304	99427	110
111	56503	68759	64988	21189	89361	01093	13458	94501	36804	75758	111
112	71015	58374	55700	17283	02822	46847	19644	56298	28363	88692	112
113	01525	06462	32199	28145	69252	56232	35744	26994	71289	25906	113
114	04748	76563	00781	89201	99705	69183	36777	66719	48022	65126	114
115	36337	76371	56419	57462	91591	85204	46652	53152	23325	27949	115
116	50516	33877	60288	52452	32220	99492	38378	77703	62410	53958	116
117	53278	71860	35327	79118	50330	72410	52210	62145	80132	79276	117
118	24396	73460	21715	15875	76225	01362	29941	96873	21765	51302	118
119	90526	85125	87761	67620	87458	53789	86249	09071	91432	93498	119
120	62016	11174	81655	95547	68586	69706	93755	08894	94045	68308	120
121	84388	10946	96056	64407	16812	86694	45320	14494	71454	33194	121
122	79364	91943	12421	13446	22397	96003	82447	57140	66739	15779	122
123	77008	24942	76020	83095	44461	47443	54642	66043	33403	37242	123
124	81770	21470	81822	91417	16685	41100	38863	69628	80463	62659	124
125	30650	00510	39786	37119	51869	57706	80670	69219	60031	57862	125
126	46808	70639	64709	79976	89899	05189	35484	86220	27698	50382	126
127	75690	09668	49236	26127	72363	38179	06614	73630	75445	50183	127
128	66841	26725	81491	47455	60061	24162	67617	61211	08660	62754	128
129	62679	76672	92854	65564	95428	61608	25213	17120	52691	10703	129
130	35996	76325	67090	92344	96575	06735	12066	09040	23352	95665	130
131	58215	30554	42593	51847	04285	91203	60282	48263	03458	92510	131
132	53846	20911	01113	41215	86563	64485	14213	99552	49607	34128	132
133	67847	22692	98162	28309	86439	29258	05728	13132	40198	54449	133
134	27984	58063	39553	84802	65095	95345	24821	39723	12232	56641	134
135	94370	02410	32366	87038	87873	13448	82265	41010	37682	64202	135
136	98352	83237	78099	55733	92336	51756	85103	57257	49016	68938	136
137	38781	77533	50745	21619	83329	64114	24668	50826	50996	35402	137
138	30910	59371	78549	00134	75070	35640	33679	62954	76462	14207	138
139	93383	68965	97175	27810	63231	38153	22138	83818	14553	00792	139
140	44771	50543	46870	84958	59666	78397	67347	45037	30278	34879	140
141	44265	18136	76765	26301	42887	58467	19346	86266	75905	83908	141
142	30942	69742	71236	82089	65346	31553	93314	69647	71233	93093	142
143	12466	26435	22634	69115	06966	12864	84535	80650	59399	80201	143
144	22246	36534	37474	97571	80532	89892	29782	55803	36494	82152	144
145	79721	95215	36459	06287	83524	06356	61171	76170	44472	91543	145
146	60028	91431	93528	17580	60218	61498	86851	05174	08201	45308	146
147	40591	53078	62739	01053	00466	30604	29595	12719	80770	64790	147
148	35984	21080	79400	27098	62417	45236	95779	14806	42222	64628	148
149	24335	70074	60121	78385	69437	23084	42324	04811	36003	35277	149
150	67517	46090	92338	98851	52758	07435	21735	99397	20332	59809	150

TABLE 10 Ten thousand random digits

	1	2	3	4	5	6	7	8	9	10	
151	13961	73549	50445	02797	77988	32480	99116	19559	92686	96188	151
152	40986	23282	40312	03108	08635	96873	37382	67014	99964	13769	152
153	63483	89965	29965	94242	02160	59656	04459	76675	90391	77608	153
154	58874	06054	93183	37271	41717	93385	50687	95813	10015	59534	154
155	22202	93835	75224	56232	29272	50386	50915	32944	84385	48293	155
156	45242	02824	49627	76875	39915	94590	04095	91482	63116	13103	156
157	36815	62809	45636	22306	70080	03830	84709	50946	28711	14240	157
158	11251	93858	66744	09068	08923	46314	16104	66114	52838	60822	158
159	03146	69101	82270	85732	18453	38142	61825	61458	59184	29726	159
160	93303	30006	37277	23975	73283	24325	04092	47153	90553	07140	160
161	06668	91453	74726	75716	16870	65097	12035	55470	57251	65855	161
162	85684	17562	53211	50234	96530	14736	89674	11447	76989	42206	162
163	78948	97463	09494	09356	61285	72337	72064	73890	21476	17576	163
164	96168	90422	35391	89622	81593	63972	37211	43715	11737	34307	164
165	37451	36052	52866	42907	92003	85818	19516	79923	08078	90027	165
166	26390	45437	02061	30574	07526	18571	66706	13281	28401	11476	166
167	43096	57671	66165	01905	36376	33569	29410	16458	14039	97225	167
168	46451	11389	22938	29517	71409	95998	62650	74830	71919	91099	168
169	24125	56973	87209	21228	27479	72709	83161	88214	98627	05266	169
170	79507	12940	01519	07579	78954	31373	11723	22783	98193	05903	170
171	98697	58698	74367	06975	25894	60973	71050	57472	15828	80313	171
172	76404	29382	20189	11287	27324	39284	70368	11591	66657	56035	172
173	43651	96897	98590	39853	90614	92444	67026	43201	01087	47095	173
174	92428	01267	47588	70879	70394	06709	49391	54079	37182	71208	174
175	48368	43909	13164	40940	63680	91273	00550	23383	10394	89394	175
176	03846	41039	99844	62562	38128	69608	06373	81183	53302	05747	176
177	35346	23964	07646	07075	55749	95942	00912	99482	18441	60600	177
178	83186	51111	97339	80202	08062	81871	87530	73461	91760	56926	178
179	58078	94427	95357	45447	69999	25025	61585	89954	38893	69334	179
180	22762	27816	74381	06738	23387	08183	96318	46138	51234	37122	180
181	65354	19914	33459	72776	34215	18749	24371	88366	38555	08983	181
182	79286	15905	28669	50696	19442	26417	60906	05173	34355	97089	182
183	35860	07775	77100	45226	76099	37087	53639	01933	46743	47394	183
184	41998	67048	53397	38732	16130	33833	87778	84418	56556	35785	184
185	08826	99500	16607	64822	13390	55106	37720	56819	46828	10690	185
186	47255	75479	72807	79522	14904	47046	33009	95298	93865	02278	186
187	35684	31325	38516	20620	42438	90223	23470	41134	12603	87511	187
188	99961	55979	83906	41570	88418	13371	31644	92267	92844	98253	188
189	31541	68322	33715	73510	01110	26036	99248	47424	38555	87958	189
190	34900	30367	23892	22642	82038	20552	35925	39647	90142	15046	190
191	94205	44990	91349	91531	64055	94886	55744	41662	72455	79836	191
192	84982	43754	09078	91460	99309	64563	61935	43214	99692	60101	192
193	04274	39885	97062	15927	12741	23963	02157	45642	02002	73102	193
194	34433	58119	70976	66159	17456	35073	85894	08935	27736	60578	194
195	97821	63898	17609	22580	42262	49780	75191	04821	87092	52186	195
196	20649	89577	45149	61825	60801	71346	67714	58156	00111	88857	196
197	64054	11178	87763	11005	09619	56917	38144	96278	47458	47826	197
198	50543	80232	69007	78996	35482	34818	43193	03514	72770	18560	198
199	84029	28230	93141	27645	21154	57594	42292	57147	46589	11512	199
200	68085	75447	30235	08190	42944	10980	35960	87461	79630	12952	200

TABLE **11** Areas of the normal curve

Arguments are furnished for the right half of the normal curve from the mean up to 3.5 standard deviations in increments of 0.01 standard deviation units, and from there to 4.9 standard deviations from the mean in increments of 0.1 standard deviation units. The quantity given is the area under the standard normal density function between the mean and the critical point. The area is generally labeled $\frac{1}{2} - \alpha$ (as shown in the figure). The proportion of the area between the mean and the critical value is given to four significant decimal places up to 3.49 standard deviations, and from there to 4.9 standard deviations to six significant decimal places.

The area between the mean and 2.21 standard deviations, for example, is found from the table to comprise 0.4864 or 48.64% of the total area of the curve. From this figure one can easily compute two other quantities. The area beyond 2.21 standard deviations is the 0.5-complement of the function just looked up. Thus, it will be $0.5000 - 0.4864 = 0.0136$, which is α, the proportion of the area of the curve to the right of 2.21 standard deviations. To find $1 - \alpha$, the total area of the curve to the left of 2.21, add 0.5 (the entire area of the left half of the curve) to the area given as a function in the table. Thus, the area to the left of 2.21 standard deviations is 0.9864.

By inverse table look-up and interpolation one can find the number of standard deviations corresponding to a given area. Thus, an area of 0.3264 represents the area between the mean and 0.94 standard deviations.

This table is of wide application in statistics. It is employed whenever one makes probability statements about normally distributed variables (Section 7.9), when normalizing a frequency distribution (Section 6.5), and in a variety of other applications.

The table was generated by using the polynomial expression given in C. Hastings, *Approximations for Digital Computers* (Princeton University Press, 1955) and shown below, which approximates the integral of the normal probability distribution with a maximum error of 7.5×10^{-8}:

$$P(X) - \tfrac{1}{2} = \tfrac{1}{2} - 1/\sqrt{2\pi}e^{-X^2/2}t(b_1 + t\{b_2 + t[b_3 + t(b_4 + tb_5)]\})$$

where $t = 1/1(1 + pX)$, $p = 0.2316419$, $b_1 = 0.319381530$, $b_2 = -0.356563782$, $b_3 = 1.781477937$, $b_4 = -1.821255978$, and $b_5 = 1.330274429$.

TABLE 11 Areas of the normal curve

Standard deviation units	0.	0.01	0.02	0.03	0.04
0.0	.0000	.0040	.0080	.0120	.0160
0.1	.0398	.0438	.0478	.0517	.0557
0.2	.0793	.0832	.0871	.0910	.0948
0.3	.1179	.1217	.1255	.1293	.1331
0.4	.1554	.1591	.1628	.1664	.1700
0.5	.1915	.1950	.1985	.2019	.2054
0.6	.2257	.2291	.2324	.2357	.2389
0.7	.2580	.2611	.2642	.2673	.2704
0.8	.2881	.2910	.2939	.2967	.2995
0.9	.3159	.3186	.3212	.3238	.3264
1.0	.3413	.3438	.3461	.3485	.3508
1.1	.3643	.3665	.3686	.3708	.3729
1.2	.3849	.3869	.3888	.3907	.3925
1.3	.4032	.4049	.4066	.4082	.4099
1.4	.4192	.4207	.4222	.4236	.4251
1.5	.4332	.4345	.4357	.4370	.4382
1.6	.4452	.4463	.4474	.4484	.4495
1.7	.4554	.4564	.4573	.4582	.4591
1.8	.4641	.4649	.4656	.4664	.4671
1.9	.4713	.4719	.4726	.4732	.4738
2.0	.4772	.4778	.4783	.4788	.4793
2.1	.4821	.4826	.4830	.4834	.4838
2.2	.4861	.4864	.4868	.4871	.4875
2.3	.4893	.4896	.4898	.4901	.4904
2.4	.4918	.4920	.4922	.4925	.4927
2.5	.4938	.4940	.4941	.4943	.4945
2.6	.4953	.4955	.4956	.4957	.4959
2.7	.4965	.4966	.4967	.4968	.4969
2.8	.4974	.4975	.4976	.4977	.4977
2.9	.4981	.4982	.4982	.4983	.4984
3.0	.4987	.4987	.4987	.4988	.4988
3.1	.4990	.4991	.4991	.4991	.4992
3.2	.4993	.4993	.4994	.4994	.4994
3.3	.4995	.4995	.4995	.4996	.4996
3.4	.4997	.4997	.4997	.4997	.4997
3.5	.499767				
3.6	.499841				
3.7	.499892				
3.8	.499928				
3.9	.499952				
4.0	.499968				
4.1	.499979				
4.2	.499987				
4.3	.499991				
4.4	.499995				
4.5	.499997				
4.6	.499998				
4.7	.499999				
4.8	.499999				
4.9	.500000				

0.05	0.06	0.07	0.08	0.09	Standard deviation units
.0199	.0239	.0279	.0319	.0359	0.0
.0596	.0636	.0675	.0714	.0753	0.1
.0987	.1026	.1064	.1103	.1141	0.2
.1368	.1406	.1443	.1480	.1517	0.3
.1736	.1772	.1808	.1844	.1879	0.4
.2088	.2123	.2157	.2190	.2224	0.5
.2422	.2454	.2486	.2517	.2549	0.6
.2734	.2764	.2794	.2823	.2852	0.7
.3023	.3051	.3078	.3106	.3133	0.8
.3289	.3315	.3340	.3365	.3389	0.9
.3531	.3554	.3577	.3599	.3621	1.0
.3749	.3770	.3790	.3810	.3830	1.1
.3944	.3962	.3980	.3997	.4015	1.2
.4115	.4131	.4147	.4162	.4177	1.3
.4265	.4279	.4292	.4306	.4319	1.4
.4394	.4406	.4418	.4429	.4441	1.5
.4505	.4515	.4525	.4535	.4545	1.6
.4599	.4608	.4616	.4625	.4633	1.7
.4678	.4686	.4693	.4699	.4706	1.8
.4744	.4750	.4756	.4761	.4767	1.9
.4798	.4803	.4808	.4812	.4817	2.0
.4842	.4846	.4850	.4854	.4857	2.1
.4878	.4881	.4884	.4887	.4890	2.2
.4906	.4909	.4911	.4913	.4916	2.3
.4929	.4931	.4932	.4934	.4936	2.4
.4946	.4948	.4949	.4951	.4952	2.5
.4960	.4961	.4962	.4963	.4964	2.6
.4970	.4971	.4972	.4973	.4974	2.7
.4978	.4979	.4979	.4980	.4981	2.8
.4984	.4985	.4985	.4986	.4986	2.9
.4989	.4989	.4989	.4990	.4990	3.0
.4992	.4992	.4992	.4993	.4993	3.1
.4994	.4994	.4995	.4995	.4995	3.2
.4996	.4996	.4996	.4996	.4997	3.3
.4997	.4997	.4997	.4997	.4998	3.4

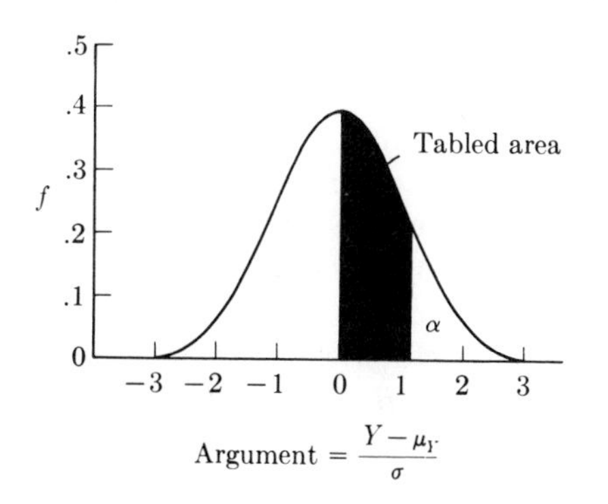

TABLE **12** Critical values of Student's t-distribution

The table furnishes critical values for Student's t-distribution for degrees of freedom $\nu = 1$ to 30 in increments of one, and for $\nu = 40, 60, 120$, and ∞. The percentage points given (corresponding to $\alpha = 0.9, 0.5, 0.4, 0.2, 0.1, 0.05, 0.02, 0.01$, and 0.001) represent the area beyond the critical values of $\pm t$ in *both* tails of the distribution, as shown in the accompanying figure. The critical values of t are given to three decimal places.

To look up the critical values of t for a given number of degrees of freedom, look up ν in the left (argument) column of the table and read off the desired values of t in that row. For example, for 22 degrees of freedom, $t_{.05} = 2.074$ and $t_{.01} = 2.819$. The last value indicates that 1% of the area of the t-distribution (with 22 degrees of freedom) is beyond $t = \pm 2.819$, with 0.5% in each tail. If a one-tailed test is desired, the probabilities at the head of the table must be halved. Thus, for a one-tailed test with 4 *df*, the critical value $t = 3.747$ delimits 0.01 of the area of the curve. For degrees of freedom $\nu > 30$, we need to interpolate between the values of the argument ν. The table is designed for harmonic interpolation. Thus, to obtain $t_{.05[43]}$, interpolate between $t_{.05[40]} = 2.021$ and $t_{.05[60]} = 2.000$, which are furnished in the table. Transform the arguments into $120/\nu = 120/43 = 2.791$ and interpolate between $120/60 = 2.000$ and $120/40 = 3.000$ by ordinary linear interpolation:

$$\begin{aligned} t_{.05[43]} &= (0.791 \times 2.021) + [(1 - 0.791) \times 2.000] \\ &= 2.017 \end{aligned}$$

When $\nu > 120$, interpolate between $120/\infty = 0$ and $120/120 = 1$. If critical values other than those furnished in this table are desired. E. S. Pearson and H. O. Hartley, *Biometrika Tables for Statisticians,* Vol. I (Cambridge University Press, 1958) quote J. B. Simaika (*Biometrika,* **32:**263–276, 1942), who recommends linear interpolation using the logarithms of the two-tailed probability values.

There are numerous applications of the t-distribution in statistics. The distribution is described in Section 7.4. Among the applications is the setting of confidence limits to means of small samples (Section 7.5), tests of difference between two means (Section 9.4), and comparisons of a single specimen with a sample (Section 9.5).

Values in this table have been taken from a more extensive one (table III) in R. A. Fisher and F. Yates, *Statistical Tables for Biological, Agricultural and Medical Research,* 5th ed. (Oliver & Boyd, Edinburgh, 1958) with permission of the authors and their publishers.

TABLE 12 Critical values of Student's t-distribution.

ν \ α	0.9	0.5	0.4	0.2	0.1	0.05	0.02	0.01	0.001	α / ν
1	.158	1.000	1.376	3.078	6.314	12.706	31.821	63.657	636.619	1
2	.142	.816	1.061	1.886	2.920	4.303	6.965	9.925	31.598	2
3	.137	.765	.978	1.638	2.353	3.182	4.541	5.841	12.924	3
4	.134	.741	.941	1.533	2.132	2.776	3.747	4.604	8.610	4
5	.132	.727	.920	1.476	2.015	2.571	3.365	4.032	6.869	5
6	.131	.718	.906	1.440	1.943	2.447	3.143	3.707	5.959	6
7	.130	.711	.896	1.415	1.895	2.365	2.998	3.499	5.408	7
8	.130	.706	.889	1.397	1.860	2.306	2.896	3.355	5.041	8
9	.129	.703	.883	1.383	1.833	2.262	2.821	3.250	4.781	9
10	.129	.700	.879	1.372	1.812	2.228	2.764	3.169	4.587	10
11	.129	.697	.876	1.363	1.796	2.201	2.718	3.106	4.437	11
12	.128	.695	.873	1.356	1.782	2.179	2.681	3.055	4.318	12
13	.128	.694	.870	1.350	1.771	2.160	2.650	3.012	4.221	13
14	.128	.692	.868	1.345	1.761	2.145	2.624	2.977	4.140	14
15	.128	.691	.866	1.341	1.753	2.131	2.602	2.947	4.073	15
16	.128	.690	.865	1.337	1.746	2.120	2.583	2.921	4.015	16
17	.128	.689	.863	1.333	1.740	2.110	2.567	2.898	3.965	17
18	.127	.688	.862	1.330	1.734	2.101	2.552	2.878	3.922	18
19	.127	.688	.861	1.328	1.729	2.093	2.539	2.861	3.883	19
20	.127	.687	.860	1.325	1.725	2.086	2.528	2.845	3.850	20
21	.127	.686	.859	1.323	1.721	2.080	2.518	2.831	3.819	21
22	.127	.686	.858	1.321	1.717	2.074	2.508	2.819	3.792	22
23	.127	.685	.858	1.319	1.714	2.069	2.500	2.807	3.767	23
24	.127	.685	.857	1.318	1.711	2.064	2.492	2.797	3.745	24
25	.127	.684	.856	1.316	1.708	2.060	2.485	2.787	3.725	25
26	.127	.684	.856	1.315	1.706	2.056	2.479	2.779	3.707	26
27	.127	.684	.855	1.314	1.703	2.052	2.473	2.771	3.690	27
28	.127	.683	.855	1.313	1.701	2.048	2.467	2.763	3.674	28
29	.127	.683	.854	1.311	1.699	2.045	2.462	2.756	3.659	29
30	.127	.683	.854	1.310	1.697	2.042	2.457	2.750	3.646	30
40	.126	.681	.851	1.303	1.684	2.021	2.423	2.704	3.551	40
60	.126	.679	.848	1.296	1.671	2.000	2.390	2.660	3.460	60
120	.126	.677	.845	1.289	1.658	1.980	2.358	2.617	3.373	120
∞	.126	.674	.842	1.282	1.645	1.960	2.326	2.576	3.291	∞

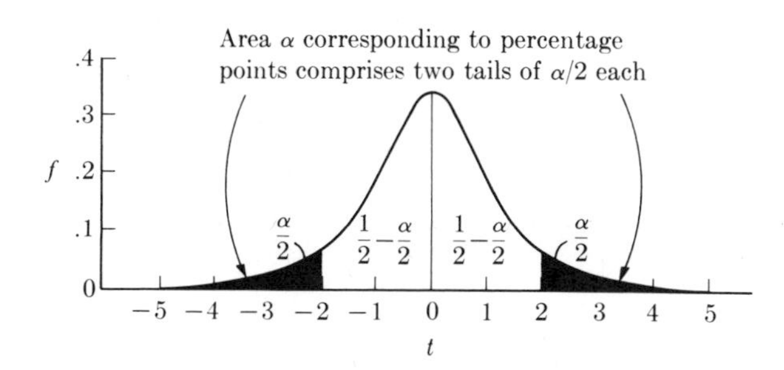

TABLE **13** Critical values of Student's t-distribution based on Šidák's multiplicative inequality

This table furnishes critical values for Student's t-distribution for unusual percentage points not found in standard t-tables such as Table **12.** These unusual percentage points α' are obtained from Šidák's multiplicative inequality; $\alpha' = 1 - (1 - \alpha)^{1/k}$ where α is the experimentwise error rate (also known as the familywise type I error rate) and k is the number of comparisons intended. Values of $t_{\alpha'}$ are furnished for degrees of freedom ν from 2 to 24 in increments of one, and for $\nu = 26, 28, 30, 40, 60, 120$, and ∞. These values of ν lend themselves to harmonic interpolation as shown for Table **12.** Arguments for k, the number of comparisons, are given from 1 to 60 in increments of one. Three levels of experimentwise α (0.10, 0.05, and 0.01) are given for each combination of ν and k in the table.

To find the critical value for a 5% experimentwise error based on a mean square with $\nu = 20$ degrees of freedom and making $k = 5$ comparisons overall, we enter the table with $k = 5$, $\nu = 20$, and $\alpha = .05$ to obtain $t_{.010206[20]} = 2.836$. This table is two-tailed. For a one-tailed test halve the experimentwise error rate α.

This table is employed for nonorthogonal multiple comparisons tests (Section 9.6) and for setting multiple confidence intervals (Section 14.10).

The table was computed directly using a modification of the Hewlett Packard 9830 Statistical Distributions Pack, volume 1. All computations were carried out to 12 decimal place accuracy. This table extends that published by P. A. Games (*J. Amer. Stat. Assoc.*, **72**:531–534, 1977) from $k = 50$ to $k = 60$. The computations were checked by comparison with values published by Games. A few values differ by 1 in the last decimal place due to rounding error.

TABLE 13 Critical values of Student's t-distribution based on Šidák's multiplicative inequality

		Degrees of freedom ν									
k	α	**2**	**3**	**4**	**5**	**6**	**7**	**8**	**9**	**10**	**11**
1	.1	2.920	2.353	2.132	2.015	1.943	1.895	1.860	1.833	1.812	1.796
	.05	4.303	3.182	2.776	2.571	2.447	2.365	2.306	2.262	2.228	2.201
	.01	9.925	5.841	4.604	4.032	3.707	3.499	3.355	3.250	3.169	3.106
2	.1	4.243	3.149	2.751	2.549	2.428	2.347	2.289	2.246	2.213	2.186
	.05	6.164	4.156	3.481	3.152	2.959	2.832	2.743	2.677	2.626	2.586
	.01	14.071	7.447	5.594	4.771	4.315	4.027	3.831	3.688	3.580	3.495
3	.1	5.243	3.690	3.150	2.882	2.723	2.618	2.544	2.488	2.446	2.412
	.05	7.582	4.826	3.941	3.518	3.274	3.115	3.005	2.923	2.860	2.811
	.01	17.248	8.565	6.248	5.243	4.695	4.353	4.120	3.952	3.825	3.726
4	.1	6.081	4.115	3.452	3.129	2.939	2.814	2.726	2.661	2.611	2.571
	.05	8.774	5.355	4.290	3.791	3.505	3.321	3.193	3.099	3.027	2.970
	.01	19.925	9.453	6.751	5.599	4.977	4.591	4.331	4.143	4.002	3.892
5	.1	6.816	4.471	3.699	3.327	3.110	2.969	2.869	2.796	2.739	2.695
	.05	9.823	5.799	4.577	4.012	3.690	3.484	3.342	3.237	3.157	3.094
	.01	22.282	10.201	7.166	5.888	5.203	4.782	4.498	4.294	4.141	4.022
6	.1	7.480	4.780	3.909	3.493	3.253	3.097	2.987	2.907	2.845	2.796
	.05	10.769	6.185	4.822	4.197	3.845	3.620	3.464	3.351	3.264	3.196
	.01	24.413	10.853	7.520	6.133	5.394	4.941	4.637	4.419	4.256	4.129
7	.1	8.090	5.055	4.093	3.638	3.376	3.206	3.088	3.001	2.934	2.881
	.05	11.639	6.529	5.036	4.358	3.978	3.736	3.569	3.448	3.355	3.283
	.01	26.372	11.436	7.832	6.346	5.559	5.078	4.756	4.526	4.354	4.221
8	.1	8.656	5.304	4.257	3.765	3.484	3.302	3.176	3.083	3.012	2.955
	.05	12.449	6.842	5.228	4.501	4.095	3.838	3.661	3.532	3.434	3.358
	.01	28.196	11.966	8.112	6.535	5.704	5.198	4.860	4.619	4.439	4.300
9	.1	9.188	5.532	4.406	3.880	3.580	3.388	3.254	3.155	3.080	3.021
	.05	13.208	7.128	5.402	4.630	4.200	3.929	3.743	3.607	3.505	3.424
	.01	29.908	12.453	8.367	6.706	5.835	5.306	4.953	4.703	4.515	4.371
10	.1	9.691	5.744	4.542	3.985	3.668	3.465	3.324	3.221	3.142	3.079
	.05	13.927	7.394	5.562	4.747	4.296	4.011	3.816	3.675	3.568	3.484
	.01	31.528	12.904	8.600	6.862	5.954	5.404	5.038	4.778	4.584	4.434
11	.1	10.169	5.941	4.668	4.081	3.748	3.535	3.388	3.280	3.197	3.133
	.05	14.610	7.642	5.710	4.855	4.383	4.086	3.883	3.736	3.625	3.538
	.01	33.068	13.326	8.817	7.006	6.063	5.493	5.115	4.846	4.646	4.492
12	.1	10.625	6.127	4.785	4.170	3.822	3.600	3.446	3.334	3.249	3.181
	.05	15.263	7.876	5.848	4.955	4.464	4.156	3.945	3.793	3.677	3.587
	.01	34.540	13.724	9.019	7.139	6.164	5.575	5.185	4.909	4.703	4.545
13	.1	11.063	6.302	4.895	4.253	3.891	3.660	3.501	3.384	3.296	3.226
	.05	15.889	8.096	5.977	5.049	4.539	4.220	4.002	3.845	3.726	3.633
	.01	35.951	14.099	9.208	7.264	6.258	5.652	5.251	4.967	4.756	4.594
14	.1	11.484	6.468	4.999	4.330	3.955	3.716	3.551	3.431	3.339	3.268
	.05	16.491	8.306	6.099	5.136	4.609	4.280	4.055	3.893	3.771	3.675
	.01	37.309	14.456	9.387	7.381	6.345	5.724	5.312	5.021	4.806	4.639
15	.1	11.890	6.627	5.097	4.403	4.015	3.768	3.598	3.474	3.380	3.306
	.05	17.072	8.505	6.214	5.219	4.675	4.336	4.105	3.939	3.813	3.715
	.01	38.620	14.796	9.556	7.491	6.428	5.791	5.370	5.072	4.852	4.682

TABLE 13 Critical values of Student's t-distribution based on Šidák's multiplicative inequality

k	α	2	3	4	5	6	7	8	9	10	11
		Degrees of freedom ν									
16	.1	12.283	6.778	5.189	4.473	4.072	3.818	3.643	3.515	3.419	3.343
	.05	17.633	8.696	6.323	5.297	4.737	4.389	4.152	3.981	3.852	3.752
	.01	39.887	15.121	9.717	7.596	6.506	5.854	5.424	5.120	4.895	4.721
17	.1	12.663	6.923	5.278	4.538	4.125	3.864	3.685	3.554	3.455	3.377
	.05	18.178	8.879	6.428	5.371	4.796	4.439	4.196	4.021	3.889	3.787
	.01	41.116	15.433	9.871	7.695	6.580	5.914	5.475	5.165	4.936	4.759
18	.1	13.032	7.062	5.363	4.601	4.176	3.908	3.724	3.590	3.489	3.409
	.05	18.706	9.054	6.527	5.441	4.852	4.486	4.238	4.059	3.924	3.820
	.01	42.308	15.733	10.018	7.790	6.651	5.971	5.523	5.208	4.974	4.795
19	.1	13.392	7.196	5.444	4.660	4.225	3.950	3.762	3.625	3.521	3.439
	.05	19.220	9.223	6.623	5.509	4.905	4.531	4.278	4.095	3.958	3.851
	.01	43.468	16.021	10.158	7.881	6.718	6.025	5.569	5.248	5.011	4.828
20	.1	13.741	7.326	5.521	4.718	4.272	3.990	3.798	3.658	3.552	3.468
	.05	19.721	9.387	6.714	5.573	4.956	4.574	4.316	4.130	3.989	3.880
	.01	44.598	16.300	10.294	7.968	6.782	6.077	5.613	5.287	5.046	4.860
21	.1	14.083	7.451	5.596	4.772	4.316	4.029	3.832	3.689	3.581	3.496
	.05	20.209	9.544	6.803	5.635	5.005	4.615	4.352	4.162	4.020	3.909
	.01	45.700	16.570	10.424	8.051	6.844	6.126	5.655	5.324	5.079	4.891
22	.1	14.416	7.572	5.668	4.825	4.359	4.065	3.865	3.719	3.609	3.522
	.05	20.686	9.697	6.888	5.695	5.052	4.655	4.386	4.194	4.049	3.936
	.01	46.776	16.831	10.549	8.131	6.903	6.174	5.696	5.359	5.111	4.920
23	.1	14.741	7.689	5.738	4.876	4.400	4.100	3.896	3.748	3.636	3.548
	.05	21.152	9.846	6.970	5.752	5.097	4.693	4.420	4.224	4.076	3.962
	.01	47.828	17.084	10.671	8.208	6.960	6.220	5.734	5.393	5.141	4.948
24	.1	15.060	7.803	5.805	4.925	4.439	4.134	3.926	3.775	3.661	3.572
	.05	21.608	9.990	7.050	5.808	5.141	4.729	4.452	4.252	4.103	3.986
	.01	48.857	17.330	10.788	8.283	7.015	6.263	5.771	5.426	5.171	4.975
25	.1	15.371	7.914	5.870	4.972	4.477	4.167	3.955	3.802	3.686	3.595
	.05	22.054	10.130	7.127	5.861	5.182	4.764	4.482	4.280	4.128	4.010
	.01	49.865	17.569	10.902	8.355	7.068	6.306	5.807	5.457	5.199	5.001
26	.1	15.677	8.022	5.934	5.017	4.514	4.198	3.983	3.827	3.710	3.617
	.05	22.492	10.266	7.201	5.913	5.223	4.798	4.512	4.307	4.153	4.033
	.01	50.853	17.802	11.012	8.425	7.119	6.347	5.842	5.487	5.226	5.026
27	.1	15.977	8.127	5.995	5.062	4.549	4.229	4.010	3.852	3.732	3.639
	.05	22.921	10.399	7.274	5.963	5.262	4.831	4.540	4.333	4.177	4.055
	.01	51.822	18.029	11.120	8.492	7.168	6.386	5.875	5.516	5.252	5.050
28	.1	16.271	8.230	6.055	5.105	4.584	4.258	4.036	3.876	3.755	3.660
	.05	23.343	10.528	7.344	6.012	5.300	4.862	4.568	4.357	4.199	4.077
	.01	52.773	18.251	11.224	8.558	7.216	6.424	5.907	5.545	5.278	5.073
29	.1	16.560	8.330	6.113	5.146	4.617	4.286	4.061	3.899	3.776	3.680
	.05	23.757	10.654	7.413	6.059	5.336	4.893	4.595	4.382	4.222	4.097
	.01	53.707	18.467	11.325	8.622	7.263	6.461	5.938	5.572	5.302	5.095
30	.1	16.845	8.427	6.169	5.187	4.649	4.314	4.086	3.921	3.796	3.699
	.05	24.163	10.778	7.480	6.105	5.372	4.923	4.621	4.405	4.243	4.117
	.01	54.626	18.678	11.424	8.684	7.308	6.497	5.969	5.598	5.326	5.117

TABLE 13 Critical values of Student's t-distribution based on Šidák's multiplicative inequality

		Degrees of freedom ν									
k	α	**2**	**3**	**4**	**5**	**6**	**7**	**8**	**9**	**10**	**11**
31	.1	17.124	8.523	6.224	5.226	4.681	4.341	4.110	3.943	3.816	3.718
	.05	24.564	10.899	7.545	6.149	5.407	4.952	4.646	4.427	4.264	4.137
	.01	55.529	18.885	11.521	8.744	7.352	6.532	5.998	5.624	5.349	5.138
32	.1	17.399	8.616	6.278	5.264	4.711	4.367	4.133	3.963	3.836	3.736
	.05	24.957	11.017	7.608	6.193	5.440	4.980	4.670	4.449	4.284	4.155
	.01	56.418	19.087	11.615	8.803	7.395	6.566	6.026	5.649	5.371	5.159
33	.1	17.670	8.708	6.330	5.302	4.741	4.392	4.155	3.984	3.855	3.754
	.05	25.345	11.133	7.670	6.235	5.473	5.007	4.694	4.471	4.303	4.174
	.01	57.293	19.285	11.707	8.860	7.436	6.599	6.054	5.673	5.393	5.178
34	.1	17.936	8.798	6.381	5.338	4.770	4.417	4.177	4.004	3.873	3.771
	.05	25.727	11.246	7.731	6.277	5.505	5.033	4.717	4.491	4.322	4.191
	.01	58.155	19.479	11.796	8.916	7.477	6.631	6.081	5.697	5.414	5.198
35	.1	18.199	8.886	6.432	5.374	4.798	4.441	4.198	4.023	3.891	3.788
	.05	26.103	11.357	7.790	6.317	5.536	5.059	4.740	4.512	4.341	4.209
	.01	59.004	19.670	11.884	8.971	7.516	6.663	6.107	5.720	5.434	5.216
36	.1	18.458	8.972	6.480	5.408	4.826	4.464	4.218	4.041	3.908	3.804
	.05	26.474	11.466	7.848	6.356	5.567	5.085	4.762	4.531	4.359	4.225
	.01	59.841	19.856	11.970	9.024	7.555	6.693	6.133	5.742	5.454	5.235
37	.1	18.713	9.056	6.528	5.442	4.853	4.487	4.238	4.060	3.925	3.820
	.05	26.839	11.573	7.905	6.395	5.596	5.109	4.783	4.550	4.376	4.242
	.01	60.667	20.040	12.054	9.076	7.593	6.723	6.158	5.764	5.474	5.252
38	.1	18.965	9.140	6.575	5.475	4.879	4.509	4.258	4.077	3.941	3.835
	.05	27.200	11.678	7.961	6.433	5.625	5.133	4.804	4.569	4.393	4.257
	.01	61.481	20.220	12.137	9.127	7.630	6.752	6.182	5.785	5.493	5.270
39	.1	19.213	9.221	6.621	5.508	4.905	4.531	4.277	4.095	3.957	3.850
	.05	27.556	11.782	8.015	6.470	5.654	5.157	4.824	4.587	4.410	4.273
	.01	62.285	20.396	12.218	9.177	7.666	6.781	6.206	5.806	5.511	5.287
40	.1	19.459	9.301	6.667	5.540	4.930	4.552	4.296	4.112	3.973	3.865
	.05	27.908	11.883	8.069	6.506	5.682	5.180	4.844	4.605	4.426	4.288
	.01	63.079	20.570	12.297	9.226	7.701	6.809	6.230	5.826	5.529	5.303
41	.1	19.701	9.380	6.711	5.571	4.954	4.573	4.314	4.128	3.988	3.879
	.05	28.255	11.983	8.121	6.541	5.709	5.202	4.863	4.622	4.442	4.303
	.01	63.863	20.741	12.375	9.274	7.735	6.836	6.252	5.846	5.547	5.319
42	.1	19.941	9.458	6.754	5.601	4.979	4.593	4.332	4.144	4.003	3.893
	.05	28.598	12.081	8.173	6.576	5.735	5.224	4.882	4.639	4.458	4.317
	.01	64.637	20.909	12.451	9.321	7.769	6.863	6.275	5.865	5.564	5.335
43	.1	20.177	9.534	6.797	5.631	5.002	4.613	4.349	4.160	4.018	3.907
	.05	28.936	12.178	8.223	6.610	5.761	5.245	4.901	4.656	4.473	4.331
	.01	65.402	21.075	12.526	9.367	7.802	6.889	6.296	5.884	5.581	5.350
44	.1	20.411	9.609	6.839	5.661	5.025	4.632	4.366	4.176	4.032	3.920
	.05	29.271	12.273	8.273	6.643	5.787	5.266	4.919	4.672	4.488	4.345
	.01	66.159	21.238	12.600	9.412	7.835	6.914	6.318	5.902	5.598	5.365
45	.1	20.642	9.683	6.880	5.689	5.048	4.651	4.383	4.191	4.046	3.933
	.05	29.603	12.366	8.322	6.676	5.812	5.287	4.937	4.688	4.502	4.358
	.01	66.906	21.398	12.672	9.457	7.867	6.939	6.339	5.921	5.614	5.380

TABLE 13 Critical values of Student's t-distribution based on Šidák's multiplicative inequality

k	α	2	3	4	5	6	7	8	9	10	11
		Degrees of freedom ν									
46	.1	20.871	9.756	6.921	5.718	5.070	4.670	4.400	4.206	4.060	3.946
	.05	29.930	12.459	8.370	6.708	5.836	5.307	4.955	4.704	4.516	4.372
	.01	67.646	21.557	12.743	9.500	7.898	6.964	6.359	5.938	5.630	5.395
47	.1	21.097	9.828	6.960	5.746	5.092	4.688	4.416	4.220	4.073	3.959
	.05	30.254	12.550	8.417	6.740	5.861	5.327	4.972	4.719	4.530	4.385
	.01	68.377	21.712	12.814	9.543	7.929	6.988	6.379	5.956	5.645	5.409
48	.1	21.321	9.899	7.000	5.773	5.113	4.706	4.432	4.234	4.086	3.971
	.05	30.574	12.639	8.463	6.771	5.884	5.347	4.989	4.734	4.544	4.397
	.01	69.101	21.866	12.883	9.585	7.959	7.012	6.399	5.973	5.661	5.423
49	.1	21.542	9.969	7.038	5.800	5.134	4.724	4.447	4.248	4.099	3.983
	.05	30.892	12.728	8.509	6.801	5.908	5.366	5.005	4.749	4.557	4.410
	.01	69.817	22.018	12.950	9.627	7.989	7.035	6.418	5.990	5.676	5.436
50	.1	21.761	10.038	7.076	5.826	5.155	4.741	4.462	4.262	4.112	3.995
	.05	31.206	12.815	8.554	6.831	5.930	5.385	5.021	4.763	4.571	4.422
	.01	70.526	22.167	13.017	9.668	8.018	7.058	6.437	6.006	5.690	5.450
51	.1	21.978	10.106	7.114	5.852	5.175	4.758	4.477	4.275	4.124	4.006
	.05	31.516	12.901	8.599	6.861	5.953	5.403	5.037	4.777	4.583	4.434
	.01	71.228	22.315	13.083	9.708	8.046	7.080	6.456	6.022	5.705	5.463
52	.1	22.193	10.173	7.150	5.878	5.195	4.775	4.492	4.289	4.136	4.018
	.05	31.824	12.986	8.642	6.890	5.975	5.421	5.053	4.791	4.596	4.446
	.01	71.923	22.460	13.148	9.747	8.074	7.102	6.474	6.038	5.719	5.475
53	.1	22.406	10.239	7.187	5.903	5.215	4.791	4.506	4.302	4.148	4.029
	.05	32.129	13.070	8.685	6.919	5.997	5.439	5.068	4.805	4.609	4.457
	.01	72.612	22.604	13.212	9.786	8.102	7.124	6.492	6.054	5.733	5.488
54	.1	22.617	10.304	7.222	5.928	5.234	4.808	4.520	4.314	4.160	4.040
	.05	32.431	13.153	8.728	6.947	6.018	5.456	5.083	4.818	4.621	4.468
	.01	73.294	22.746	13.275	9.824	8.129	7.145	6.510	6.069	5.746	5.500
55	.1	22.826	10.369	7.258	5.952	5.253	4.823	4.534	4.327	4.171	4.051
	.05	32.730	13.234	8.770	6.975	6.039	5.474	5.098	4.831	4.633	4.479
	.01	73.969	22.886	13.337	9.862	8.156	7.166	6.528	6.084	5.760	5.513
56	.1	23.033	10.433	7.292	5.976	5.272	4.839	4.548	4.339	4.183	4.061
	.05	33.027	13.315	8.811	7.002	6.060	5.491	5.113	4.844	4.645	4.490
	.01	74.639	23.025	13.399	9.899	8.183	7.187	6.545	6.099	5.773	5.525
57	.1	23.238	10.496	7.327	6.000	5.290	4.855	4.561	4.351	4.194	4.071
	.05	33.321	13.395	8.852	7.029	6.080	5.507	5.127	4.857	4.656	4.501
	.01	75.302	23.161	13.459	9.936	8.209	7.207	6.562	6.114	5.786	5.536
58	.1	23.441	10.558	7.361	6.023	5.309	4.870	4.575	4.363	4.205	4.082
	.05	33.612	13.474	8.892	7.056	6.100	5.524	5.141	4.870	4.668	4.512
	.01	75.960	23.297	13.519	9.972	8.235	7.228	6.578	6.128	5.799	5.548
59	.1	23.643	10.620	7.394	6.046	5.326	4.885	4.588	4.375	4.216	4.092
	.05	33.901	13.552	8.931	7.082	6.120	5.540	5.155	4.882	4.679	4.522
	.01	76.612	23.430	13.578	10.008	8.260	7.247	6.595	6.142	5.811	5.559
60	.1	23.843	10.681	7.427	6.069	5.344	4.899	4.600	4.386	4.226	4.102
	.05	34.187	13.629	8.971	7.108	6.140	5.556	5.169	4.894	4.690	4.532
	.01	77.259	23.563	13.636	10.043	8.285	7.267	6.611	6.156	5.824	5.571

TABLE 13 Critical values of Student's t-distribution based on Šidák's multiplicative inequality

Degrees of freedom ν

k	α	12	13	14	15	16	17	18	19	20	21
1	.1	1.782	1.771	1.761	1.753	1.746	1.740	1.734	1.729	1.725	1.721
	.05	2.179	2.160	2.145	2.131	2.120	2.110	2.101	2.093	2.086	2.080
	.01	3.055	3.012	2.977	2.947	2.921	2.898	2.878	2.861	2.845	2.831
2	.1	2.164	2.146	2.131	2.118	2.106	2.096	2.088	2.080	2.073	2.067
	.05	2.553	2.526	2.503	2.483	2.467	2.452	2.439	2.427	2.417	2.408
	.01	3.427	3.371	3.324	3.285	3.251	3.221	3.195	3.173	3.152	3.134
3	.1	2.384	2.361	2.342	2.325	2.311	2.298	2.287	2.277	2.269	2.261
	.05	2.770	2.737	2.709	2.685	2.665	2.647	2.631	2.617	2.605	2.594
	.01	3.647	3.582	3.528	3.482	3.443	3.409	3.379	3.353	3.329	3.308
4	.1	2.539	2.512	2.489	2.470	2.453	2.439	2.426	2.415	2.405	2.395
	.05	2.924	2.886	2.854	2.827	2.804	2.783	2.766	2.750	2.736	2.723
	.01	3.804	3.733	3.673	3.622	3.579	3.541	3.508	3.479	3.454	3.431
5	.1	2.658	2.628	2.603	2.582	2.563	2.547	2.532	2.520	2.508	2.498
	.05	3.044	3.002	2.967	2.937	2.911	2.889	2.869	2.852	2.836	2.822
	.01	3.927	3.850	3.785	3.731	3.684	3.644	3.609	3.578	3.550	3.525
6	.1	2.756	2.723	2.696	2.672	2.652	2.634	2.619	2.605	2.593	2.581
	.05	3.141	3.096	3.058	3.026	2.998	2.974	2.953	2.934	2.918	2.903
	.01	4.029	3.946	3.878	3.820	3.771	3.728	3.691	3.658	3.629	3.602
7	.1	2.838	2.803	2.774	2.748	2.726	2.708	2.691	2.676	2.663	2.651
	.05	3.224	3.176	3.135	3.101	3.072	3.046	3.024	3.004	2.986	2.970
	.01	4.114	4.028	3.956	3.895	3.844	3.799	3.760	3.725	3.695	3.667
8	.1	2.910	2.872	2.841	2.814	2.791	2.771	2.753	2.738	2.724	2.711
	.05	3.296	3.245	3.202	3.166	3.135	3.108	3.085	3.064	3.045	3.029
	.01	4.189	4.099	4.024	3.961	3.907	3.860	3.820	3.784	3.752	3.724
9	.1	2.973	2.933	2.900	2.872	2.848	2.826	2.808	2.791	2.777	2.764
	.05	3.359	3.306	3.261	3.224	3.191	3.163	3.138	3.116	3.097	3.080
	.01	4.256	4.162	4.084	4.019	3.963	3.914	3.872	3.835	3.802	3.773
10	.1	3.029	2.988	2.953	2.924	2.898	2.876	2.857	2.839	2.824	2.810
	.05	3.416	3.361	3.314	3.275	3.241	3.212	3.186	3.163	3.143	3.125
	.01	4.315	4.218	4.138	4.071	4.013	3.963	3.920	3.881	3.848	3.817
11	.1	3.080	3.037	3.001	2.970	2.944	2.921	2.901	2.883	2.867	2.853
	.05	3.468	3.410	3.362	3.321	3.286	3.256	3.229	3.206	3.185	3.166
	.01	4.369	4.270	4.187	4.118	4.058	4.007	3.962	3.923	3.888	3.857
12	.1	3.127	3.082	3.045	3.013	2.985	2.962	2.941	2.922	2.906	2.891
	.05	3.515	3.455	3.406	3.364	3.327	3.296	3.269	3.245	3.223	3.204
	.01	4.419	4.317	4.232	4.160	4.100	4.047	4.001	3.961	3.926	3.894
13	.1	3.170	3.124	3.085	3.052	3.024	2.999	2.977	2.958	2.941	2.926
	.05	3.558	3.497	3.446	3.402	3.365	3.333	3.305	3.280	3.258	3.238
	.01	4.465	4.360	4.273	4.200	4.138	4.084	4.037	3.996	3.960	3.927
14	.1	3.210	3.162	3.122	3.088	3.059	3.034	3.011	2.992	2.974	2.959
	.05	3.598	3.535	3.483	3.439	3.400	3.367	3.338	3.313	3.290	3.270
	.01	4.507	4.400	4.311	4.237	4.173	4.118	4.071	4.029	3.992	3.958
15	.1	3.247	3.198	3.157	3.122	3.092	3.066	3.043	3.023	3.005	2.989
	.05	3.636	3.571	3.518	3.472	3.433	3.399	3.370	3.343	3.320	3.300
	.01	4.547	4.438	4.347	4.271	4.206	4.150	4.102	4.059	4.021	3.987

TABLE 13 Critical values of Student's t-distribution based on Šidák's multiplicative inequality

k	α	12	13	14	15	16	17	18	19	20	21
		Degrees of freedom ν									
16	.1	3.281	3.231	3.189	3.153	3.122	3.096	3.072	3.052	3.033	3.017
	.05	3.671	3.605	3.550	3.503	3.464	3.429	3.399	3.372	3.348	3.327
	.01	4.584	4.473	4.381	4.303	4.237	4.180	4.131	4.087	4.049	4.014
17	.1	3.314	3.262	3.219	3.183	3.151	3.124	3.100	3.079	3.060	3.043
	.05	3.704	3.637	3.581	3.533	3.492	3.457	3.426	3.399	3.375	3.353
	.01	4.619	4.506	4.412	4.333	4.266	4.208	4.158	4.114	4.075	4.040
18	.1	3.345	3.292	3.248	3.211	3.178	3.151	3.126	3.105	3.085	3.068
	.05	3.735	3.667	3.609	3.561	3.519	3.483	3.452	3.424	3.399	3.377
	.01	4.652	4.537	4.442	4.362	4.293	4.235	4.184	4.139	4.099	4.064
19	.1	3.374	3.320	3.275	3.237	3.204	3.176	3.151	3.129	3.109	3.092
	.05	3.765	3.695	3.637	3.587	3.545	3.508	3.476	3.448	3.423	3.400
	.01	4.684	4.567	4.470	4.389	4.319	4.260	4.208	4.162	4.122	4.086
20	.1	3.402	3.347	3.301	3.262	3.228	3.199	3.174	3.152	3.132	3.114
	.05	3.793	3.722	3.662	3.612	3.569	3.532	3.499	3.470	3.445	3.422
	.01	4.714	4.595	4.497	4.414	4.344	4.284	4.231	4.185	4.144	4.108
21	.1	3.428	3.372	3.325	3.286	3.252	3.222	3.196	3.173	3.153	3.135
	.05	3.820	3.747	3.687	3.636	3.592	3.554	3.521	3.492	3.466	3.443
	.01	4.742	4.622	4.522	4.439	4.368	4.306	4.253	4.206	4.165	4.128
22	.1	3.453	3.396	3.349	3.308	3.274	3.244	3.217	3.194	3.173	3.155
	.05	3.846	3.772	3.710	3.659	3.614	3.576	3.542	3.512	3.486	3.463
	.01	4.769	4.647	4.547	4.462	4.390	4.328	4.274	4.227	4.185	4.147
23	.1	3.477	3.419	3.371	3.330	3.295	3.264	3.237	3.214	3.193	3.174
	.05	3.870	3.795	3.733	3.680	3.635	3.596	3.562	3.532	3.505	3.481
	.01	4.795	4.672	4.570	4.484	4.411	4.349	4.294	4.246	4.204	4.166
24	.1	3.500	3.441	3.392	3.351	3.315	3.284	3.257	3.233	3.211	3.192
	.05	3.894	3.818	3.754	3.701	3.655	3.616	3.581	3.551	3.524	3.499
	.01	4.821	4.695	4.592	4.506	4.432	4.368	4.313	4.265	4.222	4.184
25	.1	3.522	3.463	3.413	3.370	3.334	3.303	3.275	3.251	3.229	3.210
	.05	3.916	3.839	3.775	3.721	3.675	3.634	3.599	3.569	3.541	3.517
	.01	4.845	4.718	4.614	4.526	4.451	4.387	4.332	4.283	4.239	4.201
26	.1	3.544	3.483	3.432	3.389	3.353	3.321	3.293	3.268	3.246	3.227
	.05	3.938	3.860	3.795	3.740	3.693	3.653	3.617	3.586	3.558	3.533
	.01	4.868	4.740	4.634	4.546	4.470	4.406	4.349	4.300	4.256	4.217
27	.1	3.564	3.503	3.451	3.408	3.371	3.338	3.310	3.285	3.263	3.243
	.05	3.959	3.880	3.814	3.758	3.711	3.670	3.634	3.602	3.574	3.549
	.01	4.890	4.761	4.654	4.565	4.489	4.423	4.366	4.316	4.272	4.233
28	.1	3.584	3.522	3.470	3.426	3.388	3.355	3.326	3.301	3.279	3.259
	.05	3.979	3.899	3.832	3.776	3.728	3.687	3.650	3.618	3.590	3.565
	.01	4.912	4.781	4.673	4.583	4.506	4.440	4.383	4.332	4.288	4.248
29	.1	3.603	3.540	3.487	3.443	3.404	3.371	3.342	3.317	3.294	3.274
	.05	3.998	3.917	3.850	3.793	3.745	3.703	3.666	3.634	3.605	3.580
	.01	4.932	4.801	4.692	4.601	4.523	4.457	4.399	4.348	4.303	4.263
30	.1	3.621	3.557	3.504	3.459	3.420	3.387	3.358	3.332	3.309	3.288
	.05	4.017	3.935	3.867	3.810	3.761	3.718	3.681	3.649	3.620	3.594
	.01	4.953	4.819	4.710	4.618	4.540	4.472	4.414	4.363	4.317	4.277

TABLE 13 Critical values of Student's t-distribution based on Šidák's multiplicative inequality

Degrees of freedom ν

k	α	12	13	14	15	16	17	18	19	20	21
31	.1	3.639	3.575	3.521	3.475	3.436	3.402	3.372	3.346	3.323	3.302
	.05	4.035	3.953	3.884	3.826	3.776	3.733	3.696	3.663	3.634	3.608
	.01	4.972	4.838	4.727	4.634	4.556	4.488	4.429	4.377	4.331	4.291
32	.1	3.657	3.591	3.537	3.490	3.451	3.417	3.387	3.360	3.337	3.316
	.05	4.053	3.969	3.900	3.841	3.791	3.748	3.710	3.677	3.647	3.621
	.01	4.991	4.856	4.744	4.650	4.571	4.503	4.443	4.391	4.345	4.304
33	.1	3.673	3.607	3.552	3.505	3.465	3.431	3.401	3.374	3.350	3.329
	.05	4.070	3.986	3.916	3.857	3.806	3.762	3.724	3.690	3.661	3.634
	.01	5.009	4.873	4.760	4.666	4.586	4.517	4.457	4.405	4.358	4.317
34	.1	3.690	3.623	3.567	3.520	3.480	3.445	3.414	3.387	3.363	3.342
	.05	4.087	4.002	3.931	3.871	3.820	3.776	3.737	3.704	3.673	3.647
	.01	5.027	4.890	4.776	4.681	4.600	4.531	4.471	4.418	4.371	4.329
35	.1	3.705	3.638	3.582	3.534	3.493	3.458	3.427	3.400	3.376	3.354
	.05	4.103	4.017	3.946	3.885	3.834	3.789	3.750	3.716	3.686	3.659
	.01	5.045	4.906	4.792	4.696	4.614	4.544	4.484	4.430	4.383	4.342
36	.1	3.721	3.653	3.596	3.548	3.507	3.471	3.440	3.412	3.388	3.366
	.05	4.119	4.032	3.960	3.899	3.847	3.802	3.763	3.729	3.698	3.671
	.01	5.061	4.922	4.806	4.710	4.628	4.558	4.497	4.443	4.395	4.353
37	.1	3.736	3.667	3.610	3.561	3.520	3.484	3.452	3.424	3.400	3.378
	.05	4.134	4.047	3.974	3.913	3.860	3.815	3.775	3.741	3.710	3.682
	.01	5.078	4.937	4.821	4.724	4.641	4.571	4.509	4.455	4.407	4.365
38	.1	3.750	3.681	3.623	3.574	3.532	3.496	3.464	3.436	3.411	3.389
	.05	4.149	4.061	3.987	3.926	3.873	3.827	3.787	3.752	3.721	3.693
	.01	5.094	4.952	4.835	4.737	4.654	4.583	4.521	4.467	4.419	4.376
39	.1	3.765	3.695	3.636	3.587	3.544	3.508	3.476	3.448	3.422	3.400
	.05	4.164	4.075	4.001	3.938	3.885	3.839	3.799	3.764	3.732	3.704
	.01	5.110	4.967	4.849	4.751	4.667	4.595	4.533	4.478	4.430	4.387
40	.1	3.779	3.708	3.649	3.599	3.556	3.519	3.487	3.459	3.433	3.411
	.05	4.178	4.088	4.014	3.951	3.897	3.851	3.810	3.775	3.743	3.715
	.01	5.125	4.981	4.863	4.764	4.679	4.607	4.544	4.489	4.441	4.397
41	.1	3.792	3.721	3.661	3.611	3.568	3.531	3.498	3.470	3.444	3.421
	.05	4.192	4.101	4.026	3.963	3.909	3.862	3.821	3.786	3.754	3.725
	.01	5.140	4.995	4.876	4.776	4.691	4.619	4.556	4.500	4.451	4.408
42	.1	3.805	3.733	3.673	3.623	3.579	3.542	3.509	3.480	3.454	3.432
	.05	4.205	4.114	4.038	3.975	3.920	3.873	3.832	3.796	3.764	3.736
	.01	5.154	5.009	4.889	4.788	4.703	4.630	4.566	4.511	4.462	4.418
43	.1	3.818	3.746	3.685	3.634	3.591	3.553	3.520	3.491	3.465	3.442
	.05	4.218	4.127	4.050	3.986	3.931	3.884	3.843	3.806	3.774	3.745
	.01	5.169	5.022	4.901	4.800	4.715	4.641	4.577	4.521	4.472	4.428
44	.1	3.831	3.758	3.697	3.646	3.601	3.563	3.530	3.501	3.475	3.451
	.05	4.231	4.139	4.062	3.998	3.942	3.895	3.853	3.816	3.784	3.755
	.01	5.183	5.035	4.914	4.812	4.726	4.652	4.588	4.531	4.482	4.437
45	.1	3.843	3.770	3.708	3.656	3.612	3.574	3.540	3.511	3.484	3.461
	.05	4.244	4.151	4.074	4.009	3.953	3.905	3.863	3.826	3.794	3.765
	.01	5.196	5.048	4.926	4.824	4.737	4.662	4.598	4.541	4.491	4.447

TABLE 13 Critical values of Student's t-distribution based on Šidák's multiplicative inequality

Degrees of freedom ν

k	α	12	13	14	15	16	17	18	19	20	21
46	.1	3.855	3.781	3.719	3.667	3.623	3.584	3.550	3.520	3.494	3.470
	.05	4.256	4.163	4.085	4.019	3.963	3.915	3.873	3.836	3.803	3.774
	.01	5.209	5.060	4.938	4.835	4.748	4.673	4.608	4.551	4.501	4.456
47	.1	3.867	3.792	3.730	3.678	3.633	3.594	3.560	3.530	3.503	3.479
	.05	4.269	4.174	4.096	4.030	3.974	3.925	3.883	3.845	3.812	3.783
	.01	5.223	5.073	4.949	4.846	4.758	4.683	4.618	4.560	4.510	4.465
48	.1	3.879	3.804	3.741	3.688	3.643	3.603	3.569	3.539	3.512	3.488
	.05	4.281	4.186	4.107	4.040	3.984	3.935	3.892	3.855	3.821	3.792
	.01	5.235	5.084	4.960	4.857	4.768	4.693	4.627	4.570	4.519	4.474
49	.1	3.890	3.814	3.751	3.698	3.652	3.613	3.578	3.548	3.521	3.497
	.05	4.292	4.197	4.117	4.050	3.993	3.944	3.901	3.864	3.830	3.800
	.01	5.248	5.096	4.971	4.867	4.779	4.703	4.637	4.579	4.528	4.482
50	.1	3.901	3.825	3.761	3.708	3.662	3.622	3.587	3.557	3.530	3.505
	.05	4.304	4.207	4.128	4.060	4.003	3.954	3.910	3.872	3.839	3.809
	.01	5.260	5.108	4.982	4.877	4.788	4.712	4.646	4.588	4.536	4.491
51	.1	3.912	3.835	3.772	3.718	3.671	3.631	3.596	3.566	3.538	3.514
	.05	4.315	4.218	4.138	4.070	4.013	3.963	3.919	3.881	3.847	3.817
	.01	5.272	5.119	4.993	4.888	4.798	4.721	4.655	4.596	4.545	4.499
52	.1	3.923	3.846	3.781	3.727	3.680	3.640	3.605	3.574	3.547	3.522
	.05	4.326	4.228	4.148	4.080	4.022	3.972	3.928	3.890	3.856	3.825
	.01	5.284	5.130	5.004	4.898	4.808	4.731	4.664	4.605	4.553	4.507
53	.1	3.934	3.856	3.791	3.736	3.689	3.649	3.614	3.582	3.555	3.530
	.05	4.337	4.239	4.157	4.089	4.031	3.980	3.937	3.898	3.864	3.833
	.01	5.296	5.141	5.014	4.907	4.817	4.740	4.672	4.613	4.561	4.515
54	.1	3.944	3.866	3.800	3.745	3.698	3.658	3.622	3.591	3.563	3.538
	.05	4.347	4.249	4.167	4.098	4.040	3.989	3.945	3.906	3.872	3.841
	.01	5.307	5.152	5.024	4.917	4.826	4.748	4.681	4.622	4.569	4.523
55	.1	3.954	3.875	3.810	3.754	3.707	3.666	3.630	3.599	3.571	3.546
	.05	4.358	4.259	4.177	4.107	4.048	3.998	3.953	3.914	3.880	3.849
	.01	5.319	5.162	5.034	4.926	4.835	4.757	4.689	4.630	4.577	4.531
56	.1	3.964	3.885	3.819	3.763	3.716	3.674	3.638	3.607	3.579	3.553
	.05	4.368	4.268	4.186	4.116	4.057	4.006	3.961	3.922	3.887	3.856
	.01	5.330	5.173	5.044	4.936	4.844	4.766	4.697	4.638	4.585	4.538
57	.1	3.974	3.894	3.828	3.772	3.724	3.682	3.646	3.614	3.586	3.561
	.05	4.378	4.278	4.195	4.125	4.065	4.014	3.969	3.930	3.895	3.864
	.01	5.341	5.183	5.053	4.945	4.853	4.774	4.706	4.646	4.593	4.546
58	.1	3.983	3.903	3.837	3.780	3.732	3.690	3.654	3.622	3.594	3.568
	.05	4.388	4.287	4.204	4.134	4.074	4.022	3.977	3.937	3.902	3.871
	.01	5.351	5.193	5.063	4.954	4.861	4.782	4.713	4.653	4.600	4.553
59	.1	3.993	3.912	3.845	3.789	3.740	3.698	3.662	3.630	3.601	3.575
	.05	4.398	4.296	4.213	4.142	4.082	4.030	3.985	3.945	3.910	3.878
	.01	5.362	5.203	5.072	4.962	4.870	4.790	4.721	4.661	4.608	4.560
60	.1	4.002	3.921	3.854	3.797	3.748	3.706	3.669	3.637	3.608	3.583
	.05	4.407	4.305	4.221	4.150	4.090	4.038	3.992	3.952	3.917	3.885
	.01	5.372	5.212	5.081	4.971	4.878	4.798	4.729	4.668	4.615	4.567

TABLE 13 Critical values of Student's t-distribution based on Šidák's multiplicative inequality

k	α	22	23	24	26	28	30	40	60	120	∞
		Degrees of freedom ν									
1	.1	1.717	1.714	1.711	1.706	1.701	1.697	1.684	1.671	1.658	1.645
	.05	2.074	2.069	2.064	2.056	2.048	2.042	2.021	2.000	1.980	1.960
	.01	2.819	2.807	2.797	2.779	2.763	2.750	2.704	2.660	2.617	2.576
2	.1	2.061	2.056	2.051	2.043	2.036	2.030	2.009	1.989	1.968	1.949
	.05	2.400	2.392	2.385	2.373	2.363	2.354	2.323	2.294	2.265	2.236
	.01	3.118	3.103	3.089	3.066	3.046	3.029	2.970	2.914	2.859	2.806
3	.1	2.254	2.247	2.241	2.231	2.222	2.215	2.189	2.163	2.138	2.114
	.05	2.584	2.574	2.566	2.551	2.539	2.528	2.492	2.456	2.422	2.388
	.01	3.289	3.272	3.257	3.230	3.207	3.188	3.121	3.056	2.994	2.934
4	.1	2.387	2.380	2.373	2.361	2.351	2.342	2.312	2.283	2.254	2.226
	.05	2.712	2.701	2.692	2.675	2.661	2.649	2.608	2.568	2.529	2.491
	.01	3.410	3.392	3.375	3.345	3.320	3.298	3.225	3.155	3.087	3.022
5	.1	2.489	2.481	2.473	2.460	2.449	2.439	2.406	2.373	2.342	2.311
	.05	2.810	2.798	2.788	2.770	2.755	2.742	2.696	2.653	2.610	2.569
	.01	3.503	3.483	3.465	3.433	3.407	3.384	3.305	3.230	3.158	3.089
6	.1	2.572	2.563	2.554	2.540	2.528	2.517	2.481	2.446	2.411	2.378
	.05	2.889	2.877	2.866	2.847	2.830	2.816	2.768	2.721	2.675	2.631
	.01	3.579	3.558	3.539	3.505	3.477	3.453	3.370	3.291	3.215	3.143
7	.1	2.641	2.631	2.622	2.607	2.594	2.582	2.544	2.506	2.469	2.434
	.05	2.956	2.943	2.931	2.911	2.893	2.878	2.827	2.777	2.729	2.683
	.01	3.643	3.621	3.601	3.566	3.536	3.511	3.425	3.342	3.263	3.188
8	.1	2.700	2.690	2.680	2.664	2.650	2.638	2.597	2.558	2.519	2.481
	.05	3.014	3.000	2.988	2.966	2.948	2.932	2.878	2.826	2.776	2.727
	.01	3.698	3.675	3.654	3.618	3.587	3.561	3.472	3.386	3.304	3.226
9	.1	2.752	2.741	2.731	2.714	2.700	2.687	2.644	2.603	2.562	2.523
	.05	3.064	3.050	3.037	3.014	2.995	2.979	2.923	2.869	2.816	2.766
	.01	3.747	3.723	3.702	3.664	3.632	3.605	3.513	3.425	3.340	3.260
10	.1	2.798	2.787	2.777	2.759	2.744	2.731	2.686	2.643	2.600	2.560
	.05	3.109	3.094	3.081	3.058	3.038	3.021	2.963	2.906	2.852	2.800
	.01	3.790	3.766	3.744	3.705	3.672	3.644	3.549	3.459	3.372	3.289
11	.1	2.840	2.828	2.818	2.799	2.783	2.770	2.723	2.678	2.635	2.592
	.05	3.150	3.134	3.121	3.096	3.076	3.058	2.998	2.940	2.884	2.830
	.01	3.830	3.804	3.782	3.742	3.708	3.680	3.582	3.489	3.401	3.316
12	.1	2.878	2.866	2.855	2.835	2.819	2.805	2.757	2.711	2.666	2.622
	.05	3.187	3.171	3.157	3.132	3.111	3.092	3.031	2.971	2.913	2.858
	.01	3.865	3.840	3.816	3.776	3.741	3.712	3.612	3.517	3.427	3.340
13	.1	2.912	2.900	2.889	2.869	2.852	2.838	2.788	2.740	2.694	2.649
	.05	3.220	3.204	3.190	3.164	3.142	3.124	3.060	2.999	2.940	2.883
	.01	3.898	3.872	3.848	3.807	3.771	3.742	3.640	3.543	3.451	3.362
14	.1	2.944	2.932	2.920	2.900	2.882	2.868	2.817	2.768	2.720	2.674
	.05	3.252	3.235	3.220	3.194	3.172	3.153	3.088	3.025	2.965	2.906
	.01	3.929	3.902	3.878	3.835	3.799	3.769	3.665	3.567	3.473	3.383
15	.1	2.974	2.961	2.949	2.928	2.911	2.895	2.843	2.793	2.744	2.697
	.05	3.281	3.264	3.249	3.222	3.199	3.180	3.113	3.049	2.987	2.928
	.01	3.957	3.930	3.905	3.862	3.825	3.794	3.689	3.589	3.493	3.402

TABLE 13 Critical values of Student's t-distribution based on Šidák's multiplicative inequality

Degrees of freedom ν

k	α	22	23	24	26	28	30	40	60	120	∞
16	.1	3.002	2.989	2.976	2.955	2.937	2.921	2.868	2.816	2.767	2.718
	.05	3.308	3.291	3.275	3.248	3.224	3.205	3.137	3.071	3.008	2.948
	.01	3.983	3.956	3.931	3.887	3.850	3.818	3.711	3.609	3.512	3.419
17	.1	3.028	3.014	3.002	2.980	2.961	2.946	2.891	2.838	2.787	2.738
	.05	3.333	3.316	3.300	3.272	3.248	3.228	3.159	3.092	3.028	2.966
	.01	4.008	3.980	3.955	3.910	3.872	3.840	3.732	3.628	3.530	3.436
18	.1	3.053	3.039	3.026	3.003	2.985	2.968	2.913	2.859	2.807	2.757
	.05	3.357	3.340	3.323	3.295	3.271	3.250	3.180	3.112	3.047	2.984
	.01	4.032	4.003	3.977	3.932	3.894	3.861	3.751	3.646	3.546	3.451
19	.1	3.076	3.061	3.048	3.026	3.006	2.990	2.933	2.878	2.826	2.774
	.05	3.380	3.362	3.345	3.316	3.292	3.271	3.199	3.130	3.064	3.000
	.01	4.054	4.025	3.999	3.953	3.914	3.881	3.769	3.663	3.562	3.466
20	.1	3.098	3.083	3.070	3.047	3.027	3.010	2.952	2.897	2.843	2.791
	.05	3.402	3.383	3.366	3.337	3.312	3.291	3.218	3.148	3.081	3.016
	.01	4.075	4.046	4.019	3.972	3.933	3.900	3.787	3.679	3.577	3.479
21	.1	3.118	3.104	3.090	3.067	3.047	3.029	2.971	2.914	2.860	2.807
	.05	3.422	3.403	3.386	3.356	3.331	3.309	3.235	3.165	3.096	3.031
	.01	4.095	4.065	4.038	3.991	3.952	3.918	3.803	3.694	3.591	3.493
22	.1	3.138	3.123	3.109	3.085	3.065	3.048	2.988	2.931	2.875	2.822
	.05	3.441	3.422	3.405	3.375	3.349	3.327	3.252	3.180	3.111	3.045
	.01	4.114	4.084	4.057	4.009	3.969	3.935	3.819	3.709	3.604	3.505
23	.1	3.157	3.142	3.128	3.104	3.083	3.065	3.005	2.947	2.890	2.836
	.05	3.460	3.441	3.423	3.392	3.366	3.344	3.268	3.195	3.125	3.058
	.01	4.132	4.102	4.074	4.026	3.985	3.951	3.834	3.723	3.617	3.517
24	.1	3.175	3.160	3.146	3.121	3.100	3.082	3.021	2.962	2.904	2.849
	.05	3.478	3.458	3.440	3.409	3.383	3.360	3.283	3.210	3.139	3.071
	.01	4.150	4.119	4.091	4.042	4.001	3.966	3.848	3.736	3.629	3.528
25	.1	3.193	3.177	3.162	3.137	3.116	3.098	3.036	2.976	2.918	2.862
	.05	3.495	3.475	3.457	3.425	3.399	3.376	3.298	3.223	3.152	3.083
	.01	4.166	4.135	4.107	4.058	4.017	3.981	3.862	3.749	3.641	3.539
26	.1	3.209	3.193	3.179	3.153	3.132	3.113	3.050	2.990	2.931	2.875
	.05	3.511	3.491	3.473	3.441	3.414	3.391	3.312	3.237	3.164	3.095
	.01	4.182	4.151	4.122	4.073	4.031	3.996	3.875	3.761	3.652	3.549
27	.1	3.225	3.209	3.194	3.169	3.147	3.128	3.064	3.003	2.944	2.887
	.05	3.527	3.506	3.488	3.456	3.428	3.405	3.326	3.249	3.176	3.106
	.01	4.198	4.166	4.137	4.087	4.045	4.009	3.888	3.773	3.663	3.559
28	.1	3.240	3.224	3.209	3.183	3.161	3.142	3.078	3.016	2.956	2.898
	.05	3.542	3.521	3.503	3.470	3.442	3.419	3.339	3.261	3.187	3.116
	.01	4.213	4.181	4.152	4.101	4.059	4.023	3.900	3.784	3.673	3.569
29	.1	3.255	3.239	3.224	3.197	3.175	3.156	3.091	3.028	2.967	2.909
	.05	3.557	3.536	3.517	3.484	3.456	3.432	3.351	3.273	3.198	3.127
	.01	4.227	4.195	4.166	4.115	4.072	4.035	3.912	3.795	3.683	3.578
30	.1	3.270	3.253	3.238	3.211	3.188	3.169	3.103	3.040	2.979	2.920
	.05	3.571	3.550	3.531	3.497	3.469	3.445	3.363	3.284	3.209	3.137
	.01	4.241	4.208	4.179	4.128	4.084	4.048	3.923	3.805	3.693	3.587

TABLE 13 Critical values of Student's t-distribution based on Šidák's multiplicative inequality

Degrees of freedom ν

k	α	22	23	24	26	28	30	40	60	120	∞
31	.1	3.283	3.266	3.251	3.224	3.201	3.182	3.115	3.051	2.989	2.930
	.05	3.584	3.563	3.544	3.510	3.482	3.457	3.375	3.295	3.219	3.146
	.01	4.254	4.222	4.192	4.140	4.097	4.060	3.934	3.815	3.702	3.595
32	.1	3.297	3.280	3.264	3.237	3.214	3.194	3.127	3.062	3.000	2.940
	.05	3.597	3.576	3.557	3.523	3.494	3.469	3.386	3.306	3.229	3.156
	.01	4.267	4.234	4.204	4.152	4.108	4.071	3.945	3.825	3.711	3.603
33	.1	3.310	3.292	3.277	3.249	3.226	3.206	3.138	3.073	3.010	2.949
	.05	3.610	3.589	3.569	3.535	3.506	3.481	3.397	3.316	3.239	3.165
	.01	4.280	4.247	4.216	4.164	4.120	4.082	3.955	3.834	3.720	3.611
34	.1	3.322	3.305	3.289	3.261	3.238	3.218	3.149	3.083	3.019	2.958
	.05	3.622	3.601	3.581	3.546	3.517	3.492	3.407	3.326	3.248	3.173
	.01	4.292	4.259	4.228	4.175	4.131	4.093	3.965	3.843	3.728	3.619
35	.1	3.334	3.317	3.301	3.273	3.249	3.229	3.160	3.093	3.029	2.967
	.05	3.634	3.613	3.593	3.558	3.528	3.503	3.418	3.336	3.257	3.182
	.01	4.304	4.270	4.240	4.186	4.142	4.103	3.975	3.852	3.736	3.627
36	.1	3.346	3.329	3.312	3.284	3.260	3.240	3.170	3.103	3.038	2.976
	.05	3.646	3.624	3.604	3.569	3.539	3.514	3.427	3.345	3.266	3.190
	.01	4.316	4.282	4.251	4.197	4.152	4.114	3.984	3.861	3.744	3.634
37	.1	3.358	3.340	3.324	3.295	3.271	3.250	3.180	3.112	3.047	2.984
	.05	3.657	3.635	3.615	3.579	3.550	3.524	3.437	3.354	3.274	3.198
	.01	4.327	4.293	4.262	4.208	4.162	4.124	3.993	3.869	3.752	3.641
38	.1	3.369	3.351	3.334	3.306	3.281	3.261	3.190	3.121	3.056	2.992
	.05	3.669	3.646	3.626	3.590	3.560	3.534	3.446	3.363	3.282	3.205
	.01	4.338	4.303	4.272	4.218	4.172	4.133	4.002	3.877	3.760	3.648
39	.1	3.380	3.362	3.345	3.316	3.292	3.271	3.199	3.130	3.064	3.000
	.05	3.679	3.657	3.636	3.600	3.570	3.544	3.456	3.371	3.290	3.213
	.01	4.348	4.314	4.282	4.228	4.182	4.143	4.010	3.885	3.767	3.655
40	.1	3.390	3.372	3.355	3.326	3.301	3.280	3.208	3.139	3.072	3.008
	.05	3.690	3.667	3.646	3.610	3.579	3.553	3.464	3.379	3.298	3.220
	.01	4.359	4.324	4.292	4.237	4.191	4.152	4.019	3.893	3.774	3.661
41	.1	3.401	3.382	3.365	3.336	3.311	3.290	3.217	3.147	3.080	3.015
	.05	3.700	3.677	3.656	3.620	3.589	3.563	3.473	3.388	3.306	3.227
	.01	4.369	4.334	4.302	4.247	4.200	4.161	4.027	3.901	3.781	3.667
42	.1	3.411	3.392	3.375	3.346	3.321	3.299	3.226	3.155	3.088	3.023
	.05	3.710	3.687	3.666	3.629	3.598	3.572	3.482	3.395	3.313	3.234
	.01	4.379	4.343	4.311	4.256	4.209	4.169	4.035	3.908	3.787	3.673
43	.1	3.421	3.402	3.385	3.355	3.330	3.308	3.234	3.163	3.095	3.030
	.05	3.720	3.696	3.675	3.638	3.607	3.580	3.490	3.403	3.320	3.241
	.01	4.388	4.353	4.321	4.265	4.218	4.178	4.043	3.915	3.794	3.679
44	.1	3.430	3.411	3.394	3.364	3.339	3.317	3.243	3.171	3.103	3.037
	.05	3.729	3.706	3.684	3.647	3.616	3.589	3.498	3.411	3.327	3.247
	.01	4.398	4.362	4.330	4.274	4.226	4.186	4.050	3.922	3.800	3.685
45	.1	3.440	3.421	3.403	3.373	3.347	3.325	3.251	3.179	3.110	3.043
	.05	3.738	3.715	3.693	3.656	3.624	3.597	3.506	3.418	3.334	3.254
	.01	4.407	4.371	4.339	4.282	4.235	4.194	4.058	3.929	3.807	3.691

TABLE 13 Critical values of Student's t-distribution based on Šidák's multiplicative inequality

k	α	22	23	24	26	28	30	40	60	120	∞
		Degrees of freedom ν									
46	.1	3.449	3.430	3.412	3.382	3.356	3.334	3.259	3.186	3.117	3.050
	.05	3.747	3.724	3.702	3.664	3.633	3.606	3.513	3.425	3.341	3.260
	.01	4.416	4.380	4.347	4.290	4.243	4.202	4.065	3.935	3.813	3.697
47	.1	3.458	3.439	3.421	3.390	3.364	3.342	3.266	3.194	3.124	3.057
	.05	3.756	3.732	3.711	3.673	3.641	3.614	3.521	3.432	3.347	3.266
	.01	4.425	4.388	4.356	4.299	4.251	4.210	4.072	3.942	3.819	3.702
48	.1	3.467	3.447	3.429	3.399	3.372	3.350	3.274	3.201	3.130	3.063
	.05	3.765	3.741	3.719	3.681	3.649	3.621	3.528	3.439	3.354	3.272
	.01	4.433	4.397	4.364	4.307	4.259	4.218	4.079	3.948	3.825	3.707
49	.1	3.475	3.456	3.438	3.407	3.380	3.358	3.281	3.208	3.137	3.069
	.05	3.773	3.749	3.727	3.689	3.657	3.629	3.535	3.446	3.360	3.278
	.01	4.442	4.405	4.372	4.314	4.266	4.225	4.086	3.955	3.830	3.713
50	.1	3.484	3.464	3.446	3.415	3.388	3.366	3.289	3.214	3.143	3.075
	.05	3.782	3.757	3.735	3.697	3.664	3.637	3.542	3.452	3.366	3.283
	.01	4.450	4.413	4.380	4.322	4.274	4.232	4.093	3.961	3.836	3.718
51	.1	3.492	3.472	3.454	3.423	3.396	3.373	3.296	3.221	3.150	3.081
	.05	3.790	3.766	3.743	3.705	3.672	3.644	3.549	3.459	3.372	3.289
	.01	4.458	4.421	4.388	4.330	4.281	4.239	4.099	3.967	3.841	3.723
52	.1	3.500	3.480	3.462	3.430	3.404	3.381	3.303	3.228	3.156	3.087
	.05	3.798	3.773	3.751	3.712	3.679	3.651	3.556	3.465	3.378	3.295
	.01	4.466	4.429	4.395	4.337	4.288	4.246	4.106	3.972	3.847	3.728
53	.1	3.508	3.488	3.470	3.438	3.411	3.388	3.309	3.234	3.162	3.092
	.05	3.806	3.781	3.759	3.720	3.687	3.658	3.562	3.471	3.383	3.300
	.01	4.474	4.437	4.403	4.344	4.295	4.253	4.112	3.978	3.852	3.732
54	.1	3.516	3.496	3.477	3.445	3.418	3.395	3.316	3.240	3.168	3.098
	.05	3.814	3.789	3.766	3.727	3.694	3.665	3.569	3.477	3.389	3.305
	.01	4.481	4.444	4.410	4.351	4.302	4.260	4.118	3.984	3.857	3.737
55	.1	3.523	3.503	3.485	3.452	3.425	3.402	3.323	3.246	3.173	3.103
	.05	3.821	3.796	3.773	3.734	3.701	3.672	3.575	3.483	3.395	3.310
	.01	4.489	4.451	4.417	4.358	4.309	4.267	4.124	3.989	3.862	3.742
56	.1	3.531	3.510	3.492	3.460	3.432	3.409	3.329	3.253	3.179	3.109
	.05	3.828	3.803	3.781	3.741	3.707	3.679	3.582	3.489	3.400	3.315
	.01	4.496	4.459	4.425	4.365	4.315	4.273	4.130	3.995	3.867	3.746
57	.1	3.538	3.518	3.499	3.466	3.439	3.415	3.335	3.258	3.185	3.114
	.05	3.836	3.811	3.788	3.748	3.714	3.685	3.588	3.494	3.405	3.320
	.01	4.504	4.466	4.431	4.372	4.322	4.279	4.136	4.000	3.872	3.751
58	.1	3.545	3.525	3.506	3.473	3.446	3.422	3.342	3.264	3.190	3.119
	.05	3.843	3.818	3.795	3.754	3.721	3.692	3.594	3.500	3.411	3.325
	.01	4.511	4.473	4.438	4.379	4.328	4.286	4.141	4.005	3.877	3.755
59	.1	3.553	3.532	3.513	3.480	3.452	3.429	3.348	3.270	3.195	3.124
	.05	3.850	3.824	3.801	3.761	3.727	3.698	3.600	3.505	3.416	3.330
	.01	4.518	4.480	4.445	4.385	4.335	4.292	4.147	4.010	3.881	3.759
60	.1	3.560	3.539	3.520	3.487	3.459	3.435	3.354	3.276	3.201	3.129
	.05	3.857	3.831	3.808	3.768	3.733	3.704	3.605	3.511	3.421	3.335
	.01	4.525	4.486	4.452	4.391	4.341	4.298	4.153	4.015	3.886	3.764

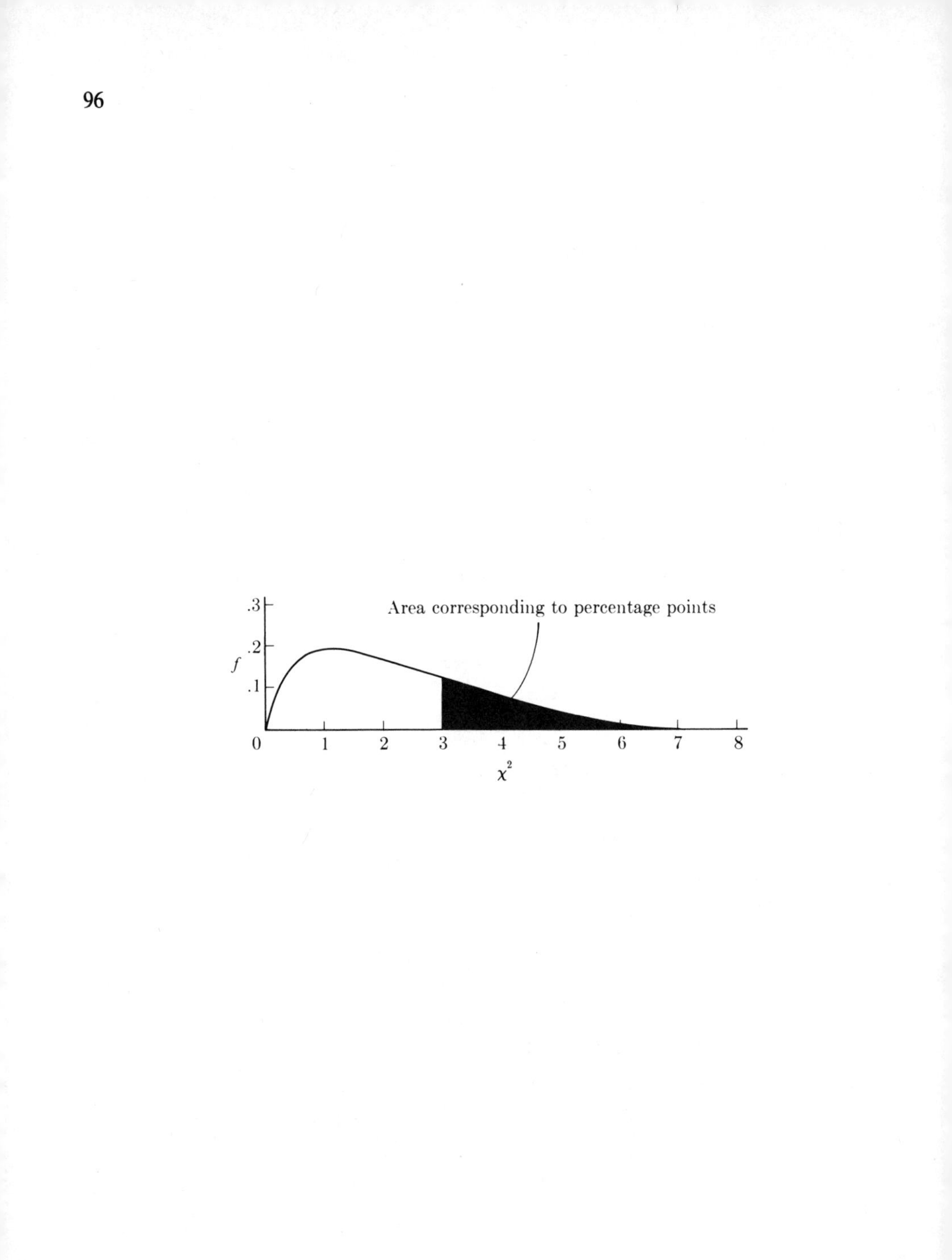
Area corresponding to percentage points
.3
.2
.1
f
0
1
2
3
4
5
6
7
8
χ^2

TABLE **14** Critical values of the chi-square distribution

This table furnishes critical values for the chi-square distribution for degrees of freedom $\nu = 1$ to 100 in increments of 1. The percentage points given (corresponding to $\alpha = 0.995, 0.975, 0.9, 0.5, 0.1, 0.05, 0.025, 0.01, 0.005$, and 0.001) represent the area to the right of the critical value of χ^2 in one tail of the distribution, as shown in the accompanying figure. The critical values of χ^2 are given to three significant decimal places, except in cases when $\chi^2 > 100$, when they are given to two significant decimal places.

To find the critical value of χ^2 for a given number of degrees of freedom, look up ν in the left (argument) column of the table and read off the desired values of χ^2 in that row. For example, for eight degrees of freedom, $\chi^2_{.05[8]} = 15.507$ and $\chi^2_{.01[8]} = 20.090$. The last value indicates that 1% of the area of the chi-square distribution (for eight degrees of freedom) is to the right of the value of $\chi^2 = 20.090$. For values of $\nu > 100$, compute approximate critical values of χ^2 by formula as follows: $\chi^2_{\alpha[\nu]} = \frac{1}{2}(t_{2\alpha[\infty]} + \sqrt{2\nu - 1})^2$, where $t_{2\alpha[\infty]}$ can be looked up in Table **12.** Thus $\chi^2_{.05[120]}$ is computed as $\frac{1}{2}(t_{.10[\infty]} + \sqrt{240 - 1})^2 = \frac{1}{2}(1.645 + \sqrt{239})^2 = \frac{1}{2}(17.10462)^2 = 146.284$. For $\alpha > 0.5$, employ $t_{1-2\alpha[\infty]}$ in the above formula. When $\alpha = 0.5$, $t_{2\alpha} = 0$.

There are numerous applications of the chi-square distribution in statistics. The distribution is described in Section 7.6. It is used to set confidence limits to sample variances (Section 7.7), to test the homogeneity of variances (Section 13.3) and of correlation coefficients (Section 15.5), and for tests of goodness of fit (Chapter 17).

Except for $\alpha = 0.001$, the values of chi-square from 1 to 30 degrees of freedom have been taken from a more extensive table by C. M. Thompson (*Biometrika* **32**:188–189, 1941) with permission of the publisher. Values between 31 and 100 degrees of freedom were approximated using the Cornish–Fisher asymptotic expansion following the account of M. Zelen and N. C. Severo, section 26.2.49 of M. Abramovitz and I. A. Stegun (eds.), *Handbook of Mathematical Functions* (U.S. National Bureau of Standards, 1964). The values of the Hermite polynomials were computed, instead of using the tables furnished in the source above. The values for $\alpha = 0.001$ were all taken from H. L. Harter, *New Tables of the Incomplete Gamman-function Ratio and of Percentage Points of the Chi-Square and Beta Distribution* (U.S. Government Printing Office, 1964).

TABLE 14 Critical values of the chi-square distribution

$\nu \backslash \alpha$	.995	.975	.9	.5	.1	.05	.025	.01	.005	.001	α / ν
1	0.000	0.000	0.016	0.455	2.706	3.841	5.024	6.635	7.879	10.828	1
2	0.010	0.051	0.211	1.386	4.605	5.991	7.378	9.210	10.597	13.816	2
3	0.072	0.216	0.584	2.366	6.251	7.815	9.348	11.345	12.838	16.266	3
4	0.207	0.484	1.064	3.357	7.779	9.488	11.143	13.277	14.860	18.467	4
5	0.412	0.831	1.610	4.351	9.236	11.070	12.832	15.086	16.750	20.515	5
6	0.676	1.237	2.204	5.348	10.645	12.592	14.449	16.812	18.548	22.458	6
7	0.989	1.690	2.833	6.346	12.017	14.067	16.013	18.475	20.278	24.322	7
8	1.344	2.180	3.490	7.344	13.362	15.507	17.535	20.090	21.955	26.124	8
9	1.735	2.700	4.168	8.343	14.684	16.919	19.023	21.666	23.589	27.877	9
10	2.156	3.247	4.865	9.342	15.987	18.307	20.483	23.209	25.188	29.588	10
11	2.603	3.816	5.578	10.341	17.275	19.675	21.920	24.725	26.757	31.264	11
12	3.074	4.404	6.304	11.340	18.549	21.026	23.337	26.217	28.300	32.910	12
13	3.565	5.009	7.042	12.340	19.812	22.362	24.736	27.688	29.819	34.528	13
14	4.075	5.629	7.790	13.339	21.064	23.685	26.119	29.141	31.319	36.123	14
15	4.601	6.262	8.547	14.339	22.307	24.996	27.488	30.578	32.801	37.697	15
16	5.142	6.908	9.312	15.338	23.542	26.296	28.845	32.000	34.267	39.252	16
17	5.697	7.564	10.085	16.338	24.769	27.587	30.191	33.409	35.718	40.790	17
18	6.265	8.231	10.865	17.338	25.989	28.869	31.526	34.805	37.156	42.312	18
19	6.844	8.907	11.651	18.338	27.204	30.144	32.852	36.191	38.582	43.820	19
20	7.434	9.591	12.443	19.337	28.412	31.410	34.170	37.566	39.997	45.315	20
21	8.034	10.283	13.240	20.337	29.615	32.670	35.479	38.932	41.401	46.797	21
22	8.643	10.982	14.042	21.337	30.813	33.924	36.781	40.289	42.796	48.268	22
23	9.260	11.688	14.848	22.337	32.007	35.172	38.076	41.638	44.181	49.728	23
24	9.886	12.401	15.659	23.337	33.196	36.415	39.364	42.980	45.558	51.179	24
25	10.520	13.120	16.473	24.337	34.382	37.652	40.646	44.314	46.928	52.620	25
26	11.160	13.844	17.292	25.336	35.563	38.885	41.923	45.642	48.290	54.052	26
27	11.808	14.573	18.114	26.336	36.741	40.113	43.194	46.963	49.645	55.476	27
28	12.461	15.308	18.939	27.336	37.916	41.337	44.461	48.278	50.993	56.892	28
29	13.121	16.047	19.768	28.336	39.088	42.557	45.722	49.588	52.336	58.301	29
30	13.787	16.791	20.599	29.336	40.256	43.773	46.979	50.892	53.672	59.703	30
31	14.458	17.539	21.434	30.336	41.422	44.985	48.232	52.191	55.003	61.098	31
32	15.134	18.291	22.271	31.336	42.585	46.194	49.480	53.486	56.329	62.487	32
33	15.815	19.047	23.110	32.336	43.745	47.400	50.725	54.776	57.649	63.870	33
34	16.501	19.806	23.952	33.336	44.903	48.602	51.966	56.061	58.964	65.247	34
35	17.192	20.569	24.797	34.336	46.059	49.802	53.203	57.342	60.275	66.619	35
36	17.887	21.336	25.643	35.336	47.212	50.998	54.437	58.619	61.582	67.985	36
37	18.586	22.106	26.492	36.335	48.363	52.192	55.668	59.892	62.884	69.346	37
38	19.289	22.878	27.343	37.335	49.513	53.384	56.896	61.162	64.182	70.703	38
39	19.996	23.654	28.196	38.335	50.660	54.572	58.120	62.428	65.476	72.055	39
40	20.707	24.433	29.051	39.335	51.805	55.758	59.342	63.691	66.766	73.402	40
41	21.421	25.215	29.907	40.335	52.949	56.942	60.561	64.950	68.053	74.745	41
42	22.138	25.999	30.765	41.335	54.090	58.124	61.777	66.206	69.336	76.084	42
43	22.859	26.785	31.625	42.335	55.230	59.304	62.990	67.459	70.616	77.419	43
44	23.584	27.575	32.487	43.335	56.369	60.481	64.202	68.710	71.893	78.750	44
45	24.311	28.366	33.350	44.335	57.505	61.656	65.410	69.957	73.166	80.077	45
46	25.042	29.160	34.215	45.335	58.641	62.830	66.617	71.201	74.437	81.400	46
47	25.775	29.956	35.081	46.335	59.774	64.001	67.821	72.443	75.704	82.720	47
48	26.511	30.755	35.949	47.335	60.907	65.171	69.023	73.683	76.969	84.037	48
49	27.249	31.555	36.818	48.335	62.038	66.339	70.222	74.919	78.231	85.351	49
50	27.991	32.357	37.689	49.335	63.167	67.505	71.420	76.154	79.490	86.661	50

TABLE 14 Critical values of the chi-square distribution

$\nu \backslash \alpha$	.995	.975	.9	.5	.1	.05	.025	.01	.005	.001	α / ν
51	28.735	33.162	38.560	50.335	64.295	68.669	72.616	77.386	80.747	87.968	51
52	29.481	33.968	39.433	51.335	65.422	69.832	73.810	78.616	82.001	89.272	52
53	30.230	34.776	40.308	52.335	66.548	70.993	75.002	79.843	83.253	90.573	53
54	30.981	35.586	41.183	53.335	67.673	72.153	76.192	81.069	84.502	91.872	54
55	31.735	36.398	42.060	54.335	68.796	73.311	77.380	82.292	85.749	93.168	55
56	32.490	37.212	42.937	55.335	69.918	74.468	78.567	83.513	86.994	94.460	56
57	33.248	38.027	43.816	56.335	71.040	75.624	79.752	84.733	88.237	95.751	57
58	34.008	38.844	44.696	57.335	72.160	76.778	80.936	85.950	89.477	97.039	58
59	34.770	39.662	45.577	58.335	73.279	77.931	82.117	87.166	90.715	98.324	59
60	35.534	40.482	46.459	59.335	74.397	79.082	83.298	88.379	91.952	99.607	60
61	36.300	41.303	47.342	60.335	75.514	80.232	84.476	89.591	93.186	100.888	61
62	37.068	42.126	48.226	61.335	76.630	81.381	85.654	90.802	94.419	102.166	62
63	37.838	42.950	49.111	62.335	77.745	82.529	86.830	92.010	95.649	103.442	63
64	38.610	43.776	49.996	63.335	78.860	83.675	88.004	93.217	96.878	104.716	64
65	39.383	44.603	50.883	64.335	79.973	84.821	89.177	94.422	98.105	105.988	65
66	40.158	45.431	51.770	65.335	81.085	85.965	90.349	95.626	99.331	107.258	66
67	40.935	46.261	52.659	66.335	82.197	87.108	91.519	96.828	100.55	108.526	67
68	41.713	47.092	53.548	67.334	83.308	88.250	92.689	98.028	101.78	109.791	68
69	42.494	47.924	54.438	68.334	84.418	89.391	93.856	99.228	103.00	111.055	69
70	43.275	48.758	55.329	69.334	85.527	90.531	95.023	100.43	104.21	112.317	70
71	44.058	49.592	56.221	70.334	86.635	91.670	96.189	101.62	105.43	113.577	71
72	44.843	50.428	57.113	71.334	87.743	92.808	97.353	102.82	106.65	114.835	72
73	45.629	51.265	58.006	72.334	88.850	93.945	98.516	104.01	107.86	116.092	73
74	46.417	52.103	58.900	73.334	89.956	95.081	99.678	105.20	109.07	117.346	74
75	47.206	52.942	59.795	74.334	91.061	96.217	100.84	106.39	110.29	118.599	75
76	47.997	53.782	60.690	75.334	92.166	97.351	102.00	107.58	111.50	119.850	76
77	48.788	54.623	61.586	76.334	93.270	98.484	103.16	108.77	112.70	121.100	77
78	49.582	55.466	62.483	77.334	94.373	99.617	104.32	109.96	113.91	122.348	78
79	50.376	56.309	63.380	78.334	95.476	100.75	105.47	111.14	115.12	123.594	79
80	51.172	57.153	64.278	79.334	96.578	101.88	106.63	112.33	116.32	124.839	80
81	51.969	57.998	65.176	80.334	97.680	103.01	107.78	113.51	117.52	126.082	81
82	52.767	58.845	66.076	81.334	98.780	104.14	108.94	114.69	118.73	127.324	82
83	53.567	59.692	66.976	82.334	99.880	105.27	110.09	115.88	119.93	128.565	83
84	54.368	60.540	67.876	83.334	100.98	106.39	111.24	117.06	121.13	129.804	84
85	55.170	61.389	68.777	84.334	102.08	107.52	112.39	118.24	122.32	131.041	85
86	55.973	62.239	69.679	85.334	103.18	108.65	113.54	119.41	123.52	132.277	86
87	56.777	63.089	70.581	86.334	104.28	109.77	114.69	120.59	124.72	133.512	87
88	57.582	63.941	71.484	87.334	105.37	110.90	115.84	121.77	125.91	134.745	88
89	58.389	64.793	72.387	88.334	106.47	112.02	116.99	122.94	127.11	135.978	89
90	59.196	65.647	73.291	89.334	107.56	113.15	118.14	124.12	128.30	137.208	90
91	60.005	66.501	74.196	90.334	108.66	114.27	119.28	125.29	129.49	138.438	91
92	60.815	67.356	75.101	91.334	109.76	115.39	120.43	126.46	130.68	139.666	92
93	61.625	68.211	76.006	92.334	110.85	116.51	121.57	127.63	131.87	140.893	93
94	62.437	69.068	76.912	93.334	111.94	117.63	122.72	128.80	133.06	142.119	94
95	63.250	69.925	77.818	94.334	113.04	118.75	123.86	129.97	134.25	143.344	95
96	64.063	70.783	78.725	95.334	114.13	119.87	125.00	131.14	135.43	144.567	96
97	64.878	71.642	79.633	96.334	115.22	120.99	126.14	132.31	136.62	145.789	97
98	65.694	72.501	80.541	97.334	116.32	122.11	127.28	133.48	137.80	147.010	98
99	66.510	73.361	81.449	98.334	117.41	123.23	128.42	134.64	138.99	148.230	99
100	67.328	74.222	82.358	99.334	118.50	124.34	129.56	135.81	140.17	149.449	100

TABLE **15** Critical values of the chi-square distribution based on Šidák's multiplicative inequality

This table furnishes critical values of the chi-square distribution for unusual percentage points not found in standard chi-square tables such as Table **14.** These unusual percentage points α' are obtained from Šidák's multiplicative inequality; $\alpha' = 1 - (1 - \alpha)^{1/k}$ where α is the experimentwise error rate (also known as the familywise type I error rate) and k is the number of comparisons intended. Values of $\chi^2_{\alpha'}$ are furnished for degrees of freedom ν from 1 to 20 in increments of one. Arguments for k, the number of comparisons, are given from 1 to 60 in increments of one. Two levels of experimentwise α (0.05 and 0.01) are given for each combination of ν and k in the table.

To find the critical value for a 5% experimentwise error based on a chi-square with $\nu = 6$ degrees of freedom and making $k = 5$ comparisons overall, we enter the table with $k = 5$, $\nu = 6$, and $\alpha = .05$ to obtain $\chi^2_{.010206[6]} = 16.760$. This table is one-tailed.

This table is employed for unplanned multiple comparisons tests for the homogeneity of replicates in replicated tests of goodness of fit (Section 17.3).

The table was computed directly using a modification of the Hewlett Packard 9830 Statistical Distributions Pack, volume 1. All computations were carried out to 12 decimal place accuracy.

TABLE 15 Critical values of the chi-square distribution based on Šidák's multiplicative inequality

		Degrees of freedom ν									
k	α	1	2	3	4	5	6	7	8	9	10
1	.05	3.841	5.991	7.815	9.488	11.070	12.592	14.067	15.507	16.919	18.307
	.01	6.635	9.210	11.345	13.277	15.086	16.812	18.475	20.090	21.666	23.209
2	.05	5.002	7.352	9.320	11.113	12.801	14.416	15.978	17.498	18.985	20.444
	.01	7.875	10.592	12.833	14.855	16.744	18.541	20.271	21.948	23.582	25.181
3	.05	5.701	8.155	10.198	12.054	13.797	15.462	17.070	18.633	20.160	21.657
	.01	8.609	11.401	13.699	15.770	17.702	19.539	21.305	23.015	24.681	26.310
4	.05	6.205	8.726	10.820	12.718	14.497	16.196	17.834	19.426	20.980	22.502
	.01	9.134	11.975	14.312	16.415	18.377	20.240	22.031	23.765	25.452	27.102
5	.05	6.599	9.169	11.301	13.230	15.037	16.760	18.422	20.035	21.608	23.150
	.01	9.542	12.421	14.787	16.915	18.898	20.781	22.591	24.342	26.046	27.711
6	.05	6.922	9.532	11.693	13.647	15.476	17.219	18.898	20.528	22.118	23.675
	.01	9.877	12.785	15.174	17.322	19.322	21.222	23.046	24.811	26.528	28.205
7	.05	7.197	9.840	12.024	13.998	15.845	17.604	19.299	20.943	22.546	24.115
	.01	10.161	13.094	15.501	17.665	19.680	21.593	23.429	25.205	26.933	28.621
8	.05	7.437	10.106	12.311	14.302	16.164	17.937	19.644	21.300	22.915	24.494
	.01	10.407	13.360	15.784	17.962	19.989	21.913	23.760	25.546	27.283	28.980
9	.05	7.648	10.341	12.563	14.569	16.445	18.230	19.948	21.614	23.238	24.827
	.01	10.624	13.596	16.034	18.223	20.262	22.195	24.051	25.846	27.591	29.295
10	.05	7.838	10.551	12.789	14.808	16.695	18.491	20.219	21.894	23.526	25.124
	.01	10.819	13.806	16.257	18.457	20.505	22.447	24.311	26.113	27.865	29.576
11	.05	8.010	10.741	12.993	15.024	16.921	18.726	20.463	22.146	23.786	25.391
	.01	10.996	13.997	16.458	18.668	20.724	22.674	24.545	26.354	28.113	29.830
12	.05	8.167	10.914	13.179	15.220	17.127	18.940	20.685	22.376	24.023	25.634
	.01	11.157	14.171	16.642	18.860	20.924	22.881	24.759	26.574	28.339	30.061
13	.05	8.312	11.074	13.350	15.401	17.316	19.137	20.889	22.586	24.240	25.857
	.01	11.305	14.331	16.811	19.037	21.108	23.072	24.955	26.776	28.546	30.273
14	.05	8.447	11.222	13.508	15.568	17.491	19.319	21.078	22.781	24.440	26.062
	.01	11.443	14.479	16.968	19.201	21.278	23.248	25.137	26.962	28.737	30.469
15	.05	8.572	11.360	13.655	15.723	17.653	19.488	21.253	22.962	24.626	26.254
	.01	11.571	14.617	17.113	19.353	21.436	23.411	25.305	27.136	28.915	30.651
16	.05	8.689	11.489	13.793	15.869	17.805	19.646	21.416	23.131	24.800	26.432
	.01	11.691	14.746	17.250	19.495	21.584	23.564	25.463	27.298	29.081	30.821
17	.05	8.800	11.610	13.922	16.005	17.948	19.794	21.569	23.289	24.963	26.599
	.01	11.804	14.867	17.377	19.629	21.723	23.708	25.611	27.450	29.237	30.981
18	.05	8.904	11.724	14.044	16.133	18.082	19.934	21.714	23.438	25.116	26.756
	.01	11.910	14.982	17.498	19.755	21.854	23.843	25.750	27.593	29.384	31.131
19	.05	9.003	11.832	14.159	16.254	18.209	20.066	21.850	23.578	25.261	26.905
	.01	12.011	15.090	17.612	19.874	21.977	23.970	25.881	27.728	29.522	31.273
20	.05	9.096	11.934	14.269	16.370	18.329	20.190	21.979	23.711	25.398	27.046
	.01	12.107	15.192	17.720	19.987	22.094	24.092	26.006	27.856	29.654	31.407

TABLE 15 Critical values of the chi-square distribution based on Šidák's multiplicative inequality

		Degrees of freedom ν									
k	α	**1**	**2**	**3**	**4**	**5**	**6**	**7**	**8**	**9**	**10**
21	.05	9.185	12.032	14.373	16.479	18.443	20.309	22.102	23.838	25.528	27.179
	.01	12.198	15.290	17.823	20.094	22.206	24.207	26.125	27.978	29.778	31.535
22	.05	9.270	12.125	14.472	16.583	18.552	20.422	22.219	23.959	25.652	27.306
	.01	12.285	15.383	17.921	20.196	22.312	24.316	26.238	28.094	29.897	31.657
23	.05	9.352	12.214	14.566	16.682	18.656	20.530	22.330	24.074	25.770	27.428
	.01	12.367	15.472	18.014	20.294	22.413	24.421	26.345	28.205	30.011	31.773
24	.05	9.430	12.299	14.657	16.778	18.755	20.633	22.437	24.183	25.883	27.544
	.01	12.447	15.557	18.104	20.388	22.510	24.521	26.448	28.310	30.119	31.884
25	.05	9.505	12.380	14.743	16.869	18.850	20.732	22.539	24.289	25.991	27.655
	.01	12.523	15.638	18.190	20.477	22.603	24.617	26.547	28.412	30.223	31.990
26	.05	9.576	12.459	14.827	16.957	18.942	20.827	22.637	24.390	26.095	27.762
	.01	12.596	15.717	18.272	20.563	22.692	24.709	26.642	28.509	30.323	32.092
27	.05	9.646	12.534	14.907	17.041	19.030	20.918	22.732	24.487	26.195	27.864
	.01	12.667	15.792	18.352	20.646	22.778	24.798	26.733	28.603	30.419	32.191
28	.05	9.712	12.607	14.984	17.122	19.114	21.006	22.823	24.581	26.291	27.963
	.01	12.735	15.865	18.428	20.726	22.861	24.883	26.821	28.693	30.512	32.285
29	.05	9.777	12.677	15.059	17.200	19.196	21.091	22.910	24.671	26.384	28.058
	.01	12.801	15.935	18.502	20.803	22.941	24.966	26.906	28.780	30.601	32.376
30	.05	9.839	12.744	15.131	17.276	19.275	21.172	22.995	24.758	26.474	28.150
	.01	12.864	16.003	18.573	20.877	23.018	25.045	26.988	28.864	30.687	32.464
31	.05	9.899	12.810	15.200	17.349	19.351	21.251	23.076	24.842	26.560	28.238
	.01	12.925	16.069	18.642	20.949	23.092	25.122	27.067	28.946	30.770	32.550
32	.05	9.958	12.873	15.268	17.420	19.425	21.328	23.155	24.924	26.644	28.324
	.01	12.985	16.132	18.709	21.018	23.164	25.196	27.144	29.024	30.851	32.632
33	.05	10.014	12.935	15.333	17.488	19.496	21.402	23.232	25.002	26.725	28.407
	.01	13.042	16.194	18.774	21.086	23.234	25.268	27.218	29.100	30.929	32.712
34	.05	10.069	12.995	15.396	17.555	19.565	21.474	23.306	25.079	26.803	28.488
	.01	13.098	16.253	18.836	21.151	23.302	25.338	27.290	29.174	31.004	32.789
35	.05	10.123	13.053	15.458	17.619	19.633	21.543	23.378	25.153	26.879	28.566
	.01	13.153	16.311	18.897	21.215	23.367	25.406	27.359	29.246	31.077	32.864
36	.05	10.175	13.109	15.517	17.682	19.698	21.611	23.448	25.225	26.953	28.642
	.01	13.205	16.368	18.956	21.276	23.431	25.472	27.427	29.315	31.149	32.937
37	.05	10.225	13.164	15.576	17.743	19.761	21.677	23.516	25.295	27.025	28.715
	.01	13.257	16.422	19.014	21.336	23.493	25.536	27.493	29.383	31.218	33.008
38	.05	10.274	13.217	15.632	17.802	19.823	21.741	23.582	25.363	27.095	28.787
	.01	13.307	16.476	19.070	21.394	23.554	25.598	27.557	29.449	31.285	33.076
39	.05	10.322	13.269	15.687	17.860	19.883	21.803	23.646	25.429	27.163	28.857
	.01	13.355	16.528	19.124	21.451	23.613	25.659	27.620	29.513	31.351	33.143
40	.05	10.369	13.319	15.741	17.916	19.942	21.864	23.709	25.494	27.230	28.925
	.01	13.403	16.578	19.178	21.507	23.670	25.718	27.680	29.575	31.415	33.209

TABLE 15 Critical values of the chi-square distribution based on Šidák's multiplicative inequality

		Degrees of freedom ν									
k	α	1	2	3	4	5	6	7	8	9	10
41	.05	10.415	13.369	15.793	17.971	19.999	21.923	23.770	25.557	27.294	28.991
	.01	13.449	16.628	19.229	21.561	23.726	25.776	27.740	29.636	31.477	33.273
42	.05	10.459	13.417	15.844	18.025	20.055	21.981	23.830	25.618	27.357	29.056
	.01	13.494	16.676	19.280	21.613	23.780	25.832	27.798	29.695	31.538	33.335
43	.05	10.502	13.464	15.894	18.077	20.109	22.037	23.888	25.678	27.419	29.119
	.01	13.539	16.723	19.329	21.665	23.834	25.887	27.854	29.753	31.597	33.395
44	.05	10.545	13.510	15.943	18.128	20.162	22.092	23.945	25.736	27.479	29.180
	.01	13.582	16.769	19.378	21.715	23.885	25.941	27.909	29.810	31.655	33.454
45	.05	10.586	13.555	15.990	18.178	20.214	22.146	24.000	25.794	27.537	29.240
	.01	13.624	16.814	19.425	21.764	23.936	25.993	27.963	29.865	31.712	33.512
46	.05	10.627	13.599	16.037	18.226	20.265	22.199	24.055	25.849	27.595	29.299
	.01	13.665	16.858	19.471	21.812	23.986	26.044	28.016	29.919	31.767	33.569
47	.05	10.667	13.642	16.082	18.274	20.315	22.250	24.108	25.904	27.651	29.356
	.01	13.706	16.901	19.516	21.859	24.035	26.094	28.067	29.972	31.821	33.624
48	.05	10.706	13.684	16.127	18.321	20.363	22.300	24.160	25.958	27.706	29.413
	.01	13.745	16.943	19.560	21.905	24.082	26.143	28.118	30.024	31.874	33.678
49	.05	10.744	13.725	16.170	18.367	20.411	22.350	24.210	26.010	27.759	29.468
	.01	13.784	16.984	19.603	21.950	24.129	26.191	28.167	30.074	31.926	33.731
50	.05	10.781	13.765	16.213	18.411	20.457	22.398	24.260	26.061	27.812	29.522
	.01	13.822	17.025	19.646	21.994	24.174	26.238	28.215	30.124	31.977	33.783
51	.05	10.818	13.805	16.255	18.455	20.503	22.445	24.309	26.111	27.864	29.574
	.01	13.859	17.064	19.687	22.037	24.219	26.284	28.263	30.172	32.026	33.834
52	.05	10.854	13.844	16.296	18.498	20.548	22.492	24.357	26.160	27.914	29.626
	.01	13.895	17.103	19.728	22.079	24.263	26.330	28.309	30.220	32.075	33.884
53	.05	10.889	13.882	16.337	18.540	20.592	22.537	24.404	26.209	27.964	29.677
	.01	13.931	17.141	19.768	22.121	24.306	26.374	28.355	30.267	32.123	33.933
54	.05	10.924	13.919	16.376	18.582	20.635	22.582	24.450	26.256	28.012	29.727
	.01	13.966	17.178	19.807	22.162	24.348	26.417	28.399	30.313	32.170	33.981
55	.05	10.958	13.956	16.415	18.622	20.677	22.625	24.495	26.302	28.060	29.775
	.01	14.001	17.215	19.846	22.202	24.389	26.460	28.443	30.358	32.216	34.028
56	.05	10.991	13.992	16.453	18.662	20.718	22.668	24.539	26.348	28.107	29.823
	.01	14.035	17.251	19.883	22.241	24.430	26.502	28.486	30.402	32.261	34.074
57	.05	11.024	14.027	16.490	18.702	20.759	22.710	24.583	26.393	28.153	29.870
	.01	14.068	17.287	19.920	22.280	24.470	26.543	28.529	30.445	32.306	34.119
58	.05	11.056	14.062	16.527	18.740	20.799	22.752	24.625	26.437	28.198	29.917
	.01	14.101	17.321	19.957	22.317	24.509	26.584	28.570	30.488	32.349	34.164
59	.05	11.088	14.096	16.563	18.778	20.838	22.793	24.667	26.480	28.242	29.962
	.01	14.133	17.355	19.993	22.355	24.548	26.623	28.611	30.530	32.392	34.207
60	.05	11.119	14.130	16.599	18.815	20.877	22.833	24.709	26.522	28.285	30.007
	.01	14.165	17.389	20.028	22.391	24.586	26.662	28.651	30.571	32.434	34.251

TABLE 15 Critical values of the chi-square distribution based on Šidák's multiplicative inequality

Degrees of freedom ν

k	α	11	12	13	14	15	16	17	18	19	20
1	.05	19.675	21.026	22.362	23.685	24.996	26.296	27.587	28.869	30.144	31.410
	.01	24.725	26.217	27.688	29.141	30.578	32.000	33.409	34.805	36.191	37.566
2	.05	21.880	23.295	24.693	26.075	27.444	28.800	30.145	31.479	32.804	34.121
	.01	26.750	28.292	29.812	31.312	32.793	34.259	35.710	37.148	38.574	39.988
3	.05	23.128	24.578	26.009	27.423	28.822	30.208	31.582	32.944	34.296	35.639
	.01	27.908	29.478	31.024	32.549	34.056	35.545	37.020	38.480	39.928	41.364
4	.05	23.998	25.471	26.924	28.360	29.780	31.185	32.578	33.959	35.330	36.691
	.01	28.719	30.308	31.872	33.415	34.938	36.444	37.934	39.410	40.873	42.323
5	.05	24.664	26.155	27.624	29.076	30.511	31.932	33.339	34.735	36.119	37.493
	.01	29.342	30.946	32.524	34.079	35.615	37.134	38.636	40.123	41.597	43.059
6	.05	25.203	26.707	28.190	29.655	31.102	32.535	33.954	35.360	36.755	38.140
	.01	29.849	31.463	33.052	34.618	36.165	37.693	39.204	40.701	42.184	43.655
7	.05	25.655	27.171	28.665	30.140	31.597	33.040	34.468	35.884	37.288	38.682
	.01	30.274	31.898	33.496	35.071	36.626	38.162	39.682	41.186	42.677	44.155
8	.05	26.045	27.570	29.073	30.557	32.023	33.474	34.911	36.334	37.746	39.148
	.01	30.641	32.273	33.879	35.462	37.024	38.567	40.093	41.604	43.101	44.586
9	.05	26.386	27.920	29.431	30.923	32.397	33.855	35.298	36.729	38.148	39.555
	.01	30.964	32.603	34.215	35.805	37.373	38.922	40.455	41.971	43.474	44.964
10	.05	26.691	28.232	29.750	31.248	32.729	34.193	35.643	37.080	38.504	39.918
	.01	31.252	32.897	34.515	36.110	37.684	39.239	40.776	42.298	43.806	45.300
11	.05	26.965	28.512	30.037	31.542	33.028	34.498	35.954	37.396	38.826	40.244
	.01	31.511	33.162	34.786	36.386	37.965	39.524	41.067	42.593	44.105	45.604
12	.05	27.214	28.768	30.298	31.808	33.300	34.775	36.236	37.683	39.117	40.540
	.01	31.748	33.403	35.032	36.637	38.220	39.784	41.331	42.861	44.377	45.880
13	.05	27.442	29.002	30.537	32.052	33.549	35.029	36.494	37.946	39.385	40.812
	.01	31.964	33.625	35.258	36.867	38.454	40.022	41.573	43.107	44.627	46.133
14	.05	27.653	29.218	30.758	32.278	33.779	35.263	36.733	38.188	39.631	41.062
	.01	32.165	33.829	35.466	37.079	38.671	40.242	41.796	43.334	44.857	46.367
15	.05	27.849	29.418	30.963	32.487	33.992	35.481	36.954	38.413	39.860	41.295
	.01	32.351	34.019	35.660	37.277	38.871	40.447	42.004	43.545	45.071	46.584
16	.05	28.032	29.605	31.154	32.682	34.191	35.684	37.160	38.623	40.073	41.511
	.01	32.525	34.197	35.841	37.461	39.059	40.637	42.198	43.742	45.271	46.787
17	.05	28.204	29.781	31.334	32.865	34.378	35.874	37.354	38.820	40.273	41.714
	.01	32.688	34.363	36.011	37.634	39.235	40.816	42.380	43.926	45.458	46.976
18	.05	28.365	29.946	31.502	33.037	34.553	36.052	37.535	39.005	40.461	41.905
	.01	32.842	34.520	36.171	37.797	39.401	40.985	42.551	44.100	45.635	47.155
19	.05	28.517	30.101	31.661	33.200	34.719	36.221	37.707	39.179	40.638	42.085
	.01	32.987	34.668	36.322	37.950	39.557	41.144	42.712	44.264	45.801	47.324
20	.05	28.661	30.249	31.812	33.353	34.876	36.380	37.869	39.344	40.806	42.255
	.01	33.124	34.808	36.465	38.096	39.705	41.294	42.865	44.419	45.958	47.484

TABLE 15 Critical values of the chi-square distribution based on Šidák's multiplicative inequality

k	α	11	12	13	14	15	16	17	18	19	20
		Degrees of freedom ν									
21	.05	28.798	30.389	31.955	33.499	35.024	36.532	38.023	39.501	40.965	42.417
	.01	33.254	34.941	36.600	38.234	39.846	41.437	43.010	44.567	46.108	47.635
22	.05	28.928	30.522	32.091	33.638	35.166	36.676	38.170	39.650	41.116	42.571
	.01	33.379	35.068	36.730	38.366	39.980	41.573	43.149	44.707	46.251	47.780
23	.05	29.053	30.649	32.221	33.771	35.301	36.813	38.310	39.792	41.261	42.717
	.01	33.497	35.189	36.853	38.491	40.107	41.703	43.281	44.841	46.386	47.918
24	.05	29.171	30.771	32.345	33.897	35.430	36.945	38.444	39.928	41.399	42.857
	.01	33.611	35.305	36.971	38.612	40.230	41.827	43.407	44.969	46.516	48.049
25	.05	29.285	30.887	32.464	34.018	35.553	37.071	38.572	40.058	41.531	42.991
	.01	33.720	35.416	37.084	38.727	40.347	41.946	43.528	45.092	46.641	48.175
26	.05	29.395	30.999	32.578	34.135	35.672	37.191	38.694	40.183	41.658	43.120
	.01	33.824	35.523	37.192	38.837	40.459	42.061	43.644	45.210	46.760	48.297
27	.05	29.499	31.106	32.687	34.247	35.786	37.307	38.812	40.303	41.779	43.244
	.01	33.924	35.625	37.297	38.943	40.567	42.170	43.755	45.323	46.875	48.413
28	.05	29.600	31.209	32.793	34.354	35.895	37.419	38.926	40.418	41.897	43.363
	.01	34.021	35.723	37.397	39.045	40.671	42.276	43.862	45.432	46.986	48.525
29	.05	29.698	31.309	32.895	34.458	36.001	37.526	39.035	40.529	42.009	43.477
	.01	34.114	35.818	37.494	39.144	40.771	42.378	43.966	45.537	47.092	48.633
30	.05	29.792	31.405	32.993	34.558	36.103	37.630	39.141	40.636	42.118	43.588
	.01	34.204	35.910	37.587	39.239	40.868	42.476	44.066	45.638	47.195	48.737
31	.05	29.883	31.498	33.087	34.654	36.201	37.730	39.243	40.740	42.224	43.695
	.01	34.291	35.999	37.678	39.331	40.961	42.571	44.162	45.736	47.294	48.838
32	.05	29.970	31.588	33.179	34.748	36.297	37.827	39.341	40.840	42.325	43.798
	.01	34.375	36.084	37.765	39.420	41.052	42.663	44.255	45.831	47.390	48.935
33	.05	30.055	31.674	33.268	34.838	36.389	37.921	39.436	40.937	42.424	43.898
	.01	34.456	36.168	37.850	39.506	41.139	42.752	44.346	45.922	47.483	49.029
34	.05	30.138	31.759	33.354	34.926	36.478	38.012	39.529	41.031	42.519	43.994
	.01	34.535	36.248	37.932	39.590	41.224	42.838	44.433	46.011	47.573	49.121
35	.05	30.218	31.840	33.437	35.011	36.565	38.10[illegible]	39.619	41.122	42.612	44.088
	.01	34.612	36.326	38.011	39.671	41.307	42.922	44.518	46.097	47.661	49.209
36	.05	30.295	31.920	33.518	35.094	36.649	38.186	39.706	41.210	42.701	44.180
	.01	34.686	36.402	38.089	39.749	41.387	43.003	44.601	46.181	47.746	49.296
37	.05	30.371	31.997	33.597	35.174	36.731	38.269	39.790	41.296	42.789	44.268
	.01	34.758	36.476	38.164	39.826	41.464	43.082	44.681	46.263	47.828	49.379
38	.05	30.444	32.072	33.673	35.252	36.810	38.350	39.872	41.380	42.873	44.354
	.01	34.829	36.547	38.237	39.900	41.540	43.159	44.759	46.342	47.908	49.461
39	.05	30.516	32.145	33.748	35.328	36.887	38.428	39.952	41.461	42.956	44.438
	.01	34.897	36.617	38.308	39.972	41.614	43.234	44.835	46.419	47.987	49.540
40	.05	30.585	32.216	33.820	35.402	36.963	38.505	40.030	41.540	43.036	44.519
	.01	34.964	36.685	38.377	40.043	41.685	43.307	44.909	46.494	48.063	49.617

TABLE 15 Critical values of the chi-square distribution based on Šidák's multiplicative inequality

		Degrees of freedom ν									
k	α	11	12	13	14	15	16	17	18	19	20
41	.05	30.653	32.285	33.891	35.474	37.036	38.580	40.106	41.618	43.115	44.599
	.01	35.029	36.752	38.445	40.112	41.755	43.378	44.981	46.567	48.137	49.692
42	.05	30.719	32.353	33.960	35.544	37.108	38.652	40.180	41.693	43.191	44.676
	.01	35.092	36.816	38.511	40.179	41.823	43.447	45.051	46.638	48.209	49.765
43	.05	30.784	32.419	34.027	35.613	37.177	38.723	40.252	41.766	43.266	44.752
	.01	35.154	36.879	38.575	40.244	41.890	43.514	45.120	46.708	48.280	49.837
44	.05	30.847	32.483	34.093	35.680	37.246	38.793	40.323	41.838	43.338	44.826
	.01	35.215	36.941	38.638	40.308	41.955	43.580	45.187	46.776	48.349	49.907
45	.05	30.908	32.546	34.157	35.745	37.312	38.860	40.392	41.908	43.409	44.898
	.01	35.274	37.001	38.699	40.370	42.018	43.645	45.252	46.842	48.416	49.975
46	.05	30.968	32.607	34.220	35.809	37.377	38.927	40.459	41.976	43.479	44.968
	.01	35.331	37.060	38.759	40.431	42.080	43.708	45.316	46.907	48.482	50.042
47	.05	31.027	32.667	34.281	35.871	37.441	38.991	40.525	42.043	43.547	45.037
	.01	35.388	37.118	38.817	40.491	42.141	43.769	45.379	46.970	48.546	50.107
48	.05	31.084	32.726	34.341	35.932	37.503	39.055	40.589	42.108	43.613	45.104
	.01	35.443	37.174	38.875	40.549	42.200	43.830	45.440	47.033	48.609	50.171
49	.05	31.141	32.783	34.400	35.992	37.564	39.117	40.652	42.172	43.678	45.170
	.01	35.497	37.229	38.931	40.606	42.258	43.889	45.500	47.093	48.671	50.233
50	.05	31.196	32.840	34.457	36.051	37.623	39.177	40.714	42.235	43.741	45.235
	.01	35.550	37.283	38.986	40.662	42.315	43.946	45.558	47.153	48.731	50.294
51	.05	31.250	32.895	34.513	36.108	37.682	39.237	40.774	42.296	43.804	45.298
	.01	35.602	37.336	39.040	40.717	42.371	44.003	45.616	47.211	48.790	50.354
52	.05	31.303	32.949	34.568	36.164	37.739	39.295	40.833	42.356	43.865	45.360
	.01	35.653	37.388	39.093	40.771	42.425	44.058	45.672	47.268	48.848	50.413
53	.05	31.355	33.002	34.623	36.219	37.795	39.352	40.891	42.415	43.925	45.421
	.01	35.703	37.439	39.144	40.824	42.479	44.113	45.727	47.324	48.905	50.470
54	.05	31.406	33.054	34.676	36.273	37.850	39.408	40.948	42.473	43.983	45.480
	.01	35.752	37.488	39.195	40.875	42.531	44.166	45.782	47.379	48.960	50.527
55	.05	31.456	33.105	34.728	36.326	37.904	39.463	41.004	42.530	44.041	45.539
	.01	35.800	37.537	39.245	40.926	42.583	44.219	45.835	47.433	49.015	50.582
56	.05	31.505	33.155	34.779	36.379	37.957	39.517	41.059	42.585	44.097	45.596
	.01	35.847	37.585	39.294	40.976	42.633	44.270	45.887	47.486	49.069	50.636
57	.05	31.553	33.204	34.829	36.430	38.009	39.570	41.113	42.640	44.153	45.652
	.01	35.893	37.633	39.342	41.024	42.683	44.320	45.938	47.538	49.121	50.690
58	.05	31.600	33.252	34.878	36.480	38.060	39.622	41.165	42.694	44.207	45.707
	.01	35.939	37.679	39.389	41.072	42.732	44.370	45.988	47.589	49.173	50.742
59	.05	31.646	33.300	34.926	36.529	38.110	39.673	41.217	42.746	44.261	45.762
	.01	35.983	37.724	39.435	41.120	42.780	44.419	46.038	47.639	49.224	50.794
60	.05	31.692	33.346	34.974	36.577	38.160	39.723	41.268	42.798	44.313	45.815
	.01	36.027	37.769	39.481	41.166	42.827	44.466	46.086	47.688	49.274	50.844

TABLE **16** Critical values of the *F*-distribution

This table furnishes critical values of the *F*-distribution for degrees of freedom $\nu_1 = 1$ to 12 in increments of 1 and for $\nu_1 = 15, 20, 24, 30, 40, 50, 60, 120$, and ∞, all arranged across the top of the table; also for degrees of freedom $\nu_2 = 1$ to 30 in increments of 1 and for $\nu_2 = 40, 60, 120$, and ∞ arranged along the left margin of the table. The percentage points given (corresponding to $\alpha = 0.75, 0.50, 0.25, 0.10, 0.05, 0.025, 0.01, 0.005$ and 0.001) represent the area to the right of the critical value of F in one tail of the distribution, as shown in the accompanying figure. The critical values of F are given to at least three significant figures and in a few cases to four figures.

To find the critical value of F for a given pair of numbers of degrees of freedom ν_1 and ν_2, look up these two arguments across the top of the table and in the leftmost column, respectively, and read off the desired value of F from the appropriate row in the block of nine values furnished for the various percentage points. For example, for $\nu_1 = 6$ and $\nu_2 = 28$, $F_{.05} = 2.45$ and $F_{.001} = 5.24$. The last value indicates that 0.1% of the area under the *F*-distribution for $\nu_1 = 6$ and $\nu_2 = 28$ degrees of freedom is to the right of the value $F = 5.24$. This is the correct value for a one-tailed test in which the alternative hypothesis is $\sigma_1^2 > \sigma_2^2$. For a two-tailed test one deliberately divides the greater variance by the smaller one and therefore for a type I error of α one has to look up the *F*-value for $\alpha/2$ in the table. Only one *F*-value for $\alpha > 0.5$ is given ($\alpha = 0.75$). For other values of $\alpha > 0.5$, make use of the relation $F_{\alpha[\nu_1,\nu_2]} = 1/F_{(1-\alpha)[\nu_2,\nu_1]}$. Thus one can obtain $F_{.95[8,24]}$ as $1/F_{.05[24,8]} = 1/3.12 = 0.3205$.

Interpolation for number of degrees of freedom not furnished in the arguments is by means of harmonic interpolation. If both ν_1 and ν_2 require interpolation, one needs to interpolate for each of these arguments in turn. Thus to obtain $F_{.05[55,80]}$ one first interpolates between $F_{.05[50,60]}$ and $F_{.05[60,60]}$ and between $F_{.05[50,120]}$ and $F_{.05[60,120]}$, to estimate $F_{.05[55,60]}$ and $F_{.05[55,120]}$, respectively. One then interpolates between these two values to obtain the desired quantity.

There are numerous applications of the *F*-distribution in statistics. Its most common use is in tests of significance in analysis of variance (Chapters 8 to 13), and it is also used to test the significance of differences between two variances (Section 8.3).

The values in this table were obtained in a variety of ways. Entries for $\alpha = 0.50, 0.25, 0.10, 0.05, 0.025, 0.01$, and 0.005; for $\nu_1 = 1$ to 10, 12, 15, 20, 24, 30, 40, 60, 120, and ∞; and for $\nu_2 = 1$ to 30, 40, 60, 120, and ∞ were copied from a table by M. Merrington and C. M. Thompson (*Biometrika,* **33**:73–88,

1943) with permission of the publisher. F-values for $\alpha = 0.001$ and the above values of ν_1 and ν_2 were copied from table 18 in E. S. Pearson and H. O. Hartley, *Biometrika Tables for Statisticians,* Vol. I. (Cambridge University Press, 1958) with permission of the publishers. Entries for $\nu_1 = 11$ and 50 and for $\nu_2 = 1$ to 11 were obtained by interpolation. F-values for $\alpha = 0.75$, for $\nu_1 =$ 1 to 10, 12, 15, 20, 24, 30, 40, 60, 120, and ∞, and for ν_2 equal to the same numbers of degrees of freedom were obtained by computation of reciprocals of other entries in the table, as shown above. All remaining values of F were computed by the Cornish–Fisher method as described in section 16.21 in M. G. Kendall and A. Stuart, *The Advanced Theory of Statistics,* Vol. I, 6th ed. (Charles Griffin, London, 1958).

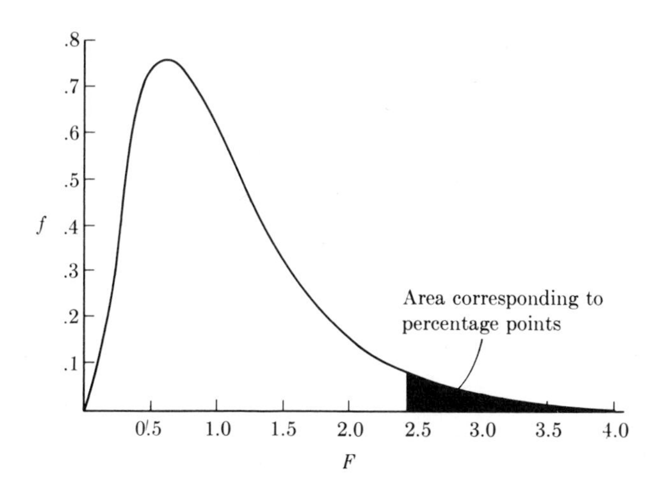

TABLE 16 Critical values of the F-distribution

ν_1 (degrees of freedom of numerator mean squares)

ν_2 (degrees of freedom of denominator mean squares)	α	1	2	3	4	5	6
1	.75	.172	.389	.494	.553	.591	.617
	.50	1.00	1.50	1.71	1.82	1.89	1.94
	.25	5.83	7.50	8.20	8.58	8.82	8.98
	.10	39.9	49.5	53.6	55.8	57.2	58.2
	.05	161	199	216	225	230	234
	.025	648	800	864	900	922	937
	.01	4050	5000	5400	5620	5760	5860
	.005	16200	20000	21600	22500	23100	23400
	.001	405300	500000	540400	562500	576400	585900
2	.75	.133	.333	.439	.500	.540	.568
	.50	.667	1.00	1.13	1.21	1.25	1.28
	.25	2.57	3.00	3.15	3.23	3.28	3.31
	.10	8.53	9.00	9.16	9.24	9.29	9.33
	.05	18.5	19.0	19.2	19.2	19.3	19.3
	.025	38.5	39.0	39.2	39.2	39.3	39.3
	.01	98.5	99.0	99.2	99.2	99.3	99.3
	.005	198	199	199	199	199	199
	.001	999	999	999	999	999	999
3	.75	.122	.317	.424	.489	.531	.561
	.50	.585	.881	1.00	1.06	1.10	1.13
	.25	2.02	2.28	2.36	2.39	2.41	2.42
	.10	5.54	5.46	5.39	5.34	5.31	5.28
	.05	10.1	9.55	9.28	9.12	9.01	8.94
	.025	17.4	16.0	15.4	15.1	14.9	14.7
	.01	34.1	30.8	29.5	28.7	28.2	27.9
	.005	55.6	49.8	47.5	46.2	45.4	44.8
	.001	167	149	141	137	135	133
4	.75	.117	.309	.418	.484	.528	.560
	.50	.549	.828	.941	1.00	1.04	1.06
	.25	1.81	2.00	2.05	2.06	2.07	2.08
	.10	4.54	4.32	4.19	4.11	4.05	4.01
	.05	7.71	6.94	6.59	6.39	6.26	6.16
	.025	12.2	10.6	9.98	9.60	9.36	9.20
	.01	21.2	18.0	16.7	16.0	15.5	15.2
	.005	31.3	26.3	24.3	23.2	22.4	22.0
	.001	74.1	61.3	56.2	53.4	51.7	50.5
5	.75	.113	.305	.415	.483	.528	.560
	.50	.528	.799	.907	.965	1.00	1.02
	.25	1.69	1.85	1.88	1.89	1.89	1.89
	.10	4.06	3.78	3.62	3.52	3.45	3.40
	.05	6.61	5.79	5.41	5.19	5.05	4.95
	.025	10.0	8.43	7.76	7.39	7.15	6.98
	.01	16.3	13.3	12.1	11.4	11.0	10.7
	.005	22.8	18.3	16.5	15.6	14.9	14.5
	.001	47.2	37.1	33.2	31.1	29.7	28.8

ν_1 (degrees of freedom of numerator mean squares)

ν_2 (degrees of freedom of denominator mean squares)	7	8	9	10	11	12	α
1	.636	.650	.661	.670	.680	.684	.75
	1.98	2.00	2.03	2.04	2.06	2.07	.50
	9.10	9.19	9.26	9.32	9.37	9.41	.25
	58.9	59.4	59.9	60.2	60.5	60.7	.10
	237	239	241	241	243	244	.05
	948	957	963	969	973	977	.025
	5930	5980	6020	6060	6080	6110	.01
	23700	23900	24100	24200	24300	24400	.005
	592900	598100	602300	605600	608400	610700	.001
2	.588	.604	.616	.626	.633	.641	.75
	1.30	1.32	1.33	1.35	1.36	1.36	.50
	3.34	3.35	3.37	3.38	3.39	3.39	.25
	9.35	9.37	9.38	9.39	9.40	9.41	.10
	19.4	19.4	19.4	19.4	19.4	19.4	.05
	39.4	39.4	39.4	39.4	39.4	39.4	.025
	99.4	99.4	99.4	99.4	99.4	99.4	.01
	199	199	199	199	199	199	.005
	999	999	999	999	999	999	.001
3	.581	.600	.613	.624	.633	.641	.75
	1.15	1.16	1.17	1.18	1.19	1.20	.50
	2.43	2.44	2.44	2.44	2.45	2.45	.25
	5.27	5.25	5.24	5.23	5.22	5.22	.10
	8.89	8.85	8.81	8.79	8.76	8.74	.05
	14.6	14.5	14.5	14.4	14.3	14.3	.025
	27.7	27.5	27.3	27.2	27.1	27.1	.01
	44.4	44.1	43.9	43.7	43.5	43.4	.005
	132	131	130	129	128	128	.001
4	.583	.601	.615	.627	.637	.645	.75
	1.08	1.09	1.10	1.11	1.12	1.13	.50
	2.08	2.08	2.08	2.08	2.08	2.08	.25
	3.98	3.95	3.94	3.92	3.91	3.90	.10
	6.09	6.04	6.00	5.96	5.93	5.91	.05
	9.07	8.98	8.90	8.84	8.79	8.75	.025
	15.0	14.8	14.7	14.5	14.4	14.4	.01
	21.6	21.4	21.1	21.0	20.8	20.7	.005
	49.7	49.0	48.5	48.1	47.7	47.4	.001
5	.584	.604	.618	.631	.641	.650	.75
	1.04	1.05	1.06	1.07	1.08	1.09	.50
	1.89	1.89	1.89	1.89	1.89	1.89	.25
	3.37	3.34	3.32	3.30	3.28	3.27	.10
	4.88	4.82	4.77	4.74	4.71	4.68	.05
	6.85	6.76	6.68	6.62	6.57	6.52	.025
	10.5	10.3	10.2	10.1	9.99	9.89	.01
	14.2	14.0	13.8	13.6	13.5	13.4	.005
	28.2	27.6	27.2	26.9	26.6	26.4	.001

TABLE 16 Critical values of the F-distribution

ν_1 (degrees of freedom of numerator mean squares)

ν_2 (degrees of freedom of denominator mean squares)

ν_2	α	1	2	3	4	5	6
6	.75	.111	.302	.413	.481	.524	.561
	.50	.515	.780	.886	.942	.977	1.00
	.25	1.62	1.76	1.78	1.79	1.79	1.78
	.10	3.78	3.46	3.29	3.18	3.11	3.05
	.05	5.99	5.14	4.76	4.53	4.39	4.28
	.025	8.81	7.26	6.60	6.23	5.99	5.82
	.01	13.7	10.9	9.78	9.15	8.75	8.47
	.005	18.6	14.5	12.9	12.0	11.5	11.1
	.001	35.5	27.0	23.7	21.9	20.8	20.0
7	.75	.110	.300	.412	.481	.528	.562
	.50	.506	.767	.871	.926	.960	.983
	.25	1.57	1.70	1.72	1.72	1.71	1.71
	.10	3.59	3.26	3.07	2.96	2.88	2.83
	.05	5.59	4.74	4.35	4.12	3.97	3.87
	.025	8.07	6.54	5.89	5.52	5.29	5.12
	.01	12.2	9.55	8.45	7.85	7.46	7.19
	.005	16.2	12.4	10.9	10.1	9.52	9.16
	.001	29.3	21.7	18.8	17.2	16.2	15.5
8	.75	.109	.298	.411	.481	.529	.563
	.50	.499	.757	.860	.915	.948	.971
	.25	1.54	1.66	1.67	1.66	1.66	1.65
	.10	3.46	3.11	2.92	2.81	2.73	2.67
	.05	5.32	4.46	4.07	3.84	3.69	3.58
	.025	7.57	6.06	5.42	5.05	4.82	4.65
	.01	11.3	8.65	7.59	7.01	6.63	6.37
	.005	14.7	11.0	9.60	8.81	8.30	7.95
	.001	25.4	18.5	15.8	14.4	13.5	12.9
9	.75	.108	.297	.410	.480	.529	.564
	.50	.494	.749	.852	.906	.939	.962
	.25	1.51	1.62	1.63	1.63	1.62	1.61
	.10	3.36	3.01	2.81	2.69	2.61	2.55
	.05	5.12	4.26	3.86	3.63	3.48	3.37
	.025	7.21	5.71	5.08	4.72	4.48	4.32
	.01	10.6	8.02	6.99	6.42	6.06	5.80
	.005	13.6	10.1	8.72	7.96	7.47	7.13
	.001	22.9	16.4	13.9	12.6	11.7	11.1
10	.75	.107	.296	.409	.480	.529	.565
	.50	.490	.743	.845	.899	.932	.954
	.25	1.49	1.60	1.60	1.59	1.59	1.58
	.10	3.29	2.92	2.73	2.61	2.52	2.46
	.05	4.96	4.10	3.71	3.48	3.33	3.22
	.025	6.94	5.46	4.83	4.47	4.24	4.07
	.01	10.0	7.56	6.55	5.99	5.64	5.39
	.005	12.8	9.43	8.08	7.34	6.87	6.54
	.001	21.0	14.9	12.5	11.3	10.5	9.92

ν_1 (degrees of freedom of numerator mean squares)

7	8	9	10	11	12	α	ν_2 (degrees of freedom of denominator mean squares)
.586	.606	.621	.635	.645	.654	.75	
1.02	1.03	1.04	1.05	1.06	1.06	.50	
1.78	1.78	1.77	1.77	1.77	1.77	.25	
3.01	2.98	2.96	2.94	2.92	2.90	.10	
4.21	4.15	4.10	4.06	4.03	4.00	.05	**6**
5.70	5.60	5.52	5.46	5.41	5.37	.025	
8.26	8.10	7.98	7.87	7.79	7.72	.01	
10.8	10.6	10.4	10.3	10.1	10.0	.005	
19.5	19.0	18.7	18.4	18.2	18.0	.001	
.588	.608	.624	.637	.649	.658	.75	
1.00	1.01	1.02	1.03	1.04	1.04	.50	
1.70	1.70	1.69	1.69	1.68	1.68	.25	
2.78	2.75	2.72	2.70	2.68	2.67	.10	
3.77	3.73	3.68	3.64	3.60	3.57	.05	**7**
4.99	4.89	4.82	4.76	4.71	4.67	.025	
6.99	6.84	6.72	6.62	6.54	6.47	.01	
8.89	8.68	8.52	8.38	8.27	8.18	.005	
15.0	14.6	14.3	14.1	13.9	13.7	.001	
.589	.610	.627	.640	.654	.661	.75	
.988	1.00	1.01	1.02	1.03	1.03	.50	
1.64	1.64	1.63	1.63	1.63	1.62	.25	
2.62	2.59	2.56	2.54	2.52	2.50	.10	
3.50	3.44	3.39	3.35	3.31	3.28	.05	**8**
4.53	4.43	4.36	4.30	4.25	4.20	.025	
6.18	6.03	5.91	5.81	5.73	5.67	.01	
7.69	7.50	7.34	7.21	7.10	7.01	.005	
12.4	12.0	11.8	11.5	11.3	11.2	.001	
.591	.612	.629	.643	.654	.664	.75	
.978	.990	1.00	1.01	1.02	1.02	.50	
1.60	1.60	1.59	1.59	1.58	1.58	.25	
2.51	2.47	2.44	2.42	2.40	2.38	.10	
3.29	3.23	3.18	3.14	3.10	3.07	.05	**9**
4.20	4.10	4.03	3.96	3.91	3.87	.025	
5.61	5.47	5.35	5.26	5.18	5.11	.01	
6.88	6.69	6.54	6.42	6.32	6.23	.005	
10.7	10.4	10.1	9.79	9.72	9.57	.001	
.592	.613	.631	.645	.657	.667	.75	
.971	.983	.992	1.00	1.01	1.01	.50	
1.57	1.56	1.56	1.55	1.54	1.54	.25	
2.41	2.38	2.35	2.32	2.30	2.28	.10	
3.14	3.07	3.02	2.98	2.94	2.91	.05	**10**
3.95	3.85	3.78	3.72	3.67	3.62	.025	
5.20	5.06	4.94	4.85	4.77	4.71	.01	
6.30	6.12	5.97	5.85	5.75	5.66	.005	
9.52	9.20	8.96	8.75	8.59	8.45	.001	

TABLE 16 Critical values of the F-distribution

ν_2 (degrees of freedom of denominator mean squares); ν_1 (degrees of freedom of numerator mean squares)

ν_2	α	1	2	3	4	5	6
11	.75	.107	.295	.408	.481	.529	.565
	.50	.486	.739	.840	.893	.926	.948
	.25	1.47	1.58	1.58	1.57	1.56	1.55
	.10	3.23	2.86	2.66	2.54	2.45	2.39
	.05	4.84	3.98	3.59	3.36	3.20	3.09
	.025	6.72	5.26	4.63	4.28	4.04	3.88
	.01	9.65	7.21	6.22	5.67	5.32	5.07
	.005	12.2	8.91	7.60	6.88	6.42	6.10
	.001	19.7	13.8	11.6	10.3	9.58	9.05
12	.75	.106	.295	.408	.480	.530	.566
	.50	.484	.735	.835	.888	.921	.943
	.25	1.46	1.56	1.56	1.55	1.54	1.53
	.10	3.18	2.81	2.61	2.48	2.39	2.33
	.05	4.75	3.89	3.49	3.26	3.11	3.00
	.025	6.55	5.10	4.47	4.12	3.89	3.73
	.01	9.33	6.93	5.95	5.41	5.06	4.82
	.005	11.8	8.51	7.23	6.52	6.07	5.76
	.001	18.6	13.0	10.8	9.63	8.89	8.38
13	.75	.106	.294	.408	.480	.530	.567
	.50	.481	.731	.832	.885	.917	.939
	.25	1.45	1.55	1.55	1.53	1.52	1.51
	.10	3.14	2.76	2.56	2.43	2.35	2.28
	.05	4.67	3.81	3.41	3.18	3.03	2.92
	.025	6.41	4.97	4.35	4.00	3.77	3.60
	.01	9.07	6.70	5.74	4.21	4.86	4.62
	.005	11.4	8.19	6.93	6.23	5.79	5.48
	.001	17.8	12.3	10.2	9.07	8.35	7.86
14	.75	.106	.294	.408	.480	.530	.567
	.50	.479	.729	.828	.881	.914	.936
	.25	1.44	1.53	1.53	1.52	1.51	1.50
	.10	3.10	2.73	2.52	2.39	2.31	2.24
	.05	4.60	3.74	3.34	3.11	2.96	2.85
	.025	6.30	4.86	4.24	3.89	3.66	3.50
	.01	8.86	6.51	5.56	5.04	4.69	4.46
	.005	11.1	7.92	6.68	6.00	5.53	5.26
	.001	17.1	11.8	9.73	8.62	7.92	7.43
15	.75	.105	.293	.407	.480	.530	.568
	.50	.478	.726	.826	.878	.911	.933
	.25	1.43	1.52	1.52	1.51	1.49	1.48
	.10	3.07	2.70	2.49	2.36	2.27	2.21
	.05	4.54	3.68	3.29	3.06	2.90	2.79
	.025	6.20	4.77	4.15	3.80	3.58	3.41
	.01	8.68	6.36	5.42	4.89	4.56	4.32
	.005	10.8	7.70	6.48	5.80	5.37	5.07
	.001	16.6	11.3	9.34	8.25	7.57	7.09

ν_2 (degrees of freedom of denominator mean squares)	ν_1 (degrees of freedom of numerator mean squares) 7	8	9	10	11	12	α
11	.592	.614	.633	.645	.658	.667	.75
	.964	.977	.986	.994	1.00	1.01	.50
	1.54	1.53	1.53	1.52	1.51	1.51	.25
	2.34	2.30	2.27	2.25	2.23	2.21	.10
	3.01	2.95	2.90	2.85	2.82	2.79	.05
	3.76	3.66	3.59	3.53	3.48	3.43	.025
	4.89	4.74	4.63	4.54	4.46	4.40	.01
	5.86	5.68	5.54	5.42	5.32	5.24	.005
	8.66	8.35	8.12	7.92	7.76	7.63	.001
12	.594	.616	.633	.649	.661	.671	.75
	.959	.972	.981	.989	.995	1.00	.50
	1.52	1.51	1.51	1.50	1.50	1.49	.25
	2.28	2.24	2.21	2.19	2.17	2.15	.10
	2.91	2.85	2.80	2.75	2.72	2.69	.05
	3.61	3.51	3.44	3.37	3.32	3.28	.025
	4.64	4.50	4.39	4.30	4.22	4.16	.01
	5.52	5.35	5.20	5.09	4.99	4.91	.005
	8.00	7.71	7.48	7.29	7.14	7.00	.001
13	.595	.617	.635	.650	.662	.673	.75
	.955	.967	.977	.984	.990	.996	.50
	1.50	1.49	1.49	1.48	1.48	1.47	.25
	2.23	2.20	2.16	2.14	2.12	2.10	.10
	2.83	2.77	2.71	2.67	2.64	2.60	.05
	3.48	3.39	3.31	3.25	3.20	3.15	.025
	4.44	4.30	4.19	4.10	4.03	3.96	.01
	5.25	5.08	4.94	4.82	4.73	4.64	.005
	7.49	7.21	6.98	6.80	6.65	6.52	.001
14	.595	.618	.636	.651	.664	.674	.75
	.952	.964	.973	.981	.987	.992	.50
	1.49	1.48	1.47	1.46	1.46	1.45	.25
	2.19	2.45	1.42	2.10	2.07	2.05	.10
	2.76	2.70	2.65	2.60	2.57	2.53	.05
	3.38	3.29	3.20	3.15	3.10	3.05	.025
	4.28	4.14	4.03	3.94	3.86	3.80	.01
	5.03	4.86	4.72	4.60	4.51	4.43	.005
	7.08	6.80	6.58	6.40	6.26	6.13	.001
15	.596	.618	.637	.652	.667	.676	.75
	.948	.960	.970	.977	.983	.989	.50
	1.47	1.46	1.46	1.45	1.44	1.44	.25
	2.16	2.12	2.09	2.06	2.04	2.02	.10
	2.71	2.64	2.59	2.54	2.51	2.48	.05
	3.29	3.20	3.12	3.06	3.01	2.96	.025
	4.14	4.00	3.89	3.80	3.73	3.67	.01
	4.85	4.67	4.54	4.42	4.33	4.25	.005
	6.74	6.47	6.26	6.08	5.94	5.81	.001

TABLE 16 Critical values of the F-distribution

ν_1 (degrees of freedom of numerator mean squares)

ν_2 (degrees of freedom of denominator mean squares)	α	1	2	3	4	5	6
	.75	.105	.293	.407	.480	.531	.568
	.50	.476	.724	.823	.876	.908	.930
	.25	1.42	1.51	1.51	1.50	1.48	1.47
	.10	3.05	2.67	2.46	2.33	2.24	2.18
16	.05	4.49	3.63	3.24	3.01	2.85	2.74
	.025	6.12	4.69	4.08	3.73	3.50	3.34
	.01	8.53	6.23	5.29	4.77	4.44	4.20
	.005	10.6	7.51	6.30	5.64	5.21	4.91
	.001	16.1	11.0	9.00	7.94	7.27	6.81
	.75	.105	.292	.407	.480	.531	.568
	.50	.475	.722	.821	.874	.906	.928
	.25	1.42	1.51	1.50	1.49	1.47	1.46
	.10	3.03	2.64	2.44	2.31	2.22	2.15
17	.05	4.45	3.59	3.20	2.96	2.91	2.70
	.025	6.04	4.62	4.01	3.66	3.44	3.28
	.01	8.40	6.11	5.19	4.67	4.34	4.10
	.005	10.4	7.35	6.16	5.50	5.07	4.78
	.001	15.7	10.7	8.73	7.68	7.02	6.56
	.75	.105	.292	.407	.480	.531	.569
	.50	.474	.721	.819	.872	.904	.926
	.25	1.41	1.50	1.49	1.48	1.46	1.45
	.10	3.01	2.62	2.42	2.29	2.20	2.13
18	.05	4.41	3.55	3.16	2.93	2.77	2.66
	.025	5.98	4.56	3.95	3.61	3.38	3.22
	.01	8.28	6.01	5.09	4.58	4.25	4.01
	.005	10.2	7.21	6.03	5.37	4.96	4.66
	.001	15.4	10.4	8.49	7.46	6.81	6.35
	.75	.104	.292	.407	.480	.531	.569
	.50	.473	.719	.818	.870	.902	.924
	.25	1.41	1.49	1.49	1.47	1.46	1.44
	.10	2.99	2.61	2.40	2.27	2.18	2.11
19	.05	4.38	3.52	3.13	2.90	2.74	2.63
	.025	5.92	4.51	3.90	3.56	3.33	3.17
	.01	8.19	5.93	5.01	4.50	4.17	3.94
	.005	10.1	7.09	5.92	5.27	4.85	4.56
	.001	15.1	10.2	8.28	7.26	6.62	6.18
	.75	.104	.292	.407	.480	.531	.569
	.50	.472	.718	.816	.868	.900	.922
	.25	1.40	1.49	1.48	1.47	1.45	1.44
	.10	2.97	2.59	2.38	2.25	2.16	2.09
20	.05	4.35	3.49	3.10	2.87	2.71	2.60
	.025	5.87	4.46	3.86	3.51	3.29	3.13
	.01	8.10	5.85	4.94	4.43	4.10	3.87
	.005	9.94	6.99	5.82	5.17	4.76	4.47
	.001	14.8	9.95	8.10	7.10	6.46	6.02

ν_1 (degrees of freedom of numerator mean squares)							
7	8	9	10	11	12	α	ν_2 (degrees of freedom of denominator mean squares)
.597	.619	.638	.653	.666	.677	.75	
.946	.958	.967	.975	.981	.986	.50	
1.46	1.45	1.44	1.44	1.43	1.43	.25	
2.13	2.09	2.06	2.03	2.01	1.99	.10	
2.66	2.59	2.54	2.49	2.46	2.42	.05	**16**
3.22	3.12	3.05	2.99	2.93	2.89	.025	
4.03	3.89	3.78	3.69	3.62	3.55	.01	
4.69	4.52	5.38	4.27	4.18	4.10	.005	
6.46	6.19	5.98	5.81	5.67	5.55	.001	
.597	.620	.639	.654	.667	.678	.75	
.943	.955	.965	.972	.978	.983	.50	
1.45	1.44	1.43	1.43	1.42	1.41	.25	
2.10	2.06	2.03	2.00	1.98	1.96	.10	
3.61	2.55	2.49	2.45	2.41	2.38	.05	**17**
3.16	3.06	2.98	2.92	2.87	2.82	.025	
3.93	3.79	3.68	3.59	3.52	3.46	.01	
4.56	4.39	4.25	4.14	4.05	3.97	.005	
6.22	5.96	5.75	5.58	5.44	5.32	.001	
.598	.621	.639	.655	.668	.679	.75	
.941	.953	.962	.970	.976	.981	.50	
1.44	1.43	1.42	1.42	1.41	1.40	.25	
2.08	2.04	2.00	1.98	1.95	1.93	.10	
2.58	2.51	2.46	2.41	2.37	2.34	.05	**18**
3.10	3.01	2.93	2.87	2.81	2.77	.025	
3.84	3.71	3.60	3.51	3.43	3.37	.01	
4.44	4.28	4.14	4.03	3.94	3.86	.005	
6.02	5.76	5.56	5.39	5.25	5.13	.001	
.598	.621	.640	.656	.669	.680	.75	
.939	.951	.961	.968	.974	.979	.50	
1.43	1.42	1.41	1.41	1.40	1.40	.25	
2.06	2.02	1.98	1.96	1.93	1.91	.10	
2.54	2.48	2.42	2.38	2.34	2.31	.05	**19**
3.05	2.76	2.88	2.82	2.76	2.72	.025	
3.77	3.63	3.52	3.43	3.56	3.30	.01	
4.34	4.18	4.04	3.93	3.84	3.76	.005	
5.85	5.59	5.39	5.22	5.08	4.97	.001	
.598	.622	.641	.656	.671	.681	.75	
.938	.950	.959	.966	.972	.977	.50	
1.43	1.42	1.41	1.40	1.39	1.39	.25	
2.04	2.00	1.96	1.94	1.91	1.89	.10	
2.51	2.45	2.39	2.35	2.31	2.28	.05	**20**
3.01	2.91	2.84	2.77	2.72	2.68	.025	
3.70	3.56	3.46	3.37	3.29	3.23	.01	
4.26	4.09	3.96	3.85	3.76	3.68	.005	
5.69	5.44	5.24	5.08	4.94	4.82	.001	

TABLE 16 Critical values of the F-distribution

ν_2 (degrees of freedom of denominator mean squares)	α	ν_1 (degrees of freedom of numerator mean squares) 1	2	3	4	5	6
21	.75	.104	.292	.407	.480	.532	.570
	.50	.471	.717	.815	.867	.899	.921
	.25	1.40	1.48	1.48	1.46	1.44	1.43
	.10	2.96	2.57	2.36	2.23	2.14	2.08
	.05	4.32	3.47	3.07	2.84	2.68	2.57
	.025	5.83	4.42	3.82	3.48	3.25	3.09
	.01	8.02	5.75	4.87	4.37	4.04	3.81
	.005	9.83	6.89	5.73	5.09	4.99	4.39
	.001	14.6	9.77	7.94	6.95	6.32	5.88
22	.75	.104	.292	.407	.481	.532	.570
	.50	.470	.715	.814	.866	.898	.919
	.25	1.40	1.48	1.47	1.45	1.44	1.42
	.10	2.95	2.56	2.35	2.22	2.13	2.06
	.05	4.30	3.44	3.05	2.82	2.66	2.55
	.025	5.79	4.38	3.78	3.44	3.22	3.05
	.01	7.95	5.72	4.82	4.31	3.99	3.76
	.005	6.73	6.81	5.65	5.02	4.61	4.32
	.001	14.4	9.61	7.80	6.81	6.19	5.76
23	.75	.104	.291	.406	.481	.532	.570
	.50	.470	.714	.813	.864	.896	.918
	.25	1.39	1.47	1.47	1.45	1.43	1.42
	.10	2.94	2.55	2.31	2.25	2.11	1.05
	.05	4.28	3.42	3.03	2.80	2.64	2.53
	.025	5.75	4.35	3.75	3.41	3.18	3.02
	.01	7.88	5.66	4.76	4.26	3.94	3.71
	.005	9.63	6.73	5.58	4.95	4.54	4.26
	.001	14.2	9.47	7.67	6.69	6.08	5.65
24	.75	.104	.291	.406	.480	.532	.570
	.50	.469	.714	.812	.863	.895	.917
	.25	1.39	1.47	1.46	1.44	1.43	1.41
	.10	2.93	2.54	2.33	2.19	2.10	2.04
	.05	4.26	3.40	3.01	2.78	2.62	2.51
	.025	5.72	4.32	3.72	3.38	3.15	2.99
	.01	7.82	5.61	4.72	4.22	3.90	3.67
	.005	9.55	6.66	5.52	4.89	4.49	4.20
	.001	14.0	9.34	7.55	6.59	5.98	5.55
25	.75	.104	.292	.407	.481	.532	.571
	.50	.468	.713	.811	.862	.984	.916
	.25	1.39	1.47	1.46	1.44	1.42	1.41
	.10	2.92	2.53	2.32	2.18	2.09	2.02
	.05	4.24	3.39	2.99	2.76	2.60	2.49
	.025	5.69	4.29	3.69	3.35	3.13	2.97
	.01	7.77	5.57	4.68	4.18	3.86	3.63
	.005	9.48	6.60	5.46	4.84	4.43	4.15
	.001	13.9	9.22	7.45	6.49	5.88	5.46

ν_1
(degrees of freedom of numerator mean squares)

7	8	9	10	11	12	α	ν_2 (degrees of freedom of denominator mean squares)
.599	.622	.641	.657	.671	.682	.75	
.936	.948	.957	.965	.971	.976	.50	
1.42	1.41	1.40	1.39	1.39	1.38	.25	
2.02	1.98	1.95	1.92	1.90	1.87	.10	
2.49	2.42	2.37	2.32	2.28	2.25	.05	**21**
2.97	2.87	2.80	2.73	2.68	2.64	.025	
3.64	3.51	3.40	3.31	3.24	3.17	.01	
4.18	4.01	3.88	3.77	3.68	3.60	.005	
5.56	5.31	5.11	4.95	4.81	4.70	.001	
.599	.623	.642	.658	.671	.683	.75	
.935	.947	.956	.963	.969	.974	.50	
1.41	1.40	1.39	1.39	1.38	1.37	.25	
2.01	1.97	1.93	1.90	1.88	1.86	.10	
2.46	2.40	2.39	2.30	2.26	2.23	.05	**22**
2.93	2.84	2.76	2.70	2.65	2.60	.025	
3.59	3.45	3.35	3.26	3.18	3.12	.01	
4.11	3.94	6.81	3.70	3.61	3.53	.005	
5.44	5.19	4.99	4.83	4.70	4.58	.001	
.600	.623	.642	.658	.672	.684	.75	
.934	.945	.955	.962	.968	.973	.50	
1.41	1.40	1.39	1.38	1.37	1.37	.25	
1.99	1.95	1.92	1.89	1.87	1.85	.10	
2.44	2.37	2.32	2.27	2.24	2.20	.05	**23**
2.90	2.81	2.73	2.67	2.62	2.57	.025	
3.54	3.41	3.30	3.21	3.14	3.07	.01	
4.05	3.88	3.75	3.64	3.55	3.47	.005	
5.33	5.09	4.89	4.73	4.59	4.48	.001	
.600	.623	.643	.659	.671	.684	.75	
.932	.944	.953	.961	.967	.972	.50	
1.40	1.39	1.38	1.37	1.37	1.36	.25	
1.98	1.94	1.91	1.88	1.85	1.83	.10	
2.42	2.36	2.30	2.25	2.22	2.18	.05	**24**
2.87	2.78	2.70	2.64	2.59	2.54	.025	
3.50	3.36	3.26	3.17	3.09	3.03	.01	
3.99	3.83	3.69	3.59	3.50	3.42	.005	
5.23	4.99	4.80	4.64	4.50	4.39	.001	
.600	.624	.643	.659	.673	.685	.75	
.931	.943	.952	.960	.966	.971	.50	
1.40	1.39	1.38	1.37	1.36	1.36	.25	
1.97	1.93	1.89	1.87	1.84	1.82	.10	
2.40	2.34	2.28	2.24	2.20	2.16	.05	**25**
2.85	2.75	2.68	2.61	2.56	2.51	.025	
3.46	3.32	3.22	3.13	3.06	2.99	.01	
3.94	3.78	3.64	3.54	3.44	3.37	.005	
5.15	4.91	4.71	4.56	4.42	4.31	.001	

TABLE 16 Critical values of the F-distribution

ν_2 (degrees of freedom of denominator mean squares); ν_1 (degrees of freedom of numerator mean squares)

ν_2	α	1	2	3	4	5	6
	.75	.104	.291	.406	.480	.532	.571
	.50	.468	.712	.810	.861	.893	.915
	.25	1.38	1.46	1.45	1.44	1.42	1.41
	.10	2.91	2.52	2.31	2.17	2.08	2.01
26	.05	4.23	3.37	2.98	2.74	2.59	2.47
	.025	5.66	4.27	3.67	3.33	3.10	2.94
	.01	7.72	5.53	4.64	4.14	3.82	3.59
	.005	9.41	6.54	5.41	4.79	4.38	4.10
	.001	13.7	9.12	7.36	6.41	5.80	5.38
	.75	.104	.291	.406	.480	.532	.571
	.50	.467	.711	.809	.861	.892	.914
	.25	1.38	1.46	1.45	1.43	1.42	1.40
	.10	2.90	2.51	2.30	2.17	2.07	2.00
27	.05	4.21	3.35	2.96	2.73	2.57	2.46
	.025	5.63	4.24	3.65	3.31	3.08	2.92
	.01	7.68	5.49	4.60	4.11	3.78	3.56
	.005	9.34	6.49	5.36	4.74	4.34	4.06
	.001	13.6	9.02	7.27	6.33	5.73	5.31
	.75	.103	.290	.406	.480	.532	.571
	.50	.467	.711	.808	.860	.892	.913
	.25	1.38	1.46	1.45	1.43	1.41	1.40
	.10	2.89	2.50	2.29	2.16	2.06	2.00
28	.05	4.20	3.34	2.95	2.71	2.56	2.45
	.025	5.61	4.22	3.63	3.29	3.06	2.90
	.01	7.64	5.45	4.57	4.07	3.75	3.53
	.005	9.28	6.44	5.32	4.70	4.30	4.02
	.001	13.5	8.93	7.19	6.25	5.66	5.24
	.75	.103	.290	.406	.480	.532	.571
	.50	.467	.710	.808	.859	.891	.912
	.25	1.38	1.45	1.45	1.43	1.41	1.40
	.10	2.89	2.50	2.28	2.15	2.06	1.99
29	.05	4.18	3.33	2.93	2.70	2.55	2.43
	.025	5.59	4.20	3.60	3.27	3.04	2.88
	.01	7.60	5.42	4.54	4.04	3.73	3.50
	.005	9.23	6.40	5.28	4.66	4.26	3.98
	.001	13.4	8.85	7.12	6.19	5.59	5.18
	.75	.103	.290	.406	.480	.532	.571
	.50	.466	.709	.807	.858	.890	.912
	.25	1.38	1.45	1.44	1.42	1.41	1.39
	.10	2.88	2.49	2.28	2.14	2.05	1.98
30	.05	4.17	3.32	2.92	2.69	2.53	2.42
	.025	5.57	4.18	3.59	3.25	3.03	2.87
	.01	7.56	5.39	4.51	4.02	3.70	3.47
	.005	9.18	6.35	5.24	4.62	4.23	3.95
	.001	13.3	8.77	7.05	6.12	5.53	5.12

ν_2 (degrees of freedom of denominator mean squares)	ν_1 (degrees of freedom of numerator mean squares) 7	8	9	10	11	12	α
26	.600	.624	.644	.660	.674	.686	.75
	.930	.942	.951	.959	.965	.970	.50
	1.39	1.38	1.37	1.37	1.36	1.35	.25
	1.96	1.92	1.88	1.86	1.83	1.81	.10
	2.39	2.32	2.27	2.22	2.18	2.15	.05
	2.82	2.73	2.65	2.59	2.54	2.49	.025
	3.42	3.29	3.18	3.09	3.02	2.96	.01
	3.89	3.73	3.60	3.49	3.40	3.33	.005
	5.07	4.83	4.64	4.48	4.35	4.24	.001
27	.601	.624	.644	.660	.674	.686	.75
	.930	.941	.950	.958	.964	.969	.50
	1.39	1.38	1.37	1.36	1.35	1.35	.25
	1.95	1.91	1.87	1.85	1.82	1.80	.10
	2.37	2.31	2.25	2.20	2.17	2.13	.05
	2.80	2.71	2.63	2.57	2.51	2.47	.025
	3.39	3.26	3.45	3.06	2.99	2.93	.01
	3.85	3.69	3.56	3.45	3.36	3.28	.005
	5.00	4.76	4.57	4.41	4.28	4.17	.001
28	.601	.625	.644	.661	.675	.687	.75
	.929	.940	.950	.957	.963	.968	.50
	1.39	1.38	1.37	1.36	1.35	1.34	.25
	1.94	1.90	1.87	1.84	1.81	1.79	.10
	2.36	2.29	2.24	2.19	2.15	2.12	.05
	2.78	2.69	2.61	2.55	2.49	2.45	.025
	3.36	3.23	3.12	3.03	2.96	2.90	.01
	3.81	3.65	3.52	3.41	3.32	3.25	.005
	4.93	4.69	4.50	4.35	4.22	4.11	.001
29	.601	.625	.645	.661	.675	.687	.75
	.928	.940	.949	.956	.962	.967	.50
	1.38	1.37	1.36	1.35	1.35	1.34	.25
	1.93	1.89	2.86	1.83	1.80	1.78	.10
	2.35	2.28	2.22	2.18	2.14	2.10	.05
	2.76	2.67	2.59	2.51	2.47	2.43	.025
	3.33	3.20	3.09	3.00	2.93	2.87	.01
	3.77	3.61	3.48	3.38	3.29	3.21	.005
	4.87	4.64	4.45	4.29	4.16	4.05	.001
30	.601	.625	.645	.661	.676	.688	.75
	.927	.939	.948	.955	.961	.966	.50
	1.38	1.37	1.36	1.35	1.34	1.34	.25
	1.93	1.88	1.85	1.82	1.79	1.77	.10
	2.33	2.27	2.21	2.16	2.13	2.09	.05
	2.75	2.65	2.57	2.51	2.46	2.41	.025
	3.30	3.17	3.07	2.98	2.90	2.84	.01
	3.74	3.58	3.45	3.34	3.25	3.18	.005
	4.82	4.58	4.39	4.24	4.11	4.00	.001

TABLE 16 Critical values of the F-distribution

ν_2 (degrees of freedom of denominator mean squares); ν_1 (degrees of freedom of numerator mean squares)

ν_2	α	1	2	3	4	5	6
40	.75	.103	.289	.404	.480	.533	.572
	.50	.463	.705	.802	.854	.885	.907
	.25	1.36	1.44	1.42	1.40	1.39	1.37
	.10	2.84	2.44	2.23	2.09	2.00	1.93
	.05	4.08	3.23	2.84	2.61	2.45	2.34
	.025	5.42	4.05	3.46	3.13	2.90	2.74
	.01	7.31	5.18	4.31	3.83	3.51	3.29
	.005	8.83	6.07	4.98	4.37	3.99	3.71
	.001	12.6	8.25	6.60	5.70	5.13	4.73
60	.75	.102	.289	.405	.480	.534	.573
	.50	.461	.701	.798	.849	.880	.901
	.25	1.35	1.42	1.41	1.38	1.37	1.35
	.10	2.79	2.39	2.18	2.04	1.95	1.87
	.05	4.00	3.15	2.76	2.53	2.37	2.25
	.025	5.29	3.93	3.34	3.01	2.79	2.63
	.01	7.08	4.98	4.13	3.65	3.34	3.12
	.005	8.49	5.79	4.73	4.14	3.76	3.49
	.001	12.0	7.76	6.17	5.31	4.76	4.37
120	.75	.102	.288	.405	.481	.534	.574
	.50	.458	.697	.793	.844	.875	.896
	.25	1.34	1.40	1.39	1.37	1.35	1.33
	.10	2.75	2.35	2.13	1.99	1.90	1.82
	.05	3.92	3.07	2.68	2.45	2.29	2.17
	.025	5.15	3.80	3.23	2.89	2.67	2.52
	.01	6.85	4.79	3.95	3.48	3.17	2.96
	.005	8.18	5.54	4.50	3.92	3.55	3.28
	.001	11.4	7.32	5.79	4.95	4.42	4.04
∞	.75	.102	.288	.404	.481	.535	.576
	.50	.455	.693	.789	.839	.870	.891
	.25	1.32	1.39	1.37	1.35	1.33	1.31
	.10	2.71	2.30	2.08	1.94	1.85	1.77
	.05	3.84	3.00	2.60	2.37	2.21	2.10
	.025	5.02	3.69	3.11	2.79	2.57	2.41
	.01	6.63	4.61	3.78	3.32	3.02	2.80
	.005	7.88	5.30	4.28	3.72	3.35	3.09
	.001	10.8	6.91	5.42	4.62	4.10	3.74

ν_1 (degrees of freedom of numerator mean squares)

7	8	9	10	11	12	α	ν_2 (degrees of freedom of denominator mean squares)
.603	.627	.647	.662	.679	.689	.75	
.922	.934	.943	.950	.956	.961	.50	
1.36	1.35	1.34	1.33	1.32	1.31	.25	
1.87	1.83	1.79	1.76	1.74	1.74	.10	
2.25	2.18	2.12	2.08	2.04	2.04	.05	**40**
2.62	2.53	2.45	2.39	2.33	2.29	.025	
3.12	2.99	2.89	2.80	2.73	2.66	.01	
3.51	3.35	3.22	3.12	3.03	2.95	.005	
4.44	4.21	4.02	3.87	3.74	3.64	.001	
.604	.629	.650	.667	.682	.695	.75	
.917	.928	.937	.945	.951	.956	.50	
1.33	1.32	1.31	1.30	1.29	1.29	.25	
1.82	1.77	1.74	1.71	1.68	1.66	.10	
2.17	2.10	2.04	1.99	1.95	1.92	.05	**60**
2.51	2.41	2.33	2.27	2.22	2.17	.025	
2.95	2.82	2.72	2.63	2.56	2.50	.01	
3.29	3.13	3.01	2.90	2.81	2.74	.005	
4.09	3.87	3.69	3.54	3.41	3.31	.001	
.606	.631	.652	.670	.686	.699	.75	
.912	.923	.932	.939	.945	.950	.50	
1.31	1.30	1.29	1.28	1.27	1.26	.25	
1.77	1.72	1.68	1.65	1.63	1.60	.10	
2.09	2.02	1.96	1.91	1.87	1.83	.05	**120**
2.39	2.30	2.22	2.16	2.10	2.05	.025	
2.79	2.66	2.56	2.47	2.40	2.34	.01	
3.09	2.93	2.81	2.71	2.62	2.54	.005	
3.77	3.55	3.38	3.24	3.11	3.02	.001	
.608	.634	.655	.674	.690	.703	.75	
.907	.918	.927	.934	.939	.945	.50	
1.29	1.28	1.27	1.25	1.24	1.24	.25	
1.72	1.67	1.63	1.60	1.57	1.55	.10	
2.01	1.94	1.88	1.83	1.79	1.75	.05	∞
2.29	2.19	2.11	2.05	1.99	1.94	.025	
2.64	2.51	2.41	2.32	2.25	2.18	.01	
2.90	2.74	2.62	2.52	2.43	2.36	.005	
3.47	3.27	3.10	2.96	2.84	2.74	.001	

TABLE 16 Critical values of the F-distribution

ν_1 (degrees of freedom of numerator mean squares)

ν_2 (degrees of freedom of denominator mean squares)	α	15	20	24	30	40	50	60	120	∞
	.75	.698	.712	.719	.727	.734	.738	.741	.749	.756
	.50	2.09	2.12	2.13	2.15	2.16	2.17	2.17	2.18	2.20
	.25	9.49	9.58	9.63	9.67	9.71	9.74	9.76	9.80	9.85
	.10	61.2	61.7	62.0	62.3	62.5	62.7	62.8	63.1	63.3
1	.05	246	248	249	250	251	252	252	253	254
	.025	985	993	997	1000	1010	1010	1010	1010	1020
	.01	6160	6210	6230	6260	6290	6300	6310	6340	6370
	.005	24630	24836	24940	25440	25148	25211	25253	25359	25465
	.001	615800	620900	623500	626100	628700	630300	631300	634000	636600
	.75	.657	.672	.680	.689	.697	.702	.705	.713	.721
	.50	1.38	1.39	1.40	1.41	1.42	1.43	1.43	1.43	1.44
	.25	3.41	3.43	3.43	3.44	3.45	3.46	3.46	3.47	3.48
	.10	9.42	9.44	9.45	9.46	9.47	9.47	9.47	9.48	9.49
2	.05	19.4	19.4	19.5	19.5	19.5	19.5	19.5	19.5	19.5
	.025	39.4	39.4	39.5	39.5	39.5	39.5	39.5	39.5	39.5
	.01	99.4	99.4	99.5	99.5	99.5	99.5	99.5	99.5	99.5
	.005	199	199	199	199	199	199	199	199	200
	.001	999	999	1000	1000	1000	1000	1000	1000	1000
	.75	.658	.675	.684	.694	.702	.708	.711	.721	.730
	.50	1.21	1.23	1.23	1.24	1.25	1.25	1.25	1.26	1.27
	.25	2.46	2.46	2.46	2.47	2.47	2.47	2.47	2.47	2.47
	.10	5.20	5.18	5.18	5.17	5.16	5.15	5.15	5.14	5.13
3	.05	8.70	8.66	8.64	8.62	8.59	8.58	8.57	8.55	8.53
	.025	14.3	14.2	14.1	14.1	14.0	14.0	14.0	13.9	13.9
	.01	26.9	26.7	26.6	26.5	26.4	26.3	26.3	26.2	26.1
	.005	43.1	42.8	42.6	42.5	42.3	42.2	42.1	42.0	41.8
	.001	127	126	126	125	125	125	124	124	124
	.75	.664	.683	.692	.702	.712	.718	.722	.733	.743
	.50	1.14	1.15	1.16	1.16	1.17	1.18	1.18	1.18	1.19
	.25	2.08	2.08	2.08	2.08	2.08	2.08	2.08	2.08	2.08
	.10	3.87	3.84	3.83	3.82	3.80	3.79	3.79	3.78	3.76
4	.05	5.86	5.80	5.77	5.75	5.72	5.70	5.69	5.66	5.63
	.025	8.66	8.56	8.51	8.46	8.41	8.38	8.36	8.31	8.26
	.01	14.2	14.0	13.9	13.8	13.7	13.7	13.7	13.6	13.5
	.005	20.4	20.2	20.0	19.9	19.8	19.7	19.6	19.5	19.3
	.001	46.8	46.1	45.8	45.4	45.1	44.9	44.8	44.4	44.0
	.75	.669	.690	.700	.711	.722	.728	.732	.743	.755
	.50	1.10	1.11	1.12	1.12	1.13	1.14	1.14	1.14	1.15
	.25	1.89	1.88	1.88	1.88	1.88	1.87	1.87	1.87	1.87
	.10	3.24	3.21	3.19	3.17	3.16	3.15	3.14	3.12	3.10
5	.05	4.62	4.56	4.53	4.50	4.46	4.44	4.43	4.40	4.36
	.025	6.43	6.33	6.28	6.23	6.18	6.14	6.12	6.07	6.02
	.01	9.72	9.55	9.47	9.38	9.29	9.24	9.20	9.11	9.02
	.005	13.1	12.9	12.8	12.7	12.5	12.4	12.4	12.3	12.1
	.001	25.9	25.4	25.1	24.9	24.6	24.4	24.3	24.1	23.7

ν_1 (degrees of freedom of numerator mean squares)

ν_2 (degrees of freedom of denominator mean squares)	15	20	24	30	40	50	60	120	∞	α
6	.675	.696	.707	.718	.729	.736	.741	.753	.765	.75
	1.07	1.08	1.09	1.10	1.10	1.11	1.11	1.12	1.12	.50
	1.76	1.76	1.75	1.75	1.75	1.74	1.74	1.74	1.74	.25
	2.87	2.84	2.82	2.80	2.78	2.77	2.76	2.74	2.72	.10
	3.94	3.87	3.84	3.81	3.77	3.75	3.74	3.70	3.67	.05
	5.27	5.17	5.12	5.07	5.01	4.98	4.96	4.90	4.85	.025
	7.56	7.40	7.31	7.23	7.14	7.09	7.06	6.97	6.88	.01
	9.81	9.59	9.47	9.36	9.24	9.17	9.12	9.00	8.88	.005
	17.6	17.1	16.9	16.7	16.4	16.3	16.2	16.0	15.8	.001
7	.679	.702	.713	.725	.737	.745	.749	.762	.775	.75
	1.05	1.07	1.07	1.08	1.08	1.09	1.09	1.10	1.10	.50
	1.68	1.67	1.67	1.66	1.66	1.65	1.65	1.65	1.65	.25
	2.63	2.59	2.58	2.56	2.54	2.52	2.51	2.49	2.47	.10
	3.51	3.44	3.41	3.38	3.34	3.32	3.30	3.27	3.23	.05
	4.57	4.47	4.42	4.36	4.31	4.27	4.25	4.20	4.14	.025
	6.31	6.16	6.07	5.99	5.91	5.86	5.82	5.74	5.65	.01
	7.97	7.75	7.65	7.53	7.42	7.35	7.31	7.19	7.08	.005
	13.3	12.9	12.7	12.5	12.3	12.2	12.1	11.9	11.7	.001
8	.684	.707	.718	.730	.743	.751	.756	.769	.783	.75
	1.04	1.05	1.06	1.07	1.08	1.07	1.08	1.08	1.09	.50
	1.62	1.61	1.60	1.60	1.59	1.59	1.59	1.58	1.58	.25
	2.46	2.42	2.40	2.38	2.36	2.35	2.34	2.32	2.29	.10
	3.22	3.15	3.12	3.08	3.04	3.02	3.01	2.97	2.93	.05
	4.10	4.00	3.95	3.89	3.84	3.80	3.78	3.73	3.67	.025
	5.52	5.36	5.28	5.20	5.12	5.07	5.03	4.95	4.86	.01
	6.81	6.61	6.50	6.40	6.29	6.22	6.18	6.06	5.95	.005
	10.8	10.5	10.3	10.1	9.9	9.8	9.7	9.5	9.3	.001
9	.687	.711	.723	.736	.749	.757	.762	.776	.791	.75
	1.03	1.04	1.05	1.05	1.06	1.07	1.07	1.07	1.08	.50
	1.57	1.56	1.56	1.55	1.55	1.54	1.54	1.53	1.53	.25
	2.34	2.30	2.28	2.25	2.23	2.22	2.21	2.18	2.16	.10
	3.01	2.94	2.90	2.86	2.83	2.81	2.79	2.75	2.71	.05
	3.77	3.67	3.61	3.56	3.51	3.47	3.45	3.39	3.33	.025
	4.96	4.81	4.73	4.65	4.57	4.52	4.48	4.40	4.31	.01
	6.03	5.83	5.73	5.62	5.52	5.45	5.41	5.30	5.19	.005
	9.24	8.90	8.72	8.55	8.37	8.26	8.19	8.00	7.81	.001
10	.691	.714	.727	.740	.754	.762	.767	.782	.797	.75
	1.02	1.03	1.04	1.05	1.05	1.06	1.06	1.06	1.07	.50
	1.53	1.52	1.52	1.51	1.51	1.50	1.50	1.49	1.48	.25
	2.24	2.20	2.18	2.16	2.13	2.12	2.11	2.08	2.06	.10
	2.85	2.77	2.74	2.70	2.66	2.64	2.62	2.58	2.54	.05
	3.52	3.42	3.37	3.31	3.26	3.22	3.20	3.14	3.08	.025
	4.56	4.41	4.33	4.25	4.17	4.12	4.08	4.00	3.91	.01
	5.47	5.27	5.17	5.07	4.97	4.90	4.86	4.75	4.64	.005
	8.13	7.80	7.64	7.47	7.30	7.19	7.12	6.94	6.76	.001

TABLE 16 Critical values of the F-distribution

ν_1 (degrees of freedom of numerator mean squares)

ν_2 (degrees of freedom of denominator mean squares)	α	15	20	24	30	40	50	60	120	∞
11	.75	.694	.719	.730	.744	.758	.767	.773	.788	.803
	.50	1.02	1.03	1.03	1.04	1.05	1.05	1.05	1.06	1.06
	.25	1.50	1.49	1.49	1.48	1.47	1.47	1.47	1.46	1.45
	.10	2.17	2.12	2.10	2.08	2.05	2.04	2.03	2.00	1.97
	.05	2.72	2.65	2.61	2.57	2.53	2.51	2.49	2.45	2.40
	.025	3.33	3.23	3.17	3.12	3.06	3.02	3.00	2.94	2.88
	.01	4.25	4.10	4.02	3.94	3.86	3.81	3.78	3.69	3.60
	.005	5.05	4.86	4.76	4.65	4.55	4.49	4.45	4.34	4.23
	.001	7.32	7.01	6.85	6.68	6.52	6.42	6.35	6.17	6.00
12	.75	.695	.721	.734	.748	.762	.771	.777	.792	.808
	.50	1.01	1.02	1.03	1.03	1.04	1.04	1.05	1.05	1.06
	.25	1.48	1.47	1.46	1.45	1.45	1.44	1.44	1.43	1.42
	.10	2.11	2.06	2.04	2.01	1.99	1.97	1.96	1.93	1.90
	.05	2.62	2.54	2.51	2.47	2.43	2.40	2.38	2.34	2.30
	.025	3.18	3.07	3.02	2.96	2.91	2.87	2.85	2.79	2.72
	.01	4.01	3.86	3.78	3.70	3.62	3.57	3.54	3.45	3.36
	.005	4.72	4.53	4.43	4.33	4.23	4.16	4.12	4.01	3.90
	.001	6.71	6.40	6.25	6.09	5.93	5.83	5.76	5.59	5.42
13	.75	.697	.723	.737	.751	.766	.775	.781	.797	.813
	.50	1.01	1.02	1.02	1.03	1.04	1.04	1.04	1.05	1.05
	.25	1.46	1.45	1.44	1.43	1.42	1.42	1.42	1.41	1.40
	.10	2.05	2.01	1.98	1.96	1.93	1.92	1.90	1.88	1.85
	.05	2.53	2.46	2.42	2.38	2.34	2.31	2.30	2.25	2.21
	.025	3.05	2.95	2.89	2.84	2.78	2.74	2.72	2.66	2.60
	.01	3.82	3.66	3.59	3.51	3.43	3.37	3.34	3.25	3.17
	.005	4.46	4.27	4.17	4.07	3.97	3.91	3.87	3.76	2.65
	.001	6.23	5.93	5.78	5.63	5.47	5.37	5.30	5.14	4.97
14	.75	.699	.726	.740	.754	.769	.778	.785	.801	.818
	.50	1.00	1.01	1.02	1.03	1.03	1.04	1.04	1.04	1.05
	.25	1.44	1.43	1.42	1.41	1.41	1.40	1.40	1.39	1.38
	.10	2.01	1.96	1.94	1.91	1.89	1.87	1.86	1.83	1.80
	.05	2.46	2.39	2.35	2.31	2.27	2.24	2.22	2.18	2.13
	.025	2.95	2.84	2.79	2.73	2.67	2.64	2.61	2.55	2.49
	.01	3.66	3.51	3.43	3.35	3.27	3.21	3.18	3.09	3.00
	.005	4.25	4.06	3.96	3.86	3.76	3.70	3.66	3.55	3.44
	.001	5.85	5.56	5.41	5.25	5.10	5.00	4.94	4.77	4.60
15	.75	.701	.728	.742	.757	.772	.782	.788	.805	.822
	.50	1.00	1.01	1.02	1.02	1.03	1.03	1.03	1.04	1.05
	.25	1.43	1.41	1.41	1.40	1.39	1.38	1.38	1.37	1.36
	.10	1.97	1.92	1.90	1.87	1.85	1.83	1.82	1.79	1.76
	.05	2.40	2.33	2.29	2.25	2.20	2.18	2.16	2.11	2.07
	.025	2.86	2.76	2.70	2.64	2.59	2.55	2.52	2.46	2.40
	.01	3.52	3.37	3.29	3.21	3.13	3.08	3.05	2.96	2.87
	.005	4.07	3.88	3.79	3.69	3.59	3.52	3.48	3.37	3.26
	.001	5.54	5.25	5.10	4.95	4.80	4.70	4.64	4.47	4.31

ν_1
(degrees of freedom of numerator mean squares)

15	20	24	30	40	50	60	120	∞	α	ν_2 (degrees of freedom of denominator mean squares)
.703	.730	.744	.759	.775	.785	.791	.808	.826	.75	
.997	1.01	1.01	1.02	1.03	1.03	1.03	1.04	1.04	.50	
1.41	1.40	1.39	1.38	1.37	1.37	1.36	1.35	1.34	.25	
1.94	1.89	1.87	1.84	1.81	1.79	1.78	1.75	1.72	.10	
2.35	2.28	2.24	2.19	2.15	2.12	2.11	2.06	2.01	.05	**16**
2.79	2.68	2.63	2.57	2.51	2.47	2.45	2.38	2.32	.025	
3.41	3.26	3.18	3.10	3.02	2.97	2.93	2.84	2.75	.01	
3.92	3.73	3.64	3.54	3.44	3.37	3.33	3.22	3.11	.005	
5.27	4.99	4.85	4.70	4.54	4.45	4.39	4.23	4.06	.001	
.704	.732	.746	.762	.777	.787	.794	.811	.830	.75	
.995	1.01	1.01	1.02	1.02	1.03	1.03	1.03	1.04	.50	
1.40	1.39	1.38	1.37	1.36	1.36	1.35	1.34	1.33	.25	
1.91	1.86	1.84	1.81	1.78	1.76	1.75	1.72	1.69	.10	
2.31	2.23	2.19	2.15	2.10	2.08	2.06	2.01	1.96	.05	**17**
2.72	2.62	2.56	2.50	2.44	2.41	2.38	2.32	2.25	.025	
3.31	3.16	3.08	3.00	2.92	2.87	2.83	2.75	2.65	.01	
3.79	3.61	3.51	3.41	3.31	3.25	3.21	3.10	2.98	.005	
5.05	4.78	4.63	4.48	4.33	4.24	4.18	4.02	3.85	.001	
.706	.733	.748	.764	.780	.790	.797	.814	.833	.75	
.992	1.00	1.01	1.02	1.02	1.02	1.03	1.03	1.04	.50	
1.39	1.38	1.37	1.36	1.35	1.34	1.34	1.33	1.32	.25	
1.89	1.84	1.81	1.78	1.75	1.74	1.72	1.69	1.66	.10	
2.27	2.19	2.15	2.11	2.06	2.04	2.02	1.97	1.92	.05	**18**
2.67	2.56	2.50	2.44	2.38	2.35	2.32	2.26	2.19	.025	
3.23	3.08	3.00	2.92	2.84	2.78	2.75	2.66	2.57	.01	
3.68	3.50	3.40	3.30	3.20	3.14	3.10	2.99	2.87	.005	
4.87	4.59	4.45	4.30	4.15	4.06	4.00	3.84	3.67	.001	
.707	.735	.750	.766	.782	.792	.799	.817	.836	.75	
.990	1.00	1.01	1.01	1.02	1.02	1.02	1.03	1.04	.50	
1.38	1.37	1.36	1.35	1.34	1.33	1.33	1.32	1.30	.25	
1.86	1.81	1.79	1.76	1.73	1.71	1.70	1.67	1.63	.10	
2.23	2.16	2.11	2.07	2.03	2.00	1.98	1.93	1.88	.05	**19**
2.62	2.51	2.45	2.39	2.33	2.30	2.27	2.20	2.13	.025	
3.15	3.00	2.92	2.84	2.76	2.71	2.67	2.58	2.49	.01	
3.59	3.40	3.31	3.21	3.11	3.04	3.00	2.89	2.78	.005	
4.70	4.43	4.29	4.14	3.99	3.90	3.84	3.68	3.51	.001	
.708	.736	.751	.767	.784	.794	.801	.820	.840	.75	
.989	1.00	1.01	1.01	1.02	1.02	1.02	1.03	1.03	.50	
1.37	1.36	1.35	1.34	1.33	1.33	1.32	1.31	1.29	.25	
1.84	1.79	1.77	1.74	1.71	1.69	1.68	1.64	1.61	.10	
2.20	2.12	2.08	2.04	1.99	1.97	1.95	1.90	1.84	.05	**20**
2.57	2.46	2.41	2.35	2.29	2.25	2.22	2.16	2.09	.025	
3.09	2.94	2.86	2.78	2.69	2.64	2.61	2.52	2.42	.01	
3.50	3.32	3.22	3.12	3.02	2.96	2.92	2.81	2.69	.005	
4.56	4.29	4.15	4.00	3.86	3.76	3.70	3.54	3.38	.001	

TABLE 16 Critical values of the F-distribution

ν_1 (degrees of freedom of numerator mean squares)

ν_2 (degrees of freedom of denominator mean squares)	α	15	20	24	30	40	50	60	120	∞
21	.75	.709	.738	.753	.769	.786	.796	.803	.822	.842
	.50	.987	.998	1.00	1.01	1.02	1.02	1.02	1.03	1.03
	.25	1.37	1.35	1.34	1.33	1.32	1.32	1.31	1.30	1.28
	.10	1.83	1.78	1.75	1.72	1.69	1.67	1.66	1.62	1.59
	.05	2.18	2.10	2.05	2.01	1.96	1.94	1.92	1.87	1.81
	.025	2.53	2.42	2.37	2.31	2.25	2.21	2.18	2.11	2.04
	.01	3.03	2.88	2.80	2.72	2.64	2.58	2.55	2.46	2.36
	.005	3.43	3.24	3.15	3.05	2.95	2.88	2.84	2.73	2.61
	.001	4.44	4.17	4.03	3.88	3.74	3.64	3.58	3.42	3.26
22	.75	.710	.739	.754	.770	.787	.798	.805	.824	.845
	.50	.986	.997	1.00	1.01	1.01	1.02	1.02	1.03	1.03
	.25	1.36	1.34	1.33	1.32	1.31	1.31	1.30	1.29	1.28
	.10	1.81	1.76	1.73	1.70	1.67	1.65	1.64	1.60	1.57
	.05	2.15	2.07	2.03	1.98	1.94	1.91	1.89	1.84	1.78
	.025	2.50	2.39	2.33	2.27	2.21	2.17	2.14	2.08	2.00
	.01	2.98	2.83	2.75	2.67	2.58	2.53	2.50	2.40	2.31
	.005	3.36	3.18	3.08	2.98	2.88	2.82	2.77	2.66	2.55
	.001	4.33	4.06	3.92	3.78	3.63	3.54	3.48	3.32	3.15
23	.75	.711	.740	.756	.772	.789	.800	.807	.827	.847
	.50	.984	.996	1.00	1.01	1.01	1.02	1.02	1.02	1.03
	.25	1.35	1.34	1.33	1.32	1.31	1.30	1.30	1.28	1.27
	.10	1.80	1.74	1.72	1.69	1.66	1.64	1.62	1.59	1.55
	.05	2.13	2.05	2.01	1.96	1.91	1.89	1.86	1.81	1.76
	.025	2.47	2.36	2.30	2.24	2.18	2.14	2.11	2.04	1.97
	.01	2.93	2.78	2.70	2.62	2.54	2.48	2.45	2.35	2.26
	.005	3.30	3.12	3.02	2.92	2.82	2.76	2.71	2.60	2.48
	.001	4.23	3.96	3.82	3.68	3.53	3.44	3.38	3.22	3.05
24	.75	.712	.741	.757	.773	.791	.802	.809	.829	.850
	.50	.983	.994	1.00	1.01	1.01	1.02	1.02	1.02	1.03
	.25	1.35	1.33	1.32	1.31	1.30	1.29	1.29	1.28	1.26
	.10	1.78	1.73	1.70	1.67	1.64	1.62	1.61	1.57	1.53
	.05	2.11	2.03	1.98	1.94	1.89	1.86	1.84	1.79	1.73
	.025	2.44	2.33	2.27	2.21	2.15	2.11	2.08	2.01	1.94
	.01	2.89	2.74	2.66	2.58	2.49	2.44	2.40	2.31	2.21
	.005	3.25	3.06	2.97	2.87	2.77	2.70	2.66	2.55	2.43
	.001	4.14	3.87	3.74	3.59	3.45	3.36	3.29	3.14	2.97
25	.75	.712	.742	.758	.775	.792	.803	.811	.831	.852
	.50	.982	.993	.999	1.00	1.01	1.01	1.02	1.02	1.03
	.25	1.34	1.33	1.32	1.31	1.29	1.29	1.28	1.27	1.25
	.10	1.77	1.72	1.69	1.66	1.63	1.61	1.59	1.56	1.52
	.05	2.09	2.01	1.96	1.92	1.87	1.84	1.82	1.77	1.71
	.025	2.41	2.30	2.24	2.18	2.12	2.08	2.05	1.98	1.91
	.01	2.85	2.70	2.62	2.54	2.45	2.40	2.36	2.27	2.17
	.005	3.20	3.01	2.92	2.82	2.72	2.65	2.61	2.50	2.38
	.001	4.06	3.97	3.66	3.52	3.37	3.28	3.22	3.06	2.89

ν_2 (degrees of freedom of denominator mean squares)	α	15	20	24	30	40	50	60	120	∞
		ν_1 (degrees of freedom of numerator mean squares)								
26	.75	.713	.743	.759	.776	.793	.805	.812	.832	.854
	.50	.981	.992	.998	1.00	1.01	1.01	1.01	1.02	1.03
	.25	1.34	1.32	1.31	1.30	1.29	1.28	1.28	1.26	1.25
	.10	1.76	1.71	1.68	1.65	1.61	1.59	1.58	1.54	1.50
	.05	2.07	1.99	1.95	1.90	1.85	1.82	1.80	1.75	1.69
	.025	2.39	2.28	2.22	2.16	2.09	2.05	2.03	1.95	1.88
	.01	2.81	2.66	2.58	2.50	2.42	2.36	2.33	2.23	2.13
	.005	3.15	2.97	2.87	2.77	2.67	2.61	2.56	2.45	2.33
	.001	3.99	3.72	3.59	3.44	3.30	3.21	3.15	2.99	2.82
27	.75	.714	.744	.760	.777	.795	.806	.814	.834	.856
	.50	.980	.991	.997	1.00	1.01	1.01	1.01	1.02	1.03
	.25	1.33	1.32	1.31	1.30	1.28	1.28	1.27	1.26	1.24
	.10	1.75	1.70	1.67	1.64	1.60	1.58	1.57	1.53	1.49
	.05	2.06	1.97	1.93	1.88	1.84	1.81	1.79	1.73	1.67
	.025	2.36	2.25	2.19	2.13	2.07	2.03	2.00	1.93	1.85
	.01	2.78	2.63	2.55	2.47	2.38	2.33	2.29	2.20	2.10
	.005	3.11	2.93	2.83	2.73	2.63	2.57	2.52	2.41	2.29
	.001	3.92	3.66	3.52	3.38	3.23	3.14	3.08	2.92	2.75
28	.75	.714	.745	.761	.778	.796	.807	.815	.856	.858
	.50	.979	.990	.996	1.00	1.01	1.01	1.01	1.02	1.02
	.25	1.33	1.31	1.30	1.29	1.28	1.27	1.27	1.25	1.24
	.10	1.74	1.69	1.66	1.63	1.59	1.57	1.56	1.52	1.48
	.05	2.04	1.96	1.91	1.87	1.82	1.79	1.77	1.71	1.65
	.025	2.34	2.23	2.17	2.11	2.05	2.01	1.98	1.91	1.83
	.01	2.75	2.60	2.52	2.44	2.35	2.30	2.26	2.17	2.06
	.005	3.07	2.89	2.79	2.69	2.59	2.53	2.48	2.37	2.25
	.001	3.86	3.60	3.46	3.32	3.18	3.09	3.02	2.86	2.69
29	.75	.715	.745	.762	.779	.797	.809	.816	.837	.860
	.50	.979	.990	.996	1.00	1.01	1.01	1.01	1.02	1.02
	.25	1.32	1.31	1.30	1.29	1.27	1.27	1.26	1.25	1.23
	.10	1.73	1.68	1.65	1.62	1.58	1.56	1.55	1.51	1.47
	.05	2.03	1.94	1.90	1.85	1.81	1.78	1.75	1.70	1.64
	.025	2.32	2.21	2.15	2.09	2.03	1.99	1.96	1.89	1.81
	.01	2.73	2.57	2.49	2.41	2.33	2.27	2.23	2.14	2.03
	.005	3.04	2.86	2.76	2.66	2.56	2.49	2.45	2.33	2.21
	.001	3.80	3.54	3.41	3.27	3.12	3.03	2.97	2.81	2.64
30	.75	.716	.746	.763	.780	.798	.810	.818	.839	.862
	.50	.978	.989	.994	1.00	1.01	1.01	1.01	1.02	1.02
	.25	1.32	1.30	1.29	1.28	1.27	1.26	1.26	1.24	1.23
	.10	1.72	1.67	1.64	1.61	1.57	1.55	1.54	1.50	1.46
	.05	2.01	1.93	1.89	1.84	1.79	1.76	1.74	1.68	1.62
	.025	2.31	2.20	2.14	2.07	2.01	1.97	1.94	1.87	1.79
	.01	2.70	2.55	2.47	2.39	2.30	2.25	2.21	2.11	2.01
	.005	3.01	2.82	2.73	2.63	2.52	2.46	2.42	2.30	2.18
	.001	3.75	3.49	3.36	3.22	3.07	2.98	2.92	2.76	2.59

TABLE 16 Critical values of the F-distribution

ν_1 (degrees of freedom of numerator mean squares)

ν_2 (degrees of freedom of denominator mean squares)

ν_2	α	15	20	24	30	40	50	60	120	∞
	.75	.720	.752	.769	.787	.806	.819	.828	.851	.877
	.50	.972	.983	.989	.994	1.00	1.00	1.01	1.01	1.02
	.25	1.30	1.28	1.26	1.25	1.24	1.23	1.22	1.21	1.19
	.10	1.66	1.61	1.57	1.54	1.51	1.48	1.47	1.42	1.38
40	.05	1.92	1.84	1.79	1.74	1.69	1.66	1.64	1.58	1.51
	.025	2.18	2.07	2.01	1.94	1.88	1.83	1.80	1.72	1.64
	.01	2.52	2.37	2.29	2.20	2.11	2.06	2.02	1.92	1.80
	.005	2.78	2.60	2.50	2.40	2.30	2.23	2.18	2.06	1.93
	.001	3.40	3.15	3.01	2.87	2.73	2.64	2.57	2.41	2.23
	.75	.725	.758	.776	.796	.816	.830	.840	.865	.896
	.50	.967	.978	.983	.989	.994	.998	1.00	1.01	1.01
	.25	1.27	1.25	1.24	1.22	1.21	1.20	1.19	1.17	1.15
	.10	1.60	1.54	1.51	1.48	1.44	1.41	1.40	1.35	1.29
60	.05	1.84	1.75	1.70	1.65	1.59	1.56	1.53	1.47	1.39
	.025	2.06	1.94	1.88	1.82	1.74	1.70	1.67	1.58	1.48
	.01	2.35	2.20	2.12	2.03	1.94	1.88	1.84	1.73	1.60
	.005	2.57	2.39	2.29	2.19	2.08	2.01	1.96	1.83	1.69
	.001	3.08	2.83	2.69	2.55	2.41	2.32	2.25	2.08	1.89
	.75	.730	.765	.784	.805	.828	.843	.853	.884	.923
	.50	.961	.972	.978	.983	.989	.992	.994	1.00	1.01
	.25	1.24	1.22	1.21	1.19	1.18	1.17	1.16	1.13	1.10
	.10	1.55	1.48	1.45	1.41	1.37	1.34	1.32	1.26	1.19
120	.05	1.75	1.66	1.61	1.55	1.50	1.46	1.43	1.35	1.25
	.025	1.95	1.82	1.76	1.69	1.61	1.56	1.53	1.43	1.31
	.01	2.19	2.03	1.95	1.86	1.76	1.70	1.66	1.53	1.38
	.005	2.37	2.19	2.09	1.98	1.87	1.80	1.75	1.61	1.43
	.001	2.78	2.53	2.40	2.26	2.11	2.02	1.95	1.76	1.54
	.75	.736	.773	.793	.816	.842	.860	.872	.910	1.00
	.50	.956	.967	.972	.978	.983	.987	.989	.994	1.00
	.25	1.22	1.19	1.18	1.16	1.14	1.13	1.12	1.08	1.00
	.10	1.49	1.42	1.38	1.34	1.30	1.26	1.24	1.17	1.00
∞	.05	1.67	1.57	1.52	1.46	1.39	1.35	1.32	1.22	1.00
	.025	1.83	1.71	1.64	1.57	1.48	1.43	1.39	1.27	1.00
	.01	2.04	1.88	1.79	1.70	1.59	1.52	1.47	1.32	1.00
	.005	2.19	2.00	1.90	1.79	1.67	1.59	1.53	1.36	1.00
	.001	2.51	2.27	2.13	1.99	1.84	1.73	1.66	1.45	1.00

TABLE **17** Critical values of F_{max}

This table furnishes critical values for the maximum F-ratio distribution. The maximum observed F-ratio is computed as s^2_{max}/s^2_{min} and compared with the critical value for a samples and degrees of freedom $\nu = n - 1$, where n is the sample size. The critical values of F_{max} are tabulated for number of samples a from 2 to 12 in increments of 1 and for degrees of freedom ν from 2 to 10 in increments of 1 and also for $\nu = 12, 15, 20, 30, 60$, and ∞. Corresponding to each value of a and ν are two critical values of F_{max} representing the upper 5% and 1% percentage points. The corresponding probabilities $\alpha = 0.05$ and 0.01 represent *one tail* of the F_{max}-distribution. The critical values are given to a varying number of decimal places, depending on the exactness of the computation. Values for $a = 2$ and $\nu = 2$ and ∞ are exact. In other places in the table the third digit may be slightly in error. The third digit for $\nu = 3$ (in parentheses) is the most uncertain value in the table.

To find the critical values of F_{max} in an example with six samples and ten items per sample, we look up $a = 6$ and $\nu = n - 1 = 9$ to obtain $F_{max} = 7.80$ and 12.1 at the 5% and 1% levels, respectively.

The use of this table for testing homogeneity of variances is explained in Section 13.3.

This table was copied from H. A. David (*Biometrika*, **39**:422–424, 1952) with the permission of the publisher.

TABLE 17 Critical values of F_{max}

$\nu \backslash a$	2	3	4	5	6	7	8	9	10	11	12
2	39.0	87.5	142.	202.	266.	333.	403.	475.	550.	626.	704.
	199.	448.	729.	1036.	1362.	1705.	2063.	2432.	2813.	3204.	3605.
3	15.4	27.8	39.2	50.7	62.0	72.9	83.5	93.9	104.	114.	124.
	47.5	85.	120.	151.	184.	21(6)	24(9)	28(1)	31(0)	33(7)	36(1)
4	9.60	15.5	20.6	25.2	29.5	33.6	37.5	41.1	44.6	48.0	51.4
	23.2	37.	49.	59.	69.	79.	89.	97.	106.	113.	120.
5	7.15	10.8	13.7	16.3	18.7	20.8	22.9	24.7	26.5	28.2	29.9
	14.9	22.	28.	33.	38.	42.	46.	50.	54.	57.	60.
6	5.82	8.38	10.4	12.1	13.7	15.0	16.3	17.5	18.6	19.7	20.7
	11.1	15.5	19.1	22.	25.	27.	30.	32.	34.	36.	37.
7	4.99	6.94	8.44	9.70	10.8	11.8	12.7	13.5	14.3	15.1	15.8
	8.89	12.1	14.5	16.5	18.4	20.	22.	23.	24.	26.	27.
8	4.43	6.00	7.18	8.12	9.03	9.78	10.5	11.1	11.7	12.2	12.7
	7.50	9.9	11.7	13.2	14.5	15.8	16.9	17.9	18.9	19.8	21.
9	4.03	5.34	6.31	7.11	7.80	8.41	8.95	9.45	9.91	10.3	10.7
	6.54	8.5	9.9	11.1	12.1	13.1	13.9	14.7	15.3	16.0	16.6
10	3.72	4.85	5.67	6.34	6.92	7.42	7.87	8.28	8.66	9.01	9.34
	5.85	7.4	8.6	9.6	10.4	11.1	11.8	12.4	12.9	13.4	13.9
12	3.28	4.16	4.79	5.30	5.72	6.09	6.42	6.72	7.00	7.25	7.48
	4.91	6.1	6.9	7.6	8.2	8.7	9.1	9.5	9.9	10.2	10.6
15	2.86	3.54	4.01	4.37	4.68	4.95	5.19	5.40	5.59	5.77	5.93
	4.07	4.9	5.5	6.0	6.4	6.7	7.1	7.3	7.5	7.8	8.0
20	2.46	2.95	3.29	3.54	3.76	3.94	4.10	4.24	4.37	4.49	4.59
	3.32	3.8	4.3	4.6	4.9	5.1	5.3	5.5	5.6	5.8	5.9
30	2.07	2.40	2.61	2.78	2.91	3.02	3.12	3.21	3.29	3.36	3.39
	2.63	3.0	3.3	3.4	3.6	3.7	3.8	3.9	4.0	4.1	4.2
60	1.67	1.85	1.96	2.04	2.11	2.17	2.22	2.26	2.30	2.33	2.36
	1.96	2.2	2.3	2.4	2.4	2.5	2.5	2.6	2.6	2.7	2.7
∞	1.00	1.00	1.00	1.00	1.00	1.00	1.00	1.00	1.00	1.00	1.00
	1.00	1.00	1.00	1.00	1.00	1.00	1.00	1.00	1.00	1.00	1.00

TABLE **18** Critical values of the studentized range

This table furnishes the critical values for the distribution of the ratio $Q_{\alpha[k,\nu]} = \text{Range}/s$, known as the studentized range. The range is computed over k variates and the standard deviation s must be independently estimated and based upon ν degrees of freedom. The critical values are tabulated for values of k, the number of items over which the range is computed, from 2 to 20 increments of 1, from 20 to 40 in increments of 2, and from 40 to 100 in increments of 10. For degrees of freedom ν, arguments are furnished from 1 to 20 in increments of 1 and for $\nu = 24, 30, 40, 60, 120$, and ∞. It will be noted that these latter values of ν lend themselves to harmonic interpolation. There are separate tables of the critical values of the studentized range representing the upper 5% and 1% percentage points, respectively. The corresponding probabilities $\alpha = 0.05$ and 0.01 represent *one tail* of the distribution of Q.

To find the 1% critical values of the studentized range in a sample of six items but with a standard deviation independently obtained and based on 18 degrees of freedom (as in an analysis of variance with six samples of four items each), enter the second part of the table ($\alpha = 0.01$) with $k = 6$ and $\nu = 18$, and find $Q_{.01[6,18]} = 5.603$.

The most frequent application of the studentized range is in unplanned multiple comparisons tests in the analysis of variance as discussed in Section 9.7.

Values in this table have been copied from a more extensive one by H. L. Harter (*Ann. Math. Stat.*, **31**:1122–1147, 1960) with permission of the publisher.

TABLE 18 Critical values of the studentized range $\alpha = \mathbf{0.05}$

ν \ k	2	3	4	5	6	7	8	9	10
1	17.97	26.98	32.82	37.08	40.41	43.12	45.40	47.36	49.07
2	6.085	8.331	9.798	10.88	11.75	12.44	13.03	13.54	13.99
3	4.501	5.910	6.825	7.502	8.037	8.478	8.853	9.177	9.462
4	3.927	5.040	5.757	6.287	6.707	7.053	7.347	7.602	7.826
5	3.635	4.602	5.218	5.673	6.033	6.330	6.582	6.802	6.995
6	3.461	4.339	4.896	5.305	5.628	5.895	6.122	6.319	6.493
7	3.344	4.165	4.681	5.060	5.359	5.606	5.815	5.998	6.158
8	3.261	4.041	4.529	4.886	5.167	5.399	5.597	5.767	5.918
9	3.199	3.949	4.415	4.756	5.024	5.244	5.432	5.595	5.739
10	3.151	3.877	4.327	4.654	4.912	5.124	5.305	5.461	5.599
11	3.113	3.820	4.256	4.574	4.823	5.028	5.202	5.353	5.487
12	3.082	3.773	4.199	4.508	4.751	4.950	5.119	5.265	5.395
13	3.055	3.735	4.151	4.453	4.690	4.885	5.049	5.192	5.318
14	3.033	3.702	4.111	4.407	4.639	4.829	4.990	5.131	5.254
15	3.014	3.674	4.076	4.367	4.595	4.782	4.940	5.077	5.198
16	2.998	3.649	4.046	4.333	4.557	4.741	4.897	5.031	5.150
17	2.984	3.628	4.020	4.303	4.524	4.705	4.858	4.991	5.108
18	2.971	3.609	3.997	4.277	4.495	4.673	4.824	4.956	5.071
19	2.960	3.593	3.977	4.253	4.469	4.645	4.794	4.924	5.038
20	2.950	3.578	3.958	4.232	4.445	4.620	4.768	4.896	5.008
24	2.919	3.532	3.901	4.166	4.373	4.541	4.684	4.807	4.915
30	2.888	3.486	3.845	4.102	4.302	4.464	4.602	4.720	4.824
40	2.858	3.442	3.791	4.039	4.232	4.389	4.521	4.635	4.735
60	2.829	3.399	3.737	3.977	4.163	4.314	4.441	4.550	4.646
120	2.800	3.356	3.685	3.917	4.096	4.241	4.363	4.468	4.560
∞	2.772	3.314	3.633	3.858	4.030	4.170	4.286	4.387	4.474

ν \ k	11	12	13	14	15	16	17	18	19
1	50.59	51.96	53.20	54.33	55.36	56.32	57.22	58.04	58.83
2	14.39	14.75	15.08	15.38	15.65	15.91	16.14	16.37	16.57
3	9.717	9.946	10.15	10.35	10.53	10.69	10.84	10.98	11.11
4	8.027	8.208	8.373	8.525	8.664	8.794	8.914	9.028	9.134
5	7.168	7.324	7.466	7.596	7.717	7.828	7.932	8.030	8.122
6	6.649	6.789	6.917	7.034	7.143	7.244	7.338	7.426	7.508
7	6.302	6.431	6.550	6.658	6.759	6.852	6.939	7.020	7.097
8	6.054	6.175	6.287	6.389	6.483	6.571	6.653	6.729	6.802
9	5.867	5.983	6.089	6.186	6.276	6.359	6.437	6.510	6.579
10	5.722	5.833	5.935	6.028	6.114	6.194	6.269	6.339	6.405
11	5.605	5.713	5.811	5.901	5.984	6.062	6.134	6.202	6.265
12	5.511	5.615	5.710	5.798	5.878	5.953	6.023	6.089	6.151
13	5.431	5.533	5.625	5.711	5.789	5.862	5.931	5.995	6.055
14	5.364	5.463	5.554	5.637	5.714	5.786	5.852	5.915	5.974
15	5.306	5.404	5.493	5.574	5.649	5.720	5.785	5.846	5.904
16	5.256	5.352	5.439	5.520	5.593	5.662	5.727	5.786	5.843
17	5.212	5.307	5.392	5.471	5.544	5.612	5.675	5.734	5.790
18	5.174	5.267	5.352	5.429	5.501	5.568	5.630	5.688	5.743
19	5.140	5.231	5.315	5.391	5.462	5.528	5.589	5.647	5.701
20	5.108	5.199	5.282	5.357	5.427	5.493	5.553	5.610	5.663
24	5.012	5.099	5.179	5.251	5.319	5.381	5.439	5.494	5.545
30	4.917	5.001	5.077	5.147	5.211	5.271	5.327	5.379	5.429
40	4.824	4.904	4.977	5.044	5.106	5.163	5.216	5.266	5.313
60	4.732	4.808	4.878	4.942	5.001	5.056	5.107	5.154	5.199
120	4.641	4.714	4.781	4.842	4.898	4.950	4.998	5.044	5.086
∞	4.552	4.622	4.685	4.743	4.796	4.845	4.891	4.934	4.974

TABLE 18 Critical values of the studentized range $\alpha = 0.05$

$\nu \backslash k$	20	22	24	26	28	30	32	34	36
1	59.56	60.91	62.12	63.22	64.23	65.15	66.01	66.81	67.56
2	16.77	17.13	17.45	17.75	18.02	18.27	18.50	18.72	18.92
3	11.24	11.47	11.68	11.87	12.05	12.21	12.36	12.50	12.63
4	9.233	9.418	9.584	9.736	9.875	10.00	10.12	10.23	10.34
5	8.208	8.368	8.512	8.643	8.764	8.875	8.979	9.075	9.165
6	7.587	7.730	7.861	7.979	8.088	8.189	8.283	8.370	8.452
7	7.170	7.303	7.423	7.533	7.634	7.728	7.814	7.895	7.972
8	6.870	6.995	7.109	7.212	7.307	7.395	7.477	7.554	7.625
9	6.644	6.763	6.871	6.970	7.061	7.145	7.222	7.295	7.363
10	6.467	6.582	6.686	6.781	6.868	6.948	7.023	7.093	7.159
11	6.326	6.436	6.536	6.628	6.712	6.790	6.863	6.930	6.994
12	6.209	6.317	6.414	6.503	6.585	6.660	6.731	6.796	6.858
13	6.112	6.217	6.312	6.398	6.478	6.551	6.620	6.684	6.744
14	6.029	6.132	6.224	6.309	6.387	6.459	6.526	6.588	6.647
15	5.958	6.059	6.149	6.233	6.309	6.379	6.445	6.506	6.564
16	5.897	5.995	6.084	6.166	6.241	6.310	6.374	6.434	6.491
17	5.842	5.940	6.027	6.107	6.181	6.249	6.313	6.372	6.427
18	5.794	5.890	5.977	6.055	6.128	6.195	6.258	6.316	6.371
19	5.752	5.846	5.932	6.009	6.081	6.147	6.209	6.267	6.321
20	5.714	5.807	5.891	5.968	6.039	6.104	6.165	6.222	6.275
24	5.594	5.683	5.764	5.838	5.906	5.968	6.027	6.081	6.132
30	5.475	5.561	5.638	5.709	5.774	5.833	5.889	5.941	5.990
40	5.358	5.439	5.513	5.581	5.642	5.700	5.753	5.803	5.849
60	5.241	5.319	5.389	5.453	5.512	5.566	5.617	5.664	5.708
120	5.126	5.200	5.266	5.327	5.382	5.434	5.481	5.526	5.568
∞	5.012	5.081	5.144	5.201	5.253	5.301	5.346	5.388	5.427

$\nu \backslash k$	38	40	50	60	70	80	90	100
1	68.26	68.92	71.73	73.97	75.82	77.40	78.77	79.98
2	19.11	19.28	20.05	20.66	21.16	21.59	21.96	22.29
3	12.75	12.87	13.36	13.76	14.08	14.36	14.61	14.82
4	10.44	10.53	10.93	11.24	11.51	11.73	11.92	12.09
5	9.250	9.330	9.674	9.949	10.18	10.38	10.54	10.69
6	8.529	8.601	8.913	9.163	9.370	9.548	9.702	9.839
7	8.043	8.110	8.400	8.632	8.824	8.989	9.133	9.261
8	7.693	7.756	8.029	8.248	8.430	8.586	8.722	8.843
9	7.428	7.488	7.749	7.958	8.132	8.281	8.410	8.526
10	7.220	7.279	7.529	7.730	7.897	8.041	8.166	8.276
11	7.053	7.110	7.352	7.546	7.708	7.847	7.968	8.075
12	6.916	6.970	7.205	7.394	7.552	7.687	7.804	7.909
13	6.800	6.854	7.083	7.267	7.421	7.552	7.667	7.769
14	6.702	6.754	6.979	7.159	7.309	7.438	7.550	7.650
15	6.618	6.669	6.888	7.065	7.212	7.339	7.449	7.546
16	6.544	6.594	6.810	6.984	7.128	7.252	7.360	7.457
17	6.479	6.529	6.741	6.912	7.054	7.176	7.283	7.377
18	6.422	6.471	6.680	6.848	6.989	7.109	7.213	7.307
19	6.371	6.419	6.626	6.792	6.930	7.048	7.152	7.244
20	6.325	6.373	6.576	6.740	6.877	6.994	7.097	7.187
24	6.181	6.226	6.421	6.579	6.710	6.822	6.920	7.008
30	6.037	6.080	6.267	6.417	6.543	6.650	6.744	6.827
40	5.893	5.934	6.112	6.255	6.375	6.477	6.566	6.645
60	5.750	5.789	5.958	6.093	6.206	6.303	6.387	6.462
120	5.607	5.644	5.802	5.929	6.035	6.126	6.205	6.275
∞	5.463	5.498	5.646	5.764	5.863	5.947	6.020	6.085

TABLE 18 Critical values of the studentized range $\alpha = \mathbf{0.01}$

$\nu \backslash k$	2	3	4	5	6	7	8	9	10
1	90.03	135.0	164.3	185.6	202.2	215.8	227.2	237.0	245.6
2	14.04	19.02	22.29	24.72	26.63	28.20	29.53	30.68	31.69
3	8.261	10.62	12.17	13.33	14.24	15.00	15.64	16.20	16.69
4	6.512	8.120	9.173	9.958	10.58	11.10	11.55	11.93	12.27
5	5.702	6.976	7.804	8.421	8.913	9.321	9.669	9.972	10.24
6	5.243	6.331	7.033	7.556	7.973	8.318	8.613	8.869	9.097
7	4.949	5.919	6.543	7.005	7.373	7.679	7.939	8.166	8.368
8	4.746	5.635	6.204	6.625	6.960	7.237	7.474	7.681	7.863
9	4.596	5.428	5.957	6.348	6.658	6.915	7.134	7.325	7.495
10	4.482	5.270	5.769	6.136	6.428	6.669	6.875	7.055	7.213
11	4.392	5.146	5.621	5.970	6.247	6.476	6.672	6.842	6.992
12	4.320	5.046	5.502	5.836	6.101	6.321	6.507	6.670	6.814
13	4.260	4.964	5.404	5.727	5.981	6.192	6.372	6.528	6.667
14	4.210	4.895	5.322	5.634	5.881	6.085	6.258	6.409	6.543
15	4.168	4.836	5.252	5.556	5.796	5.994	6.162	6.309	6.439
16	4.131	4.786	5.192	5.489	5.722	5.915	6.079	6.222	6.349
17	4.099	4.742	5.140	5.430	5.659	5.847	6.007	6.147	6.270
18	4.071	4.703	5.094	5.379	5.603	5.788	5.944	6.081	6.201
19	4.046	4.670	5.054	5.334	5.554	5.735	5.889	6.022	6.141
20	4.024	4.639	5.018	5.294	5.510	5.688	5.839	5.970	6.087
24	3.956	4.546	4.907	5.168	5.374	5.542	5.685	5.809	5.919
30	3.889	4.455	4.799	5.048	5.242	5.401	5.536	5.653	5.756
40	3.825	4.367	4.696	4.931	5.114	5.265	5.392	5.502	5.599
60	3.762	4.282	4.595	4.818	4.991	5.133	5.253	5.356	5.447
120	3.702	4.200	4.497	4.709	4.872	5.005	5.118	5.214	5.299
∞	3.643	4.120	4.403	4.603	4.757	4.882	4.987	5.078	5.157

$\nu \backslash k$	11	12	13	14	15	16	17	18	19
1	253.2	260.0	266.2	271.8	277.0	281.8	286.3	290.4	294.3
2	32.59	33.40	34.13	34.81	35.43	36.00	36.53	37.03	37.50
3	17.13	17.53	17.89	18.22	18.52	18.81	19.07	19.32	19.55
4	12.57	12.84	13.09	13.32	13.53	13.73	13.91	14.08	14.24
5	10.48	10.70	10.89	11.08	11.24	11.40	11.55	11.68	11.81
6	9.301	9.485	9.653	9.808	9.951	10.08	10.21	10.32	10.43
7	8.548	8.711	8.860	8.997	9.124	9.242	9.353	9.456	9.554
8	8.027	8.176	8.312	8.436	8.552	8.659	8.760	8.854	8.943
9	7.647	7.784	7.910	8.025	8.132	8.232	8.325	8.412	8.495
10	7.356	7.485	7.603	7.712	7.812	7.906	7.993	8.076	8.153
11	7.128	7.250	7.362	7.465	7.560	7.649	7.732	7.809	7.883
12	6.943	7.060	7.167	7.265	7.356	7.441	7.520	7.594	7.665
13	6.791	6.903	7.006	7.101	7.188	7.269	7.345	7.417	7.485
14	6.664	6.772	6.871	6.962	7.047	7.126	7.199	7.268	7.333
15	6.555	6.660	6.757	6.845	6.927	7.003	7.074	7.142	7.204
16	6.462	6.564	6.658	6.744	6.823	6.898	6.967	7.032	7.093
17	6.381	6.480	6.572	6.656	6.734	6.806	6.873	6.937	6.997
18	6.310	6.407	6.497	6.579	6.655	6.725	6.792	6.854	6.912
19	6.247	6.342	6.430	6.510	6.585	6.654	6.719	6.780	6.837
20	6.191	6.285	6.371	6.450	6.523	6.591	6.654	6.714	6.771
24	6.017	6.106	6.186	6.261	6.330	6.394	6.453	6.510	6.563
30	5.849	5.932	6.008	6.078	6.143	6.203	6.259	6.311	6.361
40	5.686	5.764	5.835	5.900	5.961	6.017	6.069	6.119	6.165
60	5.528	5.601	5.667	5.728	5.785	5.837	5.886	5.931	5.974
120	5.375	5.443	5.505	5.562	5.614	5.662	5.708	5.750	5.790
∞	5.227	5.290	5.348	5.400	5.448	5.493	5.535	5.574	5.611

TABLE 18 Critical values of the studentized range $\alpha = \mathbf{0.01}$

$\nu \backslash k$	20	22	24	26	28	30	32	34	36
1	298.0	304.7	310.8	316.3	321.3	326.0	330.3	334.3	338.0
2	37.95	38.76	39.49	40.15	40.76	41.32	41.84	42.33	42.78
3	19.77	20.17	20.53	20.86	21.16	21.44	21.70	21.95	22.17
4	14.40	14.68	14.93	15.16	15.37	15.57	15.75	15.92	16.08
5	11.93	12.16	12.36	12.54	12.71	12.87	13.02	13.15	13.28
6	10.54	10.73	10.91	11.06	11.21	11.34	11.47	11.58	11.69
7	9.646	9.815	9.970	10.11	10.24	10.36	10.47	10.58	10.67
8	9.027	9.182	9.322	9.450	9.569	9.678	9.779	9.874	9.964
9	8.573	8.717	8.847	8.966	9.075	9.177	9.271	9.360	9.443
10	8.226	8.361	8.483	8.595	8.698	8.794	8.883	8.966	9.044
11	7.952	8.080	8.196	8.303	8.400	8.491	8.575	8.654	8.728
12	7.731	7.853	7.964	8.066	8.159	8.246	8.327	8.402	8.473
13	7.548	7.665	7.772	7.870	7.960	8.043	8.121	8.193	8.262
14	7.395	7.508	7.611	7.705	7.792	7.873	7.948	8.018	8.084
15	7.264	7.374	7.474	7.566	7.650	7.728	7.800	7.869	7.932
16	7.152	7.258	7.356	7.445	7.527	7.602	7.673	7.739	7.802
17	7.053	7.158	7.253	7.340	7.420	7.493	7.563	7.627	7.687
18	6.968	7.070	7.163	7.247	7.325	7.398	7.465	7.528	7.587
19	6.891	6.992	7.082	7.166	7.242	7.313	7.379	7.440	7.498
20	6.823	6.922	7.011	7.092	7.168	7.237	7.302	7.362	7.419
24	6.612	6.705	6.789	6.865	6.936	7.001	7.062	7.119	7.173
30	6.407	6.494	6.572	6.644	6.710	6.772	6.828	6.881	6.932
40	6.209	6.289	6.362	6.429	6.490	6.547	6.600	6.650	6.697
60	6.015	6.090	6.158	6.220	6.277	6.330	6.378	6.424	6.467
120	5.827	5.897	5.959	6.016	6.069	6.117	6.162	6.204	6.244
∞	5.645	5.709	5.766	5.818	5.866	5.911	5.952	5.990	6.026

$\nu \backslash k$	38	40	50	60	70	80	90	100
1	341.5	344.8	358.9	370.1	379.4	387.3	394.1	400.1
2	43.21	43.61	45.33	46.70	47.83	48.80	49.64	50.38
3	22.39	22.59	23.45	24.13	24.71	25.19	25.62	25.99
4	16.23	16.37	16.98	17.46	17.86	18.20	18.50	18.77
5	13.40	13.52	14.00	14.39	14.72	14.99	15.23	15.45
6	11.80	11.90	12.31	12.65	12.92	13.16	13.37	13.55
7	10.77	10.85	11.23	11.52	11.77	11.99	12.17	12.34
8	10.05	10.13	10.47	10.75	10.97	11.17	11.34	11.49
9	9.521	9.594	9.912	10.17	10.38	10.57	10.73	10.87
10	9.117	9.187	9.486	9.726	9.927	10.10	10.25	10.39
11	8.798	8.864	9.148	9.377	9.568	9.732	9.875	10.00
12	8.539	8.603	8.875	9.094	9.277	9.434	9.571	9.693
13	8.326	8.387	8.648	8.859	9.035	9.187	9.318	9.436
14	8.146	8.204	8.457	8.661	8.832	8.978	9.106	9.219
15	7.992	8.049	8.295	8.492	8.658	8.800	8.924	9.035
16	7.860	7.916	8.154	8.347	8.507	8.646	8.767	8.874
17	7.745	7.799	8.031	8.219	8.377	8.511	8.630	8.735
18	7.643	7.696	7.924	8.107	8.261	8.393	8.508	8.611
19	7.553	7.605	7.828	8.008	8.159	8.288	8.401	8.502
20	7.473	7.523	7.742	7.919	8.067	8.194	8.305	8.404
24	7.223	7.270	7.476	7.642	7.780	7.900	8.004	8.097
30	6.978	7.023	7.215	7.370	7.500	7.611	7.709	7.796
40	6.740	6.782	6.960	7.104	7.225	7.328	7.419	7.500
60	6.507	6.546	6.710	6.843	6.954	7.050	7.133	7.207
120	6.281	6.316	6.467	6.588	6.689	6.776	6.852	6.919
∞	6.060	6.092	6.228	6.338	6.429	6.507	6.575	6.636

TABLE **19** Critical values for Welsch's step-up procedure

This table furnishes critical values for Welsch's step-up procedure (described by him as the GAPA procedure). The table is entered via two arguments. The first of these is k, the number of means in the set being compared, and the second is j, the "stretch" or size of the subset j in k being tested for homogeneity. Both j and k are furnished from 2 to 10 in increments of one. The values in the table are actually studentized ranges $Q_{\alpha j}$, for unusual probabilities $\alpha_j = \alpha(j/k)$ where α_j is the type I error for testing a range of j adjacent means. The values are given for 5% experimentwise error rates and are tabled for degrees of freedom ν from 5 to 16 in increments of 1, and for $\nu = 18, 20, 24, 30, 40, 60, 120$, and ∞. For degrees of freedom other than those tabled, employ harmonic interpolation.

For six means and a stretch of three, the critical value at 40 degrees of freedom can be found in the table to be 3.89. This critical value is employed in Welsch's step-up procedure for multiple comparison of means featured in Section 9.7.

This table is copied in rearranged form from a technical report by R. E. Welsch, *Tables for Stepwise Multiple Comparison Procedures* (Working paper no. 949-77. Sloan School of Management, 1977) with permission of the author.

TABLE 19 Critical values for Welsch's step-up procedure

$\nu = 5$

$j \backslash k$	2	3	4	5	6	7	8	9	10
2	3.64	3.64	4.47	4.76	5.00	5.21	5.39	5.55	5.70
3		5.38	4.60	5.37	5.64	5.86	6.06	6.24	6.41
4			5.46	5.37	5.84	6.07	6.28	6.47	6.64
5				5.72	5.84	6.21	6.42	6.61	6.79
6					6.05	6.21	6.50	6.70	6.88
7						6.34	6.50	6.75	6.94
8							6.59	6.75	6.97
9								6.80	6.97
10									7.00

$\nu = 6$

$j \backslash k$	2	3	4	5	6	7	8	9	10
2	3.46	3.46	4.20	4.44	4.65	4.83	4.98	5.12	5.24
3		4.95	4.34	5.01	5.23	5.42	5.58	5.73	5.87
4			5.10	5.01	5.42	5.62	5.80	5.95	6.09
5				5.36	5.42	5.75	5.93	6.09	6.24
6					5.65	5.75	6.02	6.19	6.33
7						5.91	6.02	6.25	6.40
8							6.13	6.25	6.44
9								6.32	6.44
10									6.50

$\nu = 7$

$j \backslash k$	2	3	4	5	6	7	8	9	10
2	3.34	3.34	4.02	4.24	4.42	4.58	4.72	4.84	4.95
3		4.69	4.17	4.76	4.96	5.13	5.27	5.40	5.52
4			4.86	4.76	5.15	5.33	5.48	5.61	5.74
5				5.11	5.15	5.46	5.61	5.75	5.88
6					5.39	5.46	5.71	5.85	5.98
7						5.62	5.71	5.92	6.04
8							5.82	5.92	6.10
9								6.00	6.10
10									6.17

$\nu = 8$

$j \backslash k$	2	3	4	5	6	7	8	9	10
2	3.26	3.26	3.89	4.10	4.27	4.41	4.53	4.65	4.75
3		4.51	4.05	4.59	4.77	4.92	5.06	5.17	5.28
4			4.69	4.59	4.96	5.12	5.26	5.38	5.49
5				4.94	4.96	5.25	5.39	5.51	5.63
6					5.19	5.25	5.48	5.61	5.72
7						5.41	5.48	5.68	5.79
8							5.60	5.68	5.85
9								5.77	5.85
10									5.92

TABLE 19 Critical values for Welsch's step-up procedure

$\nu = 9$

$j \backslash k$	2	3	4	5	6	7	8	9	10
2	3.20	3.20	3.80	3.99	4.15	4.28	4.40	4.50	4.60
3		4.37	3.95	4.47	4.63	4.77	4.90	5.01	5.10
4			4.56	4.47	4.82	4.96	5.09	5.20	5.30
5				4.81	4.82	5.09	5.22	5.34	5.44
6					5.05	5.09	5.31	5.43	5.54
7						5.26	5.31	5.50	5.61
8							5.44	5.50	5.67
9								5.60	5.67
10									5.74

$\nu = 10$

$j \backslash k$	2	3	4	5	6	7	8	9	10
2	3.15	3.15	3.73	3.91	4.06	4.19	4.30	4.39	4.48
3		4.27	3.88	4.37	4.53	4.66	4.77	4.88	4.97
4			4.47	4.37	4.71	4.84	4.96	5.07	5.16
5				4.71	4.71	4.97	5.09	5.20	5.30
6					4.94	4.97	5.18	5.30	5.39
7						5.14	5.18	5.37	5.47
8							5.32	5.37	5.52
9								5.47	5.52
10									5.60

$\nu = 11$

$j \backslash k$	2	3	4	5	6	7	8	9	10
2	3.11	3.11	3.67	3.84	3.99	4.11	4.22	4.31	4.39
3		4.19	3.83	4.29	4.44	4.57	4.68	4.77	4.86
4			4.39	4.29	4.62	4.75	4.86	4.96	5.05
5				4.63	4.62	4.87	4.99	5.09	5.18
6					4.85	4.87	5.08	5.19	5.28
7						5.04	5.08	5.26	5.35
8							5.21	5.26	5.41
9								5.36	5.41
10									5.49

$\nu = 12$

$j \backslash k$	2	3	4	5	6	7	8	9	10
2	3.08	3.08	3.62	3.79	3.93	4.05	4.15	4.24	4.32
3		4.13	3.78	4.23	4.37	4.49	4.60	4.69	4.77
4			4.32	4.23	4.55	4.67	4.78	4.88	4.96
5				4.56	4.55	4.80	4.91	5.00	5.09
6					4.78	4.80	5.00	5.10	5.18
7						4.97	5.00	5.17	5.26
8							5.13	5.17	5.31
9								5.27	5.31
10									5.40

TABLE 19 Critical values for Welsch's step-up procedure

$\nu = 13$

$j \backslash k$	2	3	4	5	6	7	8	9	10
2	3.06	3.06	3.58	3.75	3.88	4.00	4.10	4.18	4.26
3		4.07	3.74	4.18	4.32	4.43	4.53	4.62	4.70
4			4.27	4.18	4.49	4.61	4.71	4.80	4.89
5				4.51	4.49	4.73	4.84	4.93	5.01
6					4.72	4.73	4.93	5.02	5.11
7						4.90	4.93	5.09	5.18
8							5.06	5.09	5.24
9								5.20	5.24
10									5.33

$\nu = 14$

$j \backslash k$	2	3	4	5	6	7	8	9	10
2	3.03	3.03	3.55	3.71	3.84	3.95	4.05	4.14	4.21
3		4.03	3.71	4.13	4.27	4.38	4.48	4.56	4.64
4			4.23	4.13	4.44	4.56	4.66	4.74	4.82
5				4.46	4.44	4.68	4.78	4.87	4.95
6					4.67	4.68	4.87	4.96	5.04
7						4.85	4.87	5.03	5.11
8							5.00	5.03	5.17
9								5.14	5.17
10									5.26

$\nu = 15$

$j \backslash k$	2	3	4	5	6	7	8	9	10
2	3.01	3.01	3.52	3.68	3.81	3.92	4.01	4.09	4.17
3		3.99	3.68	4.10	4.23	4.34	4.43	4.52	4.59
4			4.19	4.10	4.40	4.51	4.61	4.69	4.77
5				4.42	4.40	4.63	4.73	4.82	4.89
6					4.62	4.63	4.82	4.91	4.99
7						4.80	4.82	4.98	5.06
8							4.95	4.98	5.12
9								5.09	5.12
10									5.20

$\nu = 16$

$j \backslash k$	2	3	4	5	6	7	8	9	10
2	3.00	3.00	3.50	3.65	3.78	3.89	3.98	4.06	4.13
3		3.96	3.66	4.06	4.19	4.30	4.39	4.47	4.55
4			4.16	4.06	4.36	4.47	4.57	4.65	4.72
5				4.39	4.36	4.59	4.69	4.77	4.85
6					4.59	4.59	4.78	4.86	4.94
7						4.76	4.78	4.93	5.01
8							4.91	4.93	5.07
9								5.04	5.07
10									5.16

TABLE 19 Critical values for Welsch's step-up procedure

$\nu = 18$

$j \backslash k$	2	3	4	5	6	7	8	9	10
2	2.97	2.97	3.46	3.61	3.73	3.84	3.92	4.00	4.07
3		3.91	3.62	4.01	4.13	4.24	4.33	4.40	4.47
4			4.10	4.01	4.30	4.40	4.50	4.58	4.65
5				4.33	4.30	4.52	4.62	4.70	4.77
6					4.52	4.52	4.70	4.79	4.86
7						4.69	4.70	4.86	4.93
8							4.83	4.86	4.99
9								4.96	4.99
10									5.08

$\nu = 20$

$j \backslash k$	2	3	4	5	6	7	8	9	10
2	2.95	2.95	3.43	3.58	3.70	3.80	3.88	3.96	4.02
3		3.87	3.58	3.97	4.09	4.19	4.28	4.35	4.42
4			4.06	3.97	4.25	4.35	4.44	4.52	4.59
5				4.29	4.25	4.47	4.56	4.64	4.71
6					4.47	4.47	6.65	4.73	4.80
7						4.64	4.65	4.80	4.87
8							4.78	4.80	4.92
9								4.90	4.92
10									5.01

$\nu = 24$

$j \backslash k$	2	3	4	5	6	7	8	9	10
2	2.92	2.92	3.38	3.52	3.64	3.74	3.82	3.89	3.96
3		3.80	3.54	3.91	4.02	4.12	4.20	4.27	4.34
4			4.00	3.91	4.18	4.28	4.36	4.44	4.50
5				4.22	4.18	4.39	4.48	4.55	4.62
6					4.40	4.39	4.56	4.64	4.71
7						4.56	4.56	4.71	4.78
8							4.69	4.71	4.83
9								4.81	4.83
10									4.92

$\nu = 30$

$j \backslash k$	2	3	4	5	6	7	8	9	10
2	2.89	2.89	3.34	3.48	3.59	3.68	3.76	3.83	3.89
3		3.74	3.49	3.85	3.96	4.05	4.13	4.20	4.26
4			3.94	3.85	4.11	4.21	4.29	4.36	4.42
5				4.15	4.11	4.32	4.40	4.47	4.53
6					4.33	4.32	4.48	4.55	4.62
7						4.48	4.48	4.62	4.68
8							4.61	4.62	4.74
9								4.73	4.74
10									4.83

TABLE 19 Critical values for Welsch's step-up procedure

$\nu = 40$

$j \backslash k$	2	3	4	5	6	7	8	9	10
2	2.86	2.86	3.29	3.43	3.53	3.62	3.70	3.77	3.82
3		3.69	3.45	3.79	3.89	3.98	4.06	4.12	4.18
4			3.88	3.79	4.05	4.14	4.21	4.28	4.34
5				4.09	4.05	4.24	4.32	4.39	4.45
6					4.26	4.24	4.40	4.47	4.53
7						4.40	4.40	4.54	4.60
8							4.53	4.54	4.65
9								4.64	4.65
10									4.74

$\nu = 60$

$j \backslash k$	2	3	4	5	6	7	8	9	10
2	2.83	2.83	3.25	3.38	3.48	3.57	3.64	3.71	3.76
3		3.63	3.41	3.73	3.83	3.91	3.99	4.05	4.10
4			3.82	3.74	3.98	4.07	4.14	4.20	4.26
5				4.02	3.98	4.17	4.24	4.31	4.36
6					4.19	4.17	4.33	4.39	4.45
7						4.33	4.33	4.45	4.51
8							4.45	4.45	4.56
9								4.56	4.56
10									4.65

$\nu = 120$

$j \backslash k$	2	3	4	5	6	7	8	9	10
2	2.80	2.80	3.21	3.33	3.43	3.52	3.59	3.65	3.70
3		3.58	3.36	3.68	3.77	3.85	3.92	3.98	4.03
4			3.76	3.69	3.92	4.00	4.07	4.13	4.18
5				3.96	3.92	4.10	4.17	4.23	4.28
6					4.12	4.10	4.25	4.31	4.36
7						4.26	4.25	4.37	4.43
8							4.37	4.37	4.48
9								4.48	4.48
10									4.57

$\nu = \infty$

$j \backslash k$	2	3	4	5	6	7	8	9	10
2	2.77	2.77	3.17	3.29	3.39	3.46	3.53	3.59	3.64
3		3.52	3.32	3.62	3.71	3.79	3.85	3.91	3.96
4			3.71	3.64	3.86	3.93	4.00	4.05	4.10
5				3.90	3.86	4.03	4.10	4.15	4.20
6					4.06	4.03	4.18	4.23	4.28
7						4.19	4.18	4.29	4.34
8							4.30	4.29	4.39
9								4.39	4.39
10									4.48

TABLE 20 Critical values of the studentized augmented range

This table furnishes critical values for the studentized augmented range $Q'_{\alpha[k,\nu]}$ employed in the Spjøtvoll–Stoline T′-method of multiple comparisons. Values are tabled for $k = 2$ to 8 means in increments of one and for $\nu = 5, 7,$ 10, 12, 16, 20, 24, 30, 40, 60, 120, and ∞ degrees of freedom. Three experimentwise error rates α 0.10, 0.05, and 0.01) are given for each combination of k and ν. Harmonic interpolation is recommended for degrees of freedom that are not furnished.

To find the 5% critical value of the studentized augmented range over $k = 5$ means at 20 degrees of freedom, enter the table for $k = 5$, $\nu = 20$, and $\alpha = .05$ to obtain $Q'_{.05[5,20]} = 4.233$. For $k > 8$, the studentized range furnished in Table **18** may be used since for $k > 8$ these two distributions are very similar.

This table is used for the Spjøtvoll–Stoline T′-test described in Section 9.7. It can also be employed in multiple comparisons of regression coefficients illustrated in Section 14.8. This table has been copied and rearranged from the tables furnished by M. R. Stoline (*J. Amer. Stat. Assoc.*, **73**:656–660, 1978) with permission of the publisher.

TABLE **20** Critical values of the studentized augmented range

ν	α	k = 2	3	4	5	6	7	8
5	.10	3.060	3.772	4.282	4.671	4.982	5.239	5.458
	.05	3.832	4.654	5.236	5.680	6.036	6.331	6.583
	.01	5.903	7.030	7.823	8.429	8.916	9.322	9.669
7	.10	2.848	3.491	3.943	4.285	4.556	4.781	4.972
	.05	3.486	4.198	4.692	5.064	5.360	5.606	5.816
	.01	5.063	5.947	6.551	7.008	7.374	7.679	7.939
10	.10	2.704	3.300	3.712	4.021	4.265	4.466	4.636
	.05	3.259	3.899	4.333	4.656	4.913	5.124	5.305
	.01	4.550	5.284	5.773	6.138	6.428	6.669	6.875
12	.10	2.651	3.230	3.628	3.924	4.157	4.349	4.511
	.05	3.177	3.791	4.204	4.509	4.751	4.950	5.119
	.01	4.373	5.056	5.505	5.837	6.101	6.321	6.507
16	.10	2.588	3.146	3.526	3.806	4.027	4.207	4.360
	.05	3.080	3.663	4.050	4.334	4.557	4.741	4.897
	.01	4.169	4.792	5.194	5.489	5.722	5.915	6.079
20	.10	2551	3.097	3.466	3.738	3.950	4.124	4.271
	.05	3.024	3.590	3.961	4.233	4.446	4.620	4.768
	.01	4.055	4.644	5.019	5.294	5.510	5.688	5.839
24	.10	2.527	3.065	3.427	3.693	3.901	4.070	4.213
	.05	2.988	3.542	3.904	4.167	4.373	4.541	4.684
	.01	3.982	4.549	4.908	5.169	5.374	5.542	5.685
30	.10	2.503	3.034	3.389	3.649	3.851	4.016	4.155
	.05	2.952	3.496	3.847	4.103	4.302	4.464	4.602
	.01	3.912	4.458	4.800	5.048	5.242	5.401	5.536
40	.10	2.480	3.003	3.352	3.605	3.803	3.963	4.099
	.05	2.918	3.450	3.792	4.040	4.232	4.389	4.521
	.01	3.844	4.370	4.696	4.931	5.115	5.265	5.392
60	.10	2.457	2.972	3.315	3.563	3.755	3.911	4.042
	.05	2.884	3.406	3.738	3.978	4.163	4.314	4.441
	.01	3.778	4.284	4.595	4.818	4.991	5.133	5.253
120	.10	2.434	2.943	3.278	3.520	3.707	3.859	3.987
	.05	2.851	3.362	3.686	3.917	4.096	4.241	4.363
	.01	3.714	4.201	4.497	4.709	4.872	5.005	5.118
∞	.10	2.412	2.913	3.243	3.479	3.661	3.808	3.931
	.05	2.819	3.320	3.634	3.858	4.030	4.170	4.286
	.01	3.653	4.121	4.403	4.603	4.757	4.882	4.987

TABLE **21** Critical values of the studentized maximum modulus distribution

This table furnishes values of the studentized maximum modulus $m_{\alpha[k^*,\nu]}$, which is used in Hochberg's GT2-method for multiple comparison of means. For arguments k (the number of means) ranging from $k = 3$ to $k = 20$ in increments of 1, critical values are given for degrees of freedom $\nu = 5, 7, 10, 12, 16, 20, 24, 30, 40, 60, 120$, and ∞. For k ranging from 20 to 50 in increments of 2, from 50 to 80 in increments of 10, and for $k = 100$, critical values are given for degrees of freedom ν ranging from 20 to 40 in increments of 2; from 40 to 60 in increments of 10; from 60 to 120 in increments of 20; and for 240, 480, and ∞. Three experimentwise type I error rates are furnished: $\alpha = 0.10$, 0.05, and 0.01. For a set of k means, the GT2-method permits up to $k^* = k(k - 1)/2$ pairwise comparisons. Although the proper parameter of m is k^*, the number of comparisons intended, we furnish the more convenient k. Interpolation for degrees of freedom ν in this table is harmonic. If only a subset of k^* (all possible pairwise tests) is intended, one can interpolate for the studentized maximum modulus corresponding to the intended number of comparisons k' as follows. Calculate the values of $k^* = k(k - 1)/2$ that bracket the intended number of comparisons k'. Look up the values of $m_{\alpha[k^*,\nu]}$ for the corresponding arguments of k. Interpolate linearly as well as harmonically between the bracketing values of k^* (*not* the corresponding arguments k). The desired interpolated value of m is the average of the linear and harmonic interpolate. A conservative rule would be to use the value of $m_{\alpha[k^*,\nu]}$ corresponding to the upper bracketing value of k^*.

To look up the 5% critical value of m for 24 degrees of freedom and a set of $k = 10$ means, which corresponds to $k^* = 45$, we obtain $m_{.05[45,24]} = 3.641$. These critical values are for two-tailed tests.

This table is used in the GT2-method of multiple comparisons of means described in Section 9.7. It can also be employed in multiple comparisons of regression coefficients illustrated in Section 14.8.

This table has been copied in rearranged form from tables by M. R. Stoline and H. K. Ury (*Technometrics,* **21**:87–93, 1979) and by H. K. Ury, M. R. Stoline, and B. T. Mitchell [*Communications in Statistics,* **B9**(2), 1980] with permission of the publishers.

TABLE 21 Critical values of the studentized maximum modulus distribution

k	α	5	7	10	12	16	20
		Degrees of freedom ν					
3	.10	2.769	2.555	2.410	2.357	2.293	2.255
	.05	3.399	3.055	2.829	2.747	2.650	2.594
	.01	5.106	4.296	3.801	3.631	3.434	3.323
4	.10	3.239	2.961	2.771	2.701	2.616	2.567
	.05	3.928	3.489	3.199	3.095	2.969	2.897
	.01	5.812	4.814	4.205	3.995	3.753	3.617
5	.10	3.576	3.253	3.029	2.946	2.845	2.786
	.05	4.312	3.805	3.467	3.345	3.199	3.114
	.01	6.334	5.198	4.503	4.263	3.986	3.831
6	.10	3.837	3.478	3.229	3.136	3.022	2.956
	.05	4.610	4.051	3.677	3.540	3.377	3.282
	.01	6.744	5.502	4.739	4.475	4.170	3.999
7	.10	4.048	3.661	3.391	3.290	3.166	3.093
	.05	4.853	4.252	3.848	3.700	3.522	3.419
	.01	7.079	5.752	4.933	4.650	4.322	4.137
8	.10	4.224	3.814	3.527	3.419	3.286	3.208
	.05	5.057	4.421	3.992	3.835	3.645	3.534
	.01	7.362	5.963	5.099	4.799	4.451	4.255
9	.10	4.375	3.945	3.644	3.530	3.390	3.306
	.05	5.232	4.566	4.116	3.951	3.751	3.634
	.01	7.605	6.146	5.242	4.929	4.563	4.357
10	.10	4.506	4.060	3.746	3.627	3.480	3.393
	.05	5.384	4.693	4.225	4.053	3.844	3.721
	.01	7.819	6.307	5.369	5.043	4.662	4.447
11	.10	4.623	4.162	3.836	3.713	3.560	3.469
	.05	5.520	4.806	4.322	4.143	3.926	3.799
	.01	8.008	6.450	5.482	5.145	4.750	4.528
12	.10	4.727	4.252	3.917	3.790	3.633	3.538
	.05	5.641	4.907	4.409	4.225	4.001	3.869
	.01	8.178	6.579	5.584	5.237	4.830	4.601
13	.10	4.821	4.335	3.991	3.860	3.698	3.601
	.05	5.750	4.999	4.488	4.299	4.068	3.933
	.01	8.332	6.696	5.677	5.321	4.903	4.667
14	.10	4.906	4.410	4.058	3.924	3.758	3.658
	.05	5.850	5.083	4.560	4.367	4.130	3.991
	.01	8.473	6.803	5.762	5.398	4.970	4.728
15	.10	4.985	4.479	4.119	3.983	3.813	3.710
	.05	5.942	5.160	4.627	4.429	4.187	4.045
	.01	8.602	6.902	5.841	5.469	5.032	4.785
16	.10	5.058	4.542	4.176	4.037	3.863	3.759
	.05	6.027	5.232	4.688	4.487	4.240	4.095
	.01	8.722	6.994	5.914	5.535	5.090	4.837
17	.10	5.125	4.601	4.229	4.087	3.911	3.804
	.05	6.106	5.298	4.746	4.540	4.289	4.141
	.01	8.833	7.079	5.982	5.597	5.144	4.886
18	.10	5.188	4.657	4.279	4.135	3.955	3.846
	.05	6.179	5.360	4.799	4.591	4.335	4.184
	.01	8.937	7.158	6.045	5.654	5.194	4.932
19	.10	5.247	4.708	4.325	4.179	3.996	3.886
	.05	6.248	5.418	4.849	4.638	4.379	4.225
	.01	9.034	7.233	6.105	5.708	5.242	4.975
20	.10	5.302	4.757	4.369	4.220	4.035	3.923
	.05	6.313	5.472	4.897	4.682	4.419	4.264
	.01	9.126	7.303	6.161	5.760	5.286	5.016

Degrees of freedom ν							
24	**30**	**40**	**60**	**120**	∞	k	α
2.231	2.207	2.183	2.160	2.137	2.114	3	.10
2.558	2.522	2.488	2.454	2.420	2.388		.05
3.253	3.185	3.119	3.055	2.993	2.934		.01
2.534	2.502	2.470	2.439	2.408	2.378	4	.10
2.851	2.805	2.760	2.716	2.673	2.631		.05
3.531	3.447	3.367	3.289	3.215	3.143		.01
2.747	2.709	2.671	2.633	2.596	2.560	5	.10
3.059	3.005	2.952	2.900	2.849	2.800		.05
3.732	3.637	3.545	3.456	3.371	3.289		.01
2.911	2.868	2.825	2.782	2.739	2.697	6	.10
3.220	3.160	3.100	3.041	2.984	2.928		.05
3.890	3.785	3.683	3.586	3.492	3.402		.01
3.044	2.996	2.948	2.901	2.854	2.807	7	.10
3.352	3.285	3.220	3.156	3.093	3.031		.05
4.020	3.906	3.796	3.691	3.590	3.493		.01
3.156	3.104	3.052	3.001	2.949	2.898	8	.10
3.462	3.391	3.321	3.251	3.183	3.117		.05
4.130	4.009	3.892	3.780	3.672	3.569		.01
3.251	3.196	3.141	3.086	3.031	2.976	9	.10
3.557	3.482	3.407	3.334	3.261	3.190		.05
4.225	4.098	3.975	3.857	3.743	3.634		.01
3.335	3.277	3.219	3.160	3.102	3.043	10	.10
3.641	3.562	3.483	3.406	3.329	3.254		.05
4.309	4.176	4.048	3.924	3.805	3.691		.01
3.409	3.348	3.287	3.226	3.165	3.103	11	.10
3.716	3.633	3.551	3.470	3.389	3.310		.05
4.385	4.247	4.113	3.984	3.860	3.742		.01
3.476	3.413	3.349	3.286	3.221	3.157	12	.10
3.783	3.697	3.611	3.527	3.443	3.361		.05
4.453	4.310	4.172	4.038	3.910	3.787		.01
3.536	3.471	3.405	3.339	3.272	3.205	13	.10
3.843	3.755	3.667	3.579	3.492	3.407		.05
4.515	4.368	4.225	4.088	3.955	3.829		.01
3.591	3.524	3.456	3.388	3.319	3.249	14	.10
3.899	3.808	3.717	3.627	3.537	3.449		.05
4.572	4.421	4.275	4.133	3.997	3.867		.01
3.642	3.573	3.504	3.433	3.362	3.290	15	.10
3.951	3.857	3.764	3.671	3.578	3.487		.05
4.625	4.470	4.320	4.175	4.035	3.901		.01
3.689	3.619	3.547	3.475	3.402	3.327	16	.10
3.998	3.902	3.807	3.712	3.617	3.523		.05
4.674	4.516	4.362	4.214	4.070	3.934		.01
3.733	3.661	3.588	3.514	3.438	3.362	17	.10
4.043	3.945	3.847	3.750	3.652	3.556		.05
4.720	4.559	4.402	4.250	4.103	3.963		.01
3.774	3.700	3.626	3.550	3.473	3.394	18	.10
4.084	3.984	3.885	3.785	3.685	3.587		.05
4.763	4.599	4.439	4.284	4.134	3.991		.01
3.812	3.737	3.661	3.584	3.505	3.425	19	.10
4.123	4.022	3.920	3.818	3.717	3.615		.05
4.804	4.636	4.473	4.316	4.163	4.018		.01
3.848	3.772	3.695	3.616	3.536	3.453	20	.10
4.160	4.057	3.953	3.850	3.746	3.643		.05
4.842	4.672	4.506	4.346	4.191	4.043		.01

TABLE 21 Critical values of the studentized maximum modulus distribution

Degrees of freedom ν

k	α	20	22	24	26	28	30	32	34	36	38
20	.10	3.923	3.882	3.848	3.819	3.794	3.772	3.753	3.736	3.721	3.707
	.05	4.264	4.207	4.160	4.120	4.086	4.057	4.031	4.008	3.988	3.970
	.01	5.016	4.921	4.842	4.776	4.720	4.672	4.630	4.593	4.561	4.532
22	.10	3.992	3.950	3.914	3.884	3.859	3.836	3.816	3.799	3.783	3.769
	.05	4.334	4.276	4.228	4.187	4.152	4.121	4.095	4.071	4.050	4.031
	.01	5.092	4.993	4.912	4.845	4.787	4.737	4.694	4.657	4.623	4.594
24	.10	4.054	4.010	3.974	3.943	3.917	3.894	3.874	3.855	3.839	3.825
	.05	4.398	4.339	4.289	4.247	4.211	4.180	4.152	4.128	4.107	4.087
	.01	5.160	5.059	4.976	4.907	4.848	4.797	4.753	4.714	4.680	4.649
26	.10	4.110	4.066	4.029	3.997	3.970	3.946	3.926	3.907	3.891	3.876
	.05	4.456	4.395	4.345	4.302	4.265	4.233	4.205	4.180	4.158	4.138
	.01	5.222	5.119	5.035	4.964	4.903	4.851	4.806	4.766	4.731	4.700
28	.10	4.161	4.116	4.079	4.046	4.019	3.995	3.973	3.954	3.938	3.922
	.05	4.509	4.447	4.396	4.352	4.314	4.282	4.253	4.228	4.205	4.185
	.01	5.280	5.175	5.088	5.016	4.954	4.901	4.855	4.815	4.779	4.747
30	.10	4.209	4.163	4.124	4.092	4.064	4.039	4.017	3.998	3.981	3.965
	.05	4.559	4.496	4.443	4.398	4.360	4.327	4.298	4.272	4.249	4.228
	.01	5.333	5.226	5.138	5.064	5.002	4.948	4.901	4.859	4.823	4.790
32	.10	4.253	4.206	4.167	4.134	4.105	4.080	4.058	4.039	4.021	4.005
	.05	4.604	4.540	4.487	4.441	4.402	4.369	4.339	4.313	4.289	4.268
	.01	5.382	5.274	5.184	5.110	5.046	4.991	4.943	4.901	4.864	4.831
34	.10	4.294	4.246	4.207	4.173	4.144	4.119	4.096	4.076	4.058	4.042
	.05	4.647	4.582	4.528	4.482	4.442	4.408	4.377	4.351	4.327	4.306
	.01	5.428	5.318	5.228	5.152	5.087	5.031	4.982	4.940	4.902	4.868
36	.10	4.332	4.284	4.244	4.210	4.180	4.154	4.132	4.112	4.094	4.077
	.05	4.687	4.621	4.566	4.519	4.479	4.444	4.414	4.386	4.362	4.341
	.01	5.471	5.360	5.268	5.191	5.125	5.069	5.020	4.976	4.938	4.904
38	.10	4.368	4.320	4.279	4.244	4.214	4.188	4.165	4.145	4.126	4.110
	.05	4.724	4.658	4.602	4.555	4.514	4.479	4.448	4.420	4.396	4.374
	.01	5.512	5.400	5.307	5.228	5.162	5.104	5.054	5.011	4.972	4.937

Degrees of freedom ν

k	α	20	22	24	26	28	30	32	34	36	38
40	.10	4.402	4.353	4.312	4.277	4.247	4.220	4.197	4.176	4.157	4.141
	.05	4.760	4.692	4.636	4.588	4.547	4.511	4.480	4.452	4.427	4.405
	.01	5.551	5.437	5.343	5.264	5.196	5.138	5.087	5.043	5.004	4.968
42	.10	4.434	4.385	4.343	4.308	4.277	4.250	4.227	4.206	4.187	4.170
	.05	4.793	4.725	4.668	4.620	4.578	4.542	4.510	4.482	4.457	4.434
	.01	5.587	5.472	5.377	5.297	5.229	5.170	5.119	5.074	5.034	4.998
44	.10	4.464	4.415	4.373	4.337	4.306	4.279	4.255	4.234	4.215	4.198
	.05	4.825	4.756	4.698	4.650	4.607	4.571	4.539	4.510	4.485	4.462
	.01	5.622	5.505	5.409	5.329	5.260	5.200	5.149	5.103	5.063	5.027
46	.10	4.493	4.443	4.401	4.364	4.333	4.306	4.282	4.260	4.241	4.224
	.05	4.855	4.786	4.727	4.678	4.636	4.599	4.566	4.537	4.512	4.489
	.01	5.655	5.537	5.440	5.359	5.289	5.229	5.177	5.131	5.090	5.054
48	.10	4.521	4.470	4.427	4.391	4.359	4.332	4.307	4.286	4.266	4.249
	.05	4.884	4.814	4.755	4.705	4.662	4.625	4.592	4.563	4.537	4.514
	.01	5.686	5.568	5.470	5.387	5.317	5.257	5.204	5.157	5.116	5.080
50	.10	4.547	4.496	4.453	4.416	4.384	4.356	4.332	4.310	4.290	4.273
	.05	4.911	4.841	4.781	4.731	4.688	4.650	4.617	4.588	4.562	4.538
	.01	5.716	5.597	5.498	5.415	5.344	5.283	5.230	5.183	5.141	5.104
60	.10	4.663	4.610	4.565	4.527	4.494	4.465	4.440	4.417	4.397	4.378
	.05	5.033	4.959	4.898	4.846	4.801	4.762	4.727	4.697	4.670	4.645
	.01	5.849	5.725	5.623	5.537	5.463	5.400	5.344	5.296	5.252	5.214
70	.10	4.759	4.705	4.659	4.619	4.585	4.556	4.529	4.506	4.485	4.466
	.05	5.133	5.058	4.995	4.941	4.895	4.854	4.819	4.787	4.759	4.734
	.01	5.960	5.833	5.727	5.639	5.563	5.497	5.440	5.390	5.345	5.305
80	.10	4.841	4.786	4.738	4.698	4.663	4.633	4.606	4.582	4.560	4.541
	.05	5.219	5.142	5.078	5.023	4.975	4.934	4.897	4.865	4.836	4.810
	.01	6.055	5.925	5.817	5.726	5.648	5.581	5.522	5.470	5.424	5.383
100	.10	4.976	4.918	4.869	4.827	4.791	4.759	4.731	4.706	4.684	4.664
	.05	5.361	5.281	5.214	5.157	5.107	5.064	5.026	4.993	4.962	4.935
	.01	6.211	6.076	5.964	5.870	5.789	5.719	5.658	5.604	5.556	5.513

TABLE 21 Critical values of the studentized maximum modulus distribution

		Degrees of freedom ν								
k	α	40	50	60	80	100	120	240	480	∞
20	.10	3.695	3.648	3.616	3.576	3.552	3.535	3.494	3.474	3.453
	.05	3.953	3.891	3.850	3.798	3.767	3.746	3.694	3.668	3.643
	.01	4.506	4.409	4.346	4.267	4.221	4.191	4.116	4.079	4.042
22	.10	3.756	3.708	3.675	3.633	3.608	3.591	3.549	3.527	3.505
	.05	4.015	3.950	3.907	3.854	3.821	3.800	3.746	3.719	3.693
	.01	4.567	4.467	4.401	4.320	4.273	4.241	4.164	4.126	4.088
24	.10	3.812	3.762	3.728	3.685	3.659	3.642	3.597	3.575	3.552
	.05	4.070	4.004	3.960	3.904	3.871	3.849	3.793	3.765	3.738
	.01	4.622	4.519	4.452	4.368	4.319	4.287	4.207	4.168	4.129
26	.10	3.862	3.811	3.776	3.732	3.705	3.687	3.642	3.618	3.595
	.05	4.120	4.053	4.007	3.950	3.916	3.893	3.836	3.807	3.778
	.01	4.672	4.567	4.497	4.412	4.362	4.328	4.247	4.206	4.167
28	.10	3.909	3.856	3.820	3.775	3.748	3.729	3.682	3.658	3.634
	.05	4.167	4.097	4.051	3.992	3.957	3.934	3.875	3.845	3.816
	.01	4.718	4.611	4.540	4.452	4.401	4.367	4.283	4.242	4.201
30	.10	3.951	3.898	3.861	3.815	3.787	3.768	3.720	3.695	3.670
	.05	4.210	4.139	4.091	4.031	3.995	3.971	3.911	3.881	3.850
	.01	4.761	4.651	4.579	4.490	4.437	4.402	4.316	4.274	4.233
32	.10	3.991	3.936	3.899	3.852	3.823	3.804	3.754	3.729	3.704
	.05	4.249	4.177	4.128	4.067	4.031	4.006	3.944	3.913	3.882
	.01	4.801	4.689	4.615	4.524	4.470	4.435	4.347	4.304	4.262
34	.10	4.028	3.972	3.934	3.886	3.857	3.837	3.786	3.761	3.735
	.05	4.286	4.213	4.163	4.101	4.064	4.038	3.975	3.944	3.912
	.01	4.838	4.724	4.649	4.557	4.502	4.466	4.376	4.333	4.289
36	.10	4.063	4.006	3.967	3.918	3.888	3.868	3.817	3.790	3.764
	.05	4.321	4.246	4.196	4.133	4.095	4.069	4.005	3.972	3.940
	.01	4.873	4.757	4.681	4.587	4.531	4.494	4.404	4.359	4.315
38	.10	4.095	4.038	3.999	3.949	3.918	3.898	3.845	3.818	3.791
	.05	4.354	4.278	4.227	4.163	4.124	4.098	4.032	3.999	3.966
	.01	4.906	4.788	4.711	4.616	4.559	4.521	4.429	4.384	4.339

Degrees of freedom ν

k	α	40	50	60	80	100	120	240	480	∞
40	.10	4.126	4.067	4.028	3.977	3.946	3.925	3.872	3.845	3.817
	.05	4.385	4.308	4.256	4.191	4.151	4.125	4.058	4.025	3.991
	.01	4.937	4.818	4.739	4.643	4.585	4.547	4.453	4.407	4.362
42	.10	4.155	4.096	4.056	4.004	3.973	3.951	3.897	3.869	3.841
	.05	4.414	4.336	4.283	4.217	4.177	4.150	4.082	4.048	4.014
	.01	4.966	4.846	4.766	4.668	4.610	4.571	4.476	4.430	4.384
44	.10	4.182	4.122	4.082	4.030	3.998	3.976	3.921	3.893	3.864
	.05	4.441	4.363	4.309	4.242	4.202	4.174	4.106	4.071	4.037
	.01	4.994	4.872	4.792	4.692	4.633	4.594	4.498	4.451	4.404
46	.10	4.208	4.148	4.107	4.054	4.022	4.000	3.944	3.915	3.886
	.05	4.468	4.388	4.334	4.266	4.225	4.197	4.128	4.093	4.058
	.01	5.021	4.898	4.816	4.716	4.656	4.616	4.519	4.471	4.424
48	.10	4.233	4.172	4.131	4.077	4.044	4.022	3.966	3.937	3.907
	.05	4.493	4.412	4.358	4.289	4.247	4.219	4.149	4.113	4.078
	.01	5.047	4.922	4.840	4.738	4.677	4.637	4.538	4.490	4.442
50	.10	4.257	4.195	4.153	4.099	4.066	4.044	3.986	3.957	3.927
	.05	4.517	4.435	4.381	4.311	4.269	4.240	4.169	4.133	4.097
	.01	5.071	4.945	4.862	4.759	4.698	4.657	4.557	4.508	4.460
60	.10	4.362	4.297	4.254	4.197	4.162	4.139	4.078	4.046	4.015
	.05	4.623	4.538	4.480	4.407	4.363	4.333	4.258	4.220	4.181
	.01	5.179	5.047	4.960	4.852	4.788	4.746	4.641	4.589	4.538
70	.10	4.449	4.382	4.337	4.278	4.242	4.217	4.154	4.121	4.087
	.05	4.711	4.623	4.564	4.488	4.441	4.410	4.331	4.291	4.251
	.01	5.269	5.133	5.043	4.931	4.864	4.819	4.710	4.656	4.603
80	.10	4.523	4.455	4.408	4.347	4.310	4.284	4.218	4.184	4.149
	.05	4.787	4.696	4.635	4.556	4.509	4.476	4.394	4.353	4.311
	.01	5.347	5.206	5.113	4.998	4.929	4.883	4.769	4.714	4.659
100	.10	4.645	4.574	4.525	4.461	4.422	4.395	4.325	4.288	4.251
	.05	4.911	4.816	4.752	4.670	4.619	4.585	4.498	4.454	4.409
	.01	5.474	5.328	5.230	5.108	5.036	4.987	4.868	4.809	4.750

TABLE **22** Shortest unbiased confidence limits for the variance

This table furnishes special multiplication factors f_1 and f_2, which greatly simplify the computation of shortest unbiased confidence limits for the variance with confidence coefficients 0.95 and 0.99. The coefficients of the factors are provided for degrees of freedom ν from 2 to 30 in increments of 1 and from $\nu = 40$ to 100 in increments of 10.

For a sample variance based on ten items ($\nu = 9$) and a 0.95 confidence coefficient, the factors shown in the table are $f_1 = 0.4432$ and $f_2 = 3.048$; for a confidence coefficient of 0.99, the factors are found to be $f_1 = 0.3585$ and $f_2 = 4.720$. For degrees of freedom > 100 use the ordinary (equal tails) method for computing confidence limits of a variance (Section 7.7). When the number of degrees of freedom is large, the two methods yield very similar results.

These factors are employed in the setting of confidence limits to variances as discussed in Section 7.7.

The factors in this table have been obtained by dividing the quantity $n - 1$ by the values found in a table prepared by D. V. Lindley, D. A. East, and P. A. Hamilton (*Biometrika,* **47**:433–437, 1960).

TABLE 22 Shortest unbiased confidence limits for the variance

ν	Confidence coefficients 0.95	0.99	ν	Confidence coefficients 0.95	0.99	ν	Confidence coefficients 0.95	0.99
2	.2099 23.605	.1505 114.489	14	.5135 2.354	.4289 3.244	26	.6057 1.825	.5261 2.262
3	.2681 10.127	.1983 29.689	15	.5242 2.276	.4399 3.091	27	.6110 1.802	.5319 2.223
4	.3125 6.590	.2367 15.154	16	.5341 2.208	.4502 2.961	28	.6160 1.782	.5374 2.187
5	.3480 5.054	.2685 10.076	17	.5433 2.149	.4598 2.848	29	.6209 1.762	.5427 2.153
6	.3774 4.211	.2956 7.637	18	.5520 2.097	.4689 2.750	30	.6255 1.744	.5478 2.122
7	.4025 3.679	.3192 6.238	19	.5601 2.050	.4774 2.664	40	.6636 1.608	.5900 1.896
8	.4242 3.314	.3400 5.341	20	.5677 2.008	.4855 2.588	50	.6913 1.523	.6213 1.760
9	.4432 3.048	.3585 4.720	21	.5749 1.971	.4931 2.519	60	.7128 1.464	.6458 1.668
10	.4602 2.844	.3752 4.265	22	.5817 1.936	.5004 2.458	70	.7300 1.421	.6657 1.607
11	.4755 2.683	.3904 3.919	23	.5882 1.905	.5073 2.402	80	.7443 1.387	.6824 1.549
12	.4893 2.553	.4043 3.646	24	.5943 1.876	.5139 2.351	90	.7564 1.360	.6966 1.508
13	.5019 2.445	.4171 3.426	25	.6001 1.850	.5201 2.305	100	.7669 1.338	.7090 1.475

TABLE **23** Confidence limits for percentages

This table furnishes confidence limits for percentages based on the binomial distribution.

The first part of the table furnishes limits for samples up to size $n = 30$. The arguments are Y, number of items in the sample that exhibit a given property, and n, sample size. Argument Y is tabled for integral values between 0 and 15, which yield percentages up to 50%. For each sample size n and number of items Y with the given property, three lines of numerical values are shown. The first line of values gives 95% confidence limits for the percentage, the second line lists the observed percentage incidence of the property, and the third line of values furnishes the 99% confidence limits for the percentage. For example, for $Y = 8$ individuals showing the property out of a sample of $n = 20$, the second line indicates that this represents an incidence of the property of 40.00%, the first line yields the 95% confidence limits of this percentage as 19.10% to 63.95%, and the third line gives the 99% limits as 14.60% to 70.10%.

Interpolate in this table (up to $n = 49$) by dividing L_1^- and L_2^-, the lower and upper confidence limits at the next lower tabled sample size n^-, by desired sample size n, and multiply them by the next lower tabled sample size n^-. Thus, for example, to obtain the confidence limits of the percentage corresponding to 8 individuals showing the given property in a sample of 22 individuals (which corresponds to 36.36% of the individuals showing the property), compute the lower confidence limit $L_1 = L_1^- n^-/n = (19.10)20/22 = 17.36\%$ and the upper confidence limit $L_2 = L_2^- n^-/n = (63.95)20/22 = 58.14\%$.

The second half of the table is for larger sample sizes ($n = 50$, 100, 200, 500, and 1,000). The arguments along the left margin of the table are now percentages from 0 to 50% in increments of 1%, rather than counts. The 95% and 99% confidence limits corresponding to a given percentage incidence p and sample size n are the functions given in two lines in the body of the table. For instance, the 99% confidence limits of an observed incidence of 12% in a sample of 500 are found to be 8.56–16.19%, in the second of the two lines. Interpolation in this table between the furnished sample sizes can be achieved by means of the following formula for the lower limit:

$$L_1 = \frac{L_1^- n^-(n^+ - n) + L_1^+ n^+(n - n^-)}{n(n^+ - n^-)}$$

In the above expression, n is the size of the observed sample, n^- and n^+ are the next lower and upper tabled sample sizes, respectively, L_1^- and L_1^+ are corresponding tabled confidence limits for these sample sizes, and L_1 is the lower confidence limit to be found by interpolation. The upper confidence limit,

L_2, can be obtained by a corresponding formula by substituting 2 for the subscript 1. By way of an example we shall illustrate setting 95% confidence limits to an observed percentage of 25% in a sample size of 80. The tabled 95% limits for $n = 50$ are 13.84–39.27%. For $n = 100$ the corresponding tabled limits are 16.88–34.66%. When we substitute the values for the lower limits in the above formula we obtain

$$L_1 = \frac{(13.84)(50)(100 - 80) + (16.88)(100)(80 - 50)}{80(100 - 50)} = 16.12\%$$

for the lower confidence limit. Similarly, for the upper confidence limit we compute

$$L_2 = \frac{(39.27)(50)(100 - 80) + (34.66)(100)(80 - 50)}{80(100 - 50)} = 35.81\%$$

The tabled values in parentheses are limits for percentages that could not be obtained in any real sampling problem (for example, 25% in 50 items), but are necessary for purposes of interpolation. For percentages greater than 50% look up the complementary percentage as the argument. The complements of the tabled binomial confidence limits are the desired limits.

Confidence limits for percentages are used in many different applications, for example, in the sign test (Section 13.12) and in a variety of problems with frequencies in Chapters 17 and 18.

These tables have been extracted from more extensive ones in D. Mainland, L. Herrera, and M. I. Sutcliffe, *Tables for Use with Binomial Samples* (Dept. of Medical Statistics, New York University Coll. of Medicine, 1956) with permission of the publisher. The interpolation formulas cited are also due to these authors. Confidence limits of odd percentages up to 13% for $n = 50$ were computed by interpolation. For $Y = 0$ one sided $(1 - \alpha)100\%$ confidence limits were computed as $L_2 = 1 - \alpha^{1/n}$ with $L_1 = 0$.

TABLE 23 Confidence limits for percentages

Y	Confidence coefficients	n = 5	n = 10	n = 15
0	95	0.00 - 45.07	0.00 - 25.89	0.00 - 18.10
		0.00	0.00	0.00
	99	0.00 - 60.19	0.00 - 36.90	0.00 - 26.44
1	95	0.51 - 71.60	0.25 - 44.50	0.17 - 32.00
		20.00	10.00	6.67
	99	0.10 - 81.40	0.05 - 54.4	0.03 - 40.27
2	95	5.28 - 85.34	2.52 - 55.60	1.66 - 40.49
		40.00	20.00	13.33
	99	2.28 - 91.72	1.08 - 64.80	0.71 - 48.71
3	95		6.67 - 65.2	4.33 - 48.07
			30.00	20.00
	99		3.70 - 73.50	2.39 - 56.07
4	95		12.20 - 73.80	7.80 - 55.14
			40.00	26.67
	99		7.68 - 80.91	4.88 - 62.78
5	95		18.70 - 81.30	11.85 - 61.62
			50.00	33.33
	99		12.80 - 87.20	8.03 - 68.89
6	95			16.33 - 67.74
				40.00
	99			11.67 - 74.40
7	95			21.29 - 73.38
				46.67
	99			15.87 - 79.54
8	95			
	99			
9	95			
	99			
10	95			
	99			
11	95			
	99			
12	95			
	99			
13	95			
	99			
14	95			
	99			
15	95			
	99			

n = 20	n = 25	n = 30	Confidence coefficients	Y
0.00 - 13.91	0.00 - 11.29	0.00 - 9.50	95	
0.00	0.00	0.00		0
0.00 - 20.57	0.00 - 16.82	0.00 - 14.23	99	
0.13 - 24.85	0.10 - 20.36	0.08 - 17.23	95	
5.00	4.00	3.33		1
0.02 - 31.70	0.02 - 26.24	0.02 - 22.33	99	
1.24 - 31.70	0.98 - 26.05	0.82 - 22.09	95	
10.00	8.00	6.67		2
0.53 - 38.70	0.42 - 32.08	0.35 - 27.35	99	
3.21 - 37.93	2.55 - 31.24	2.11 - 26.53	95	
15.00	12.00	10.00		3
1.77 - 45.05	1.40 - 37.48	1.16 - 32.03	99	
5.75 - 43.65	4.55 - 36.10	3.77 - 30.74	95	
20.00	16.00	13.33		4
3.58 - 50.65	2.83 - 42.41	2.34 - 36.39	99	
8.68 - 49.13	6.84 - 40.72	5.64 - 34.74	95	
25.00	20.00	16.67		5
5.85 - 56.05	4.60 - 47.00	3.79 - 40.44	99	
11.90 - 54.30	9.35 - 45.14	7.70 - 38.56	95	
30.00	24.00	20.00		6
8.45 - 60.95	6.62 - 51.38	5.43 - 44.26	99	
15.38 - 59.20	12.06 - 49.38	9.92 - 42.29	95	
35.00	28.00	23.33		7
11.40 - 65.70	8.90 - 55.56	7.29 - 48.01	99	
19.10 - 63.95	14.96 - 53.50	12.29 - 45.89	95	
40.00	32.00	26.67		8
14.60 - 70.10	11.36 - 59.54	9.30 - 51.58	99	
23.05 - 68.48	17.97 - 57.48	14.73 - 49.40	95	
45.00	36.00	30.00		9
18.08 - 74.30	14.01 - 63.36	11.43 - 55.00	99	
27.20 - 72.80	21.12 - 61.32	17.29 - 52.80	95	
50.00	40.00	33.33		10
21.75 - 78.25	16.80 - 67.04	13.69 - 58.35	99	
	24.41 - 65.06	19.93 - 56.13	95	
	44.00	36.67		11
	19.75 - 70.55	16.06 - 61.57	99	
	27.81 - 68.69	22.66 - 59.39	95	
	48.00	40.00		12
	22.84 - 73.93	18.50 - 64.69	99	
		25.46 - 62.56	95	
		43.33		13
		21.07 - 67.72	99	
		28.35 - 65.66	95	
		46.67		14
		23.73 - 70.66	99	
		31.30 - 68.70	95	
		50.00		15
		26.47 - 73.53	99	

TABLE 23 Confidence limits for percentages

%	Confidence coefficients	n: 50	100	200	500	1000
0	95	.00- 5.82	.00- 2.95	.00- 1.49	.00- 0.60	.00- 0.30
	99	.00- 8.80	.00- 4.50	.00- 2.28	.00- 0.92	.00- 0.46
1	95	(.02- 8.88)	.02- 5.45	.12- 3.57	.32- 2.32	.48- 1.83
	99	(.00-12.02)	.00- 7.21	.05- 4.55	.22- 2.80	.37- 2.13
2	95	.05-10.66	.24- 7.04	.55- 5.04	1.06- 3.56	1.29- 3.01
	99	.01-13.98	.10- 8.94	.34- 6.17	.87- 4.12	1.13- 3.36
3	95	(.27-12.19)	.62- 8.53	1.11- 6.42	1.79- 4.81	2.11- 4.19
	99	(.16-15.60)	.34-10.57	.78- 7.65	1.52- 5.44	1.88- 4.59
4	95	.49-13.72	1.10- 9.93	1.74- 7.73	2.53- 6.05	2.92- 5.36
	99	.21-17.21	.68-12.08	1.31- 9.05	2.17- 6.75	2.64- 5.82
5	95	(.88-15.14)	1.64-11.29	2.43- 9.00	3.26- 7.29	3.73- 6.54
	99	(.45-18.76)	1.10-13.53	1.89-10.40	2.83- 8.07	3.39- 7.05
6	95	1.26-16.57	2.24-12.60	3.18-10.21	4.11- 8.43	4.63- 7.64
	99	.69-20.32	1.56-14.93	2.57-11.66	3.63- 9.24	4.25- 8.18
7	95	(1.74-17.91)	2.86-13.90	3.88-11.47	4.96- 9.56	5.52- 8.73
	99	(1.04-21.72)	2.08-16.28	3.17-12.99	4.43-10.42	5.12- 9.31
8	95	2.23-19.25	3.51-15.16	4.70-12.61	5.81-10.70	6.42- 9.83
	99	1.38-23.13	2.63-17.61	3.93-14.18	5.23-11.60	5.98-10.43
9	95	(2.78-20.54)	4.20-16.40	5.46-13.82	6.66-11.83	7.32-10.93
	99	(1.80-24.46)	3.21-18.92	4.61-15.44	6.04-12.77	6.84-11.56
10	95	3.32-21.82	4.90-17.62	6.22-15.02	7.51-12.97	8.21-12.03
	99	2.22-25.80	3.82-20.20	5.29-16.70	6.84-13.95	7.70-12.69
11	95	(3.93-23.06)	5.65-18.80	7.05-16.16	8.41-14.06	9.14-13.10
	99	(2.70-27.11)	4.48-21.42	6.06-17.87	7.70-15.07	8.60-13.78
12	95	4.54-24.31	6.40-19.98	7.87-17.30	9.30-15.16	10.06-14.16
	99	3.18-28.42	5.15-22.65	6.83-19.05	8.56-16.19	9.51-14.86
13	95	(5.18-27.03)	7.11-21.20	8.70-18.44	10.20-16.25	10.99-15.23
	99	(3.72-29.67)	5.77-23.92	7.60-20.23	9.42-17.31	10.41-15.95
14	95	5.82-26.75	7.87-22.37	9.53-19.58	11.09-17.34	11.92-16.30
	99	4.25-30.92	6.46-25.13	8.38-21.40	10.28-18.43	11.31-17.04
15	95	(6.50-27.94)	8.64-23.53	10.36-20.72	11.98-18.44	12.84-17.37
	99	(4.82-32.14)	7.15-26.33	9.15-22.58	11.14-19.55	12.21-18.13

TABLE 23 Confidence limits for percentages

%	Confidence coefficients	n = 50	100	200	500	1000
16	95	7.17-29.12	9.45-24.66	11.22-21.82	12.90-19.50	13.79-18.42
	99	5.40-33.36	7.89-27.49	9.97-23.71	12.03-20.63	13.14-19.19
17	95	(7.88-30.28)	10.25-25.79	12.09-22.92	13.82-20.57	14.73-19.47
	99	(6.00-34.54)	8.63-28.65	10.79-24.84	12.92-21.72	14.07-20.25
18	95	8.58-31.44	11.06-26.92	12.96-24.02	14.74-21.64	15.67-20.52
	99	6.60-35.73	9.37-29.80	11.61-25.96	13.81-22.81	14.99-21.32
19	95	(9.31-32.58)	11.86-28.06	13.82-25.12	15.66-22.71	16.62-21.57
	99	(7.23-36.88)	10.10-30.96	12.43-27.09	14.71-23.90	15.92-22.38
20	95	10.04-33.72	12.66-29.19	14.69-26.22	16.58-23.78	17.56-22.62
	99	7.86-38.04	10.84-32.12	13.26-28.22	15.60-24.99	16.84-23.45
21	95	(10.79-34.84)	13.51-30.28	15.58-27.30	17.52-24.83	18.52-23.65
	99	(8.53-39.18)	11.63-33.24	14.11-29.31	16.51-26.05	17.78-24.50
22	95	11.54-35.95	14.35-31.37	16.48-28.37	18.45-25.88	19.47-24.69
	99	9.20-40.32	12.41-34.35	14.97-30.40	17.43-27.12	18.72-25.55
23	95	(12.30-37.06)	15.19-32.47	17.37-29.45	19.39-26.93	20.43-25.73
	99	(9.88-41.44)	13.60-34.82	15.83-31.50	18.34-28.18	19.67-26.59
24	95	13.07-38.17	16.03-33.56	18.27-30.52	20.33-27.99	21.39-26.77
	99	10.56-42.56	13.98-36.57	16.68-32.59	19.26-29.25	20.61-27.64
25	95	(13.84-39.27)	16.88-34.66	19.16-31.60	21.26-29.04	22.34-27.81
	99	(11.25-43.65)	14.77-37.69	17.54-33.68	20.17-30.31	21.55-28.69
26	95	14.63-40.34	17.75-35.72	20.08-32.65	22.21-30.08	23.31-28.83
	99	11.98-44.73	15.59-38.76	18.43-34.75	21.10-31.36	22.50-29.73
27	95	(15.45-41.40)	18.62-36.79	20.99-33.70	23.16-31.11	24.27-29.86
	99	(12.71-45.79)	16.42-39.84	19.31-35.81	22.04-32.41	23.46-30.76
28	95	16.23-42.48	19.50-37.85	21.91-34.76	24.11-32.15	25.24-30.89
	99	13.42-46.88	17.25-40.91	20.20-36.88	22.97-33.46	24.41-31.80
29	95	(17.06-43.54)	20.37-38.92	22.82-35.81	25.06-33.19	26.21-31.92
	99	(14.18-47.92)	18.07-41.99	21.08-37.94	23.90-34.51	25.37-32.84
30	95	17.87-44.61	21.24-39.98	23.74-36.87	26.01-34.23	27.17-32.95
	99	14.91-48.99	18.90-43.06	21.97-39.01	24.83-35.55	26.32-33.87

TABLE 23 Confidence limits for percentages

%	Confidence coefficients	n: 50	100	200	500	1000
31	95	(18.71-45.65)	22.14-41.02	24.67-37.90	26.97-35.25	28.15-33.97
	99	(15.68-50.02)	19.76-44.11	22.88-40.05	25.78-36.59	27.29-34.90
32	95	19.55-46.68	23.04-42.06	25.61-38.94	27.93-36.28	29.12-34.99
	99	16.46-51.05	20.61-45.15	23.79-41.09	26.73-37.62	28.25-35.92
33	95	(20.38-47.72)	23.93-43.10	26.54-39.97	28.90-37.31	30.09-36.01
	99	(17.23-52.08)	21.47-46.19	24.69-42.13	27.68-38.65	29.22-36.95
34	95	21.22-48.76	24.83-44.15	27.47-41.01	29.86-38.33	31.07-37.03
	99	18.01-53.11	22.33-47.24	25.60-43.18	28.62-39.69	30.18-37.97
35	95	(22.06-49.80)	25.73-45.19	28.41-42.04	30.82-39.36	32.04-38.05
	99	(18.78-54.14)	23.19-48.28	26.51-44.22	29.57-40.72	31.14-39.00
36	95	22.93-50.80	26.65-46.20	29.36-43.06	31.79-40.38	33.02-39.06
	99	19.60-55.13	24.08-49.30	27.44-45.24	30.53-41.74	32.12-40.02
37	95	(23.80-51.81)	27.57-47.22	30.31-44.08	32.76-41.39	34.00-40.07
	99	(20.42-56.12)	24.96-50.31	28.37-46.26	31.49-42.76	33.09-41.03
38	95	24.67-52.81	28.49-48.24	31.25-45.10	33.73-42.41	34.98-41.09
	99	21.23-57.10	25.85-51.32	29.30-47.29	32.45-43.78	34.07-42.05
39	95	(25.54-53.82)	29.41-49.26	32.20-46.12	34.70-43.43	35.97-42.10
	99	(22.05-58.09)	26.74-52.34	30.23-48.31	33.42-44.80	35.04-43.06
40	95	26.41-54.82	30.33-50.28	33.15-47.14	35.68-44.44	36.95-43.11
	99	22.87-59.08	27.63-53.35	31.16-49.33	34.38-45.82	36.02-44.08
41	95	(27.31-55.80)	31.27-51.28	34.12-48.15	36.66-45.45	37.93-44.12
	99	(23.72-60.04)	28.54-54.34	32.11-50.33	35.35-46.83	37.00-45.09
42	95	28.21-56.78	32.21-52.28	35.08-49.16	37.64-46.46	38.92-45.12
	99	24.57-60.99	29.45-55.33	33.06-51.33	36.32-47.83	37.98-46.10
43	95	(29.10-57.76)	33.15-53.27	36.05-50.16	38.62-47.46	39.91-46.13
	99	(25.42-61.95)	30.37-56.32	34.01-52.34	37.29-48.84	38.96-47.10
44	95	30.00-58.74	34.09-54.27	37.01-51.17	39.60-48.47	40.90-47.14
	99	26.27-62.90	31.28-57.31	34.95-53.34	38.27-49.85	39.95-48.11
45	95	(30.90-59.71)	35.03-55.27	37.97-52.17	40.58-49.48	41.89-48.14
	99	(27.12-63.86)	32.19-58.30	35.90-54.34	39.24-50.86	40.93-49.12
46	95	31.83-60.67	35.99-56.25	38.95-53.17	41.57-50.48	42.88-49.14
	99	28.00-64.78	33.13-59.26	36.87-55.33	40.22-51.85	41.92-50.12
47	95	(32.75-61.62)	36.95-57.23	39.93-54.16	42.56-51.48	43.87-50.14
	99	(28.89-65.69)	34.07-60.22	37.84-56.31	41.21-52.85	42.91-51.12
48	95	33.68-62.57	37.91-58.21	40.91-55.15	43.55-52.47	44.87-51.14
	99	29.78-66.61	35.01-61.19	38.80-57.30	42.19-53.85	43.90-52.12
49	95	(34.61-63.52)	38.87-59.19	41.89-56.14	44.54-53.47	45.86-52.14
	99	(30.67-67.53)	35.95-62.15	39.77-58.28	43.18-54.84	44.89-53.12
50	95	35.53-64.47	39.83-60.17	42.86-57.14	45.53-54.47	46.85-53.15
	99	31.55-68.45	36.89-63.11	40.74-59.26	44.16-55.84	45.89-54.11

TABLE 24 Relative expected frequencies for individual terms of the Poisson distribution

This table furnishes expected frequencies for individual terms of the Poisson distribution. These are given for parametric values of the mean number of observations μ ranging from 0.1 to 1.0 in increments of 0.1, from 1.0 to 2.0 in increments of 0.2, from 2 to 7 in increments of 0.5, and from 7 to 12 in increments of 1.0. The function is given as a proportion to six significant decimal places for all counts Y that have a minimum expected probability of occurrence of 0.000001.

To determine the probability of the occurrence of a count of $Y = 4$ events in a Poisson distribution with mean $\mu = 1.9$, we enter the table in the column for $\mu = 1.6$ and find the row for 4. The probability of occurrence is 0.05513. Linear interpolation will generally be adequate for values of $\mu > 1.0$. For values of $\mu < 1.0$, it is best to work out the expectations as shown in Section 5.3 or to consult more extensive tables, such as table 9.3 in D. B. Owen, *Handbook of Statistical Tables* (Addison-Wesley, Reading, Mass., 1962). This is a cumulative distribution table, and to obtain the individual terms one needs differences between successive cumulative frequencies. It may be easier to work out the expectations for values of μ intermediate between tabled arguments on a calculator rather than use interpolation.

This table is used for computing expected frequencies of the Poisson distribution as shown in Section 5.3.

The table was computed by means of a FORTRAN IV program written to evaluate expression $e^{-\mu}\mu^{Y}/Y!$ for the terms of the Poisson distribution with mean μ. FORTRAN IV library functions and single precision arithmetic were employed.

TABLE 24 Relative expected frequencies for individual terms of the Poisson distribution

Y \ μ	0.1	0.2	0.3	0.4	0.5	0.6	0.7	0.8	0.9	1.0
0	0.90484	0.81873	0.74082	0.67032	0.60653	0.54881	0.49659	0.44933	0.40657	0.36788
1	0.09048	0.16375	0.22225	0.26813	0.30327	0.32929	0.34761	0.35946	0.36591	0.36788
2	0.00452	0.01637	0.03334	0.05363	0.07582	0.09879	0.12166	0.14379	0.16466	0.18394
3	0.00015	0.00109	0.00333	0.00715	0.01264	0.01976	0.02839	0.03834	0.04940	0.06131
4	0.00000	0.00005	0.00025	0.00072	0.00158	0.00296	0.00497	0.00767	0.01111	0.01533
5		0.00000	0.00002	0.00006	0.00016	0.00036	0.00070	0.00123	0.00200	0.00307
6			0.00000	0.00000	0.00001	0.00004	0.00008	0.00016	0.00030	0.00051
7					0.00000	0.00000	0.00001	0.00002	0.00004	0.00007
8							0.00000	0.00000	0.00000	0.00001
9										0.00000

Y \ μ	1.2	1.4	1.6	1.8	2	2.5	3	3.5	4	4.5
0	0.30119	0.24660	0.20190	0.16530	0.13534	0.08209	0.04979	0.03020	0.01832	0.01111
1	0.36143	0.34524	0.32303	0.29754	0.27067	0.20521	0.14936	0.10569	0.07326	0.04999
2	0.21686	0.24166	0.25843	0.26778	0.27067	0.25652	0.22404	0.18496	0.14653	0.11248
3	0.08674	0.11278	0.13783	0.16067	0.18045	0.21376	0.22404	0.21579	0.19537	0.16872
4	0.02602	0.03947	0.05513	0.07230	0.09022	0.13360	0.16803	0.18881	0.19537	0.18981
5	0.00625	0.01105	0.01764	0.02603	0.03609	0.06680	0.10082	0.13217	0.15629	0.17083
6	0.00125	0.00258	0.00470	0.00781	0.01203	0.02783	0.05041	0.07710	0.10420	0.12812
7	0.00021	0.00052	0.00108	0.00201	0.00344	0.00994	0.02160	0.03855	0.05954	0.08236
8	0.00003	0.00009	0.00022	0.00045	0.00086	0.00311	0.00810	0.01687	0.02977	0.04633
9	0.00000	0.00001	0.00004	0.00009	0.00019	0.00086	0.00270	0.00656	0.01323	0.02316
10		0.00000	0.00001	0.00002	0.00004	0.00022	0.00081	0.00230	0.00529	0.01042
11			0.00000	0.00000	0.00001	0.00005	0.00022	0.00073	0.00192	0.00426
12					0.00000	0.00001	0.00006	0.00021	0.00064	0.00160
13						0.00000	0.00001	0.00006	0.00020	0.00055
14							0.00000	0.00001	0.00006	0.00018
15								0.00000	0.00002	0.00005
16									0.00000	0.00002
17										0.00000

Y \ μ	5	5.5	6	6.5	7	8	9	10	11	12
0	0.00674	0.00409	0.00248	0.00150	0.00091	0.00034	0.00012	0.00005	0.00002	0.00001
1	0.03369	0.02248	0.01487	0.00977	0.00638	0.00268	0.00111	0.00045	0.00018	0.00007
2	0.08422	0.06181	0.04462	0.03176	0.02234	0.01073	0.00500	0.00227	0.00101	0.00044
3	0.14037	0.11332	0.08924	0.06881	0.05213	0.02863	0.01499	0.00757	0.00370	0.00177
4	0.17547	0.15582	0.13385	0.11182	0.09123	0.05725	0.03374	0.01892	0.01019	0.00531
5	0.17547	0.17140	0.16062	0.14537	0.12772	0.09160	0.06073	0.03783	0.02242	0.01274
6	0.14622	0.15712	0.16062	0.15748	0.14900	0.12214	0.09109	0.06306	0.04109	0.02548
7	0.10444	0.12345	0.13768	0.14623	0.14900	0.13959	0.11712	0.09008	0.06458	0.04368
8	0.06528	0.08487	0.10326	0.11882	0.13038	0.13959	0.13176	0.11260	0.08879	0.06552
9	0.03627	0.05187	0.06884	0.08581	0.10140	0.12408	0.13176	0.12511	0.10853	0.08736
10	0.01813	0.02853	0.04130	0.05578	0.07098	0.09926	0.11858	0.12511	0.11938	0.10484
11	0.00824	0.01426	0.02253	0.03296	0.04517	0.07219	0.09702	0.11374	0.11938	0.11437
12	0.00343	0.00654	0.01126	0.01785	0.02635	0.04813	0.07276	0.09478	0.10943	0.11437
13	0.00132	0.00277	0.00520	0.00893	0.01419	0.02962	0.05038	0.07291	0.09259	0.10557
14	0.00047	0.00109	0.00223	0.00414	0.00709	0.01692	0.03238	0.05208	0.07275	0.09049
15	0.00016	0.00040	0.00089	0.00180	0.00331	0.00903	0.01943	0.03472	0.05335	0.07239
16	0.00005	0.00014	0.00033	0.00073	0.00145	0.00451	0.01093	0.02170	0.03668	0.05429
17	0.00001	0.00004	0.00012	0.00028	0.00060	0.00212	0.00579	0.01276	0.02373	0.03832
18	0.00000	0.00001	0.00004	0.00010	0.00023	0.00094	0.00289	0.00709	0.01450	0.02555
		0.00000	0.00001	0.00003	0.00009	0.00040	0.00137	0.00373	0.00840	0.01614
			0.00000	0.00001	0.00003	0.00016	0.00062	0.00187	0.00462	0.00968
				0.00000	0.00001	0.00006	0.00026	0.00089	0.00242	0.00553
					0.00000	0.00002	0.00011	0.00040	0.00121	0.00302
						0.00001	0.00004	0.00018	0.00058	0.00157
						0.00000	0.00002	0.00007	0.00027	0.00079
							0.00001	0.00003	0.00012	0.00038
							0.00000	0.00001	0.00005	0.00017
								0.00000	0.00002	0.00008
									0.00001	0.00003
									0.00000	0.00001
										0.00000

TABLE **25** Critical values for correlation coefficients

This table furnishes 0.05 and 0.01 critical values for product–moment correlation coefficients r and multiple correlation coefficients R. These values are given for every degree of freedom between $\nu = 1$ and $\nu = 30$ and selected degrees of freedom between $\nu = 30$ and $\nu = 1000$. This table is used to test the null hypothesis that the correlation coefficient of the population from which the sample has been taken is zero. Under such conditions the following t-test for significance of the product–moment correlation coefficient r applies:

$$t = r/\sqrt{(1 - r^2)/(n - 2)} = r\sqrt{(n - 2)/(1 - r^2)} \qquad \nu = n - 2$$

where n is the sample size (number of pairs of variates). The critical values in the table are computed by entering the correct values of $t_{\alpha[\nu]}$ and n in the above equation and solving for r. The critical values for the multiple correlation coefficients are based on a different formula involving the F-distribution,

$$R_{Y \cdot 1 \ldots k(\alpha)} = \left[\frac{k\,F_{\alpha[k,\, n - m]}}{n - m + k\,F_{\alpha[k,\, n - m]}} \right]^{1/2}$$

where $R_{Y \cdot 1 \ldots k}$ is the multiple correlation coefficient of Y with k other variables, n is the sample size, m equals $k + 1$, and $F_{\alpha[k,\, n-m]}$ is the critical value of the F-distribution as shown in Table **16**.

To test the significance of a correlation coefficient, the sample size n upon which it is based must be known. Enter the table for $\nu = n - 2$ degrees of freedom and consult the first column of values headed "number of independent variables." For example, for a sample size of $n = 28$ and $\nu = 28 - 2 = 26$, the critical values of r are found to be 0.374 at the 5% level and 0.478 at the 1% level. Thus, for an observed correlation coefficient $r = 0.31$ in a sample of 28 paired observations, one would be led to conclude that the correlation between the variables concerned is not significantly different from zero. Negative correlations are considered as positive for purposes of this test. More details on significance testing of correlation coefficients are given in Section 15.5 and Box 15.3.

The other three columns in the table give critical values for a multiple correlation coefficient involving 2, 3, and 4 independent variables. Degrees of freedom for such a problem are $\nu = n - m$, where n is the sample size and m is the number of variables, both dependent and independent. Thus, for a sample value of $R = 0.42$ based on a sample of 50 items and measurements of four variables (one dependent plus three independent), one would conclude that it is significant at $P = 0.05$ but not at $P = 0.01$. The appropriate degrees of freedom are $50 - 4 = 46$, requiring interpolation, but since these conclusions are true for both $\nu = 45$ and $\nu = 50$, bracketing the correct value for the degrees of freedom, one need not interpolate. Further details are given in Section 16.4.

TABLE 25 Critical values for correlation coefficients

ν	α	k = 1	k = 2	k = 3	k = 4	ν	α	k = 1	k = 2	k = 3	k = 4
		Number of independent variables						Number of independent variables			
1	.05	.997	.999	.999	.999	24	.05	.388	.470	.523	.562
	.01	1.000	1.000	1.000	1.000		.01	.496	.565	.609	.642
2	.05	.950	.975	.983	.987	25	.05	.381	.462	.514	.553
	.01	.990	.995	.997	.998		.01	.487	.555	.600	.633
3	.05	.878	.930	.950	.961	26	.05	.374	.454	.506	.545
	.01	.959	.976	.983	.987		.01	.478	.546	.590	.624
4	.05	.811	.881	.912	.930	27	.05	.367	.446	.498	.536
	.01	.917	.949	.962	.970		.01	.470	.538	.582	.615
5	.05	.754	.836	.874	.898	28	.05	.361	.439	.490	.529
	.01	.874	.917	.937	.949		.01	.463	.530	.573	.606
6	.05	.707	.795	.839	.867	29	.05	.355	.432	.482	.521
	.01	.834	.886	.911	.927		.01	.456	.522	.565	.598
7	.05	.666	.758	.807	.838	30	.05	.349	.426	.476	.514
	.01	.798	.855	.885	.904		.01	.449	.514	.558	.591
8	.05	.632	.726	.777	.811	35	.05	.325	.397	.445	.482
	.01	.765	.827	.860	.882		.01	.418	.481	.523	.556
9	.05	.602	.697	.750	.786	40	.05	.304	.373	.419	.455
	.01	.735	.800	.836	.861		.01	.393	.454	.494	.526
10	.05	.576	.671	.726	.763	45	.05	.288	.353	.397	.432
	.01	.708	.776	.814	.840		.01	.372	.430	.470	.501
11	.05	.553	.648	.703	.741	50	.05	.273	.336	.379	.412
	.01	.684	.753	.793	.821		.01	.354	.410	.449	.479
12	.05	.532	.627	.683	.722	60	.05	.250	.308	.348	.380
	.01	.661	.732	.773	.802		.01	.325	.377	.414	.442
13	.05	.514	.608	.664	.703	70	.05	.232	.286	.324	.354
	.01	.641	.712	.755	.785		.01	.302	.351	.386	.413
14	.05	.497	.590	.646	.686	80	.05	.217	.269	.304	.332
	.01	.623	.694	.737	.768		.01	.283	.330	.362	.389
15	.05	.482	.574	.630	.670	90	.05	.205	.254	.288	.315
	.01	.606	.677	.721	.752		.01	.267	.312	.343	.368
16	.05	.468	.559	.615	.655	100	.05	.195	.241	.274	.300
	.01	.590	.662	.706	.738		.01	.254	.297	.327	.351
17	.05	.456	.545	.601	.641	125	.05	.174	.216	.246	.269
	.01	.575	.647	.691	.724		.01	.228	.266	.294	.316
18	.05	.444	.532	.587	.628	150	.05	.159	.198	.225	.247
	.01	.561	.633	.678	.710		.01	.208	.244	.270	.290
19	.05	.433	.520	.575	.615	200	.05	.138	.172	.196	.215
	.01	.549	.620	.665	.698		.01	.181	.212	.234	.253
20	.05	.423	.509	.563	.604	300	.05	.113	.141	.160	.176
	.01	.537	.608	.652	.685		.01	.148	.174	.192	.208
21	.05	.413	.498	.522	.592	400	.05	.098	.122	.139	.153
	.01	.526	.596	.641	.674		.01	.128	.151	.167	.180
22	.05	.404	.488	.542	.582	500	.05	.088	.109	.124	.137
	.01	.515	.585	.630	.663		.01	.115	.135	.150	.162
23	.05	.396	.479	.532	.572	1,000	.05	.062	.077	.088	.097
	.01	.505	.574	.619	.652		.01	.081	.096	.106	.115

TABLE **26** Mean ranges of samples from a normal distribution

This table features the expected values (means) of the ranges of samples of varying sizes n from a standard normal distribution ($\mu = 0$, $\sigma = 1$, items normally distributed). The means are given, exact to three decimal places, for sample sizes $n = 2$ to $n = 200$ in increments of 1, and for sample sizes $n = 200$ to $n = 1000$ in increments of 10. The function is in standard deviation units.

To look up the mean range for a sample of size n, find the function corresponding to the argument in the margin. For example, the mean range for a sample of $n = 173$ is 5.395, and for a sample of $n = 713$ it is 6.286 by linear interpolation between 6.283 and 6.292, the values for $n = 710$ and 720, respectively. Note that the columns after $n = 200$ represent increments of 10. The table of mean ranges is used in the estimation of population standard deviations from sample ranges (see Section 4.9).

Values for this table have been rounded and taken from the more extensive table 27 (given to five decimal places) in E. S. Pearson and H. O. Hartley, *Biometrika Tables for Statisticians,* Vol. I (Cambridge University Press, 1958) with permission of the publishers.

TABLE 26 Mean ranges of samples from a normal distribution

n	0	1	2	3	4	5	6	7	8	9
0			1.128	1.693	2.059	2.326	2.534	2.704	2.847	2.970
10	3.078	3.173	3.258	3.336	3.407	3.472	3.532	3.588	3.640	3.689
20	3.735	3.778	3.819	3.858	3.895	3.931	3.964	3.997	4.027	4.057
30	4.086	4.113	4.139	4.165	4.189	4.213	4.236	4.259	4.280	4.301
40	4.322	4.341	4.361	4.379	4.398	4.415	4.433	4.450	4.466	4.482
50	4.498	4.514	4.529	4.543	4.558	4.572	4.586	4.599	4.613	4.626
60	4.639	4.651	4.663	4.676	4.687	4.699	4.711	4.722	4.733	4.744
70	4.755	4.765	4.776	4.786	4.796	4.806	4.816	4.825	4.835	4.844
80	4.854	4.863	4.872	4.881	4.889	4.898	4.906	4.915	4.923	4.931
90	4.939	4.947	4.955	4.963	4.971	4.978	4.986	4.993	5.001	5.008
100	5.015	5.022	5.029	5.036	5.043	5.050	5.057	5.063	5.070	5.076
110	5.083	5.089	5.096	5.102	5.108	5.114	5.120	5.126	5.132	5.138
120	5.144	5.150	5.156	5.161	5.167	5.173	5.178	5.184	5.189	5.195
130	5.200	5.205	5.211	5.216	5.221	5.226	5.231	5.236	5.241	5.246
140	5.251	5.256	5.261	5.266	5.271	5.275	5.280	5.285	5.289	5.294
150	5.298	5.303	5.308	5.312	5.316	5.321	5.325	5.330	5.334	5.338
160	5.342	5.347	5.351	5.355	5.359	5.363	5.367	5.371	5.375	5.379
170	5.383	5.387	5.391	5.395	5.399	5.403	5.407	5.411	5.414	5.418
180	5.422	5.426	5.429	5.433	5.437	5.440	5.444	5.447	5.451	5.454
190	5.458	5.461	5.465	5.468	5.472	5.475	5.479	5.482	5.485	5.489

n	0	10	20	30	40	50	60	70	80	90
200	5.492	5.524	5.555	5.584	5.612	5.638	5.664	5.688	5.771	5.734
300	5.756	5.776	5.797	5.816	5.835	5.853	5.871	5.888	5.904	5.921
400	5.936	5.952	5.967	5.981	5.995	6.009	6.023	6.036	6.049	6.061
500	6.073	6.085	6.097	6.109	6.120	6.131	6.142	6.153	6.163	6.173
600	6.183	6.193	6.203	6.213	6.222	6.231	6.240	6.249	6.258	6.267
700	6.275	6.283	6.292	6.300	6.308	6.316	6.324	6.331	6.339	6.346
800	6.354	6.361	6.368	6.375	6.382	6.389	6.396	6.402	6.409	6.416
900	6.422	6.429	6.435	6.441	6.447	6.453	6.459	6.465	6.471	6.477
1000	6.483									

TABLE **27** Rankits (normal order statistics)

This table lists the expected values of ranked normal deviates, for each position in the rank order, of samples ranging in size from $n = 2$ to $n = 50$. These values, also known as rankits, can be thought to have been obtained as follows. If one samples repeatedly n items from a standard normal population ($\mu = 0$, $\sigma = 1$, items normally distributed) and ranks the items in each sample by order of magnitude, the first rankit value would be the mean of the first item from every sample. The second rankit would be the mean of the second item from every sample, and so forth. The rankits, in population standard deviation units, are given to three significant decimal places for sample sizes up to $n = 20$, and to two decimal places beyond that number.

Rankits are symmetrical about their median. The table shows only the rankits greater than their median. For odd-numbered samples the median rankit is zero and is also shown. For example, the rankits for sample size $n = 3$ are (from the left to the right tail of the distribution) -0.846, 0.000, 0.846, and for $n = 4$ they are -1.029, -0.297, 0.297, 1.029. Thus, the second largest item in a normally distributed sample of 4 items is on the average 0.297 standard deviations above the mean. However, in a sample of 40 items the second largest item will be 1.75 standard deviations above the mean.

Rankits are used in the graphic analysis of small samples (tests for normality and estimation of mean and standard deviation; see Section 6.7 and Box 6.4).

Values in this table for sample sizes $n = 2$ to $n = 20$ were taken from table 28 in E. S. Pearson and H. O. Hartley, *Biometrika Tables for Statisticians,* Vol. I (Cambridge University Press, 1958) with permission of the publishers. The values between $n = 21$ and $n = 50$ are reproduced from table XX in R. A. Fisher and F. Yates, *Statistical Tables for Biological, Agricultural and Medical Research,* 5th ed. (Oliver & Boyd, Edinburgh, 1958) with permission of the authors and their publishers.

TABLE 27 Rankits (normal order statistics)

Rank order \ n	2	3	4	5	6	7	8	9	10
1	0.564	0.846	1.029	1.163	1.267	1.352	1.424	1.485	1.539
2		.000	.297	.495	.642	.757	.852	.932	1.001
3				.000	.202	.353	.473	.572	.656
4						.000	.153	.275	.376
5								.000	.123

Rank order \ n	11	12	13	14	15	16	17	18	19	20
1	1.586	1.629	1.668	1.703	1.736	1.766	1.794	1.820	1.844	1.867
2	1.062	1.116	1.164	1.208	1.248	1.285	1.319	1.350	1.380	1.408
3	.729	.793	.850	.901	.948	.990	1.029	1.066	1.099	1.131
4	.462	.537	.603	.662	.715	.763	.807	.848	.886	.921
5	.225	.312	.388	.456	.516	.570	.619	.665	.707	.745
6	.000	.103	.190	.267	.335	.396	.451	.502	.548	.590
7			.000	.088	.165	.234	.295	.351	.402	.448
8					.000	.077	.146	.208	.264	.315
9							.000	.069	.131	.187
10									.000	.062

Rank order \ n	21	22	23	24	25	26	27	28	29	30
1	1.89	1.91	1.93	1.95	1.97	1.98	2.00	2.01	2.03	2.04
2	1.43	1.46	1.48	1.50	1.52	1.54	1.56	1.58	1.60	1.62
3	1.16	1.19	1.21	1.24	1.26	1.29	1.31	1.33	1.35	1.36
4	.95	.98	1.01	1.04	1.07	1.09	1.11	1.14	1.16	1.18
5	.78	.82	.85	.88	.91	.93	.96	.98	1.00	1.03
6	.63	.67	.70	.73	.76	.79	.82	.85	.87	.89
7	.49	.53	.57	.60	.64	.67	.70	.73	.75	.78
8	.36	.41	.45	.48	.52	.55	.58	.61	.64	.67
9	.24	.29	.33	.37	.41	.44	.48	.51	.54	.57
10	.12	.17	.22	.26	.30	.34	.38	.41	.44	.47
11	.00	.06	.11	.16	.20	.24	.28	.32	.35	.38
12			.00	.05	.10	.14	.19	.22	.26	.29
13					.00	.05	.09	.13	.17	.21
14							.00	.04	.09	.12
15									.00	.04

TABLE 27 Rankits (normal order statistics)

Rank order \ n	31	32	33	34	35	36	37	38	39	40
1	2.06	2.07	2.08	2.09	2.11	2.12	2.13	2.14	2.15	2.16
2	1.63	1.65	1.66	1.68	1.69	1.70	1.72	1.73	1.74	1.75
3	1.38	1.40	1.42	1.43	1.45	1.46	1.48	1.49	1.50	1.52
4	1.20	1.22	1.23	1.25	1.27	1.28	1.30	1.32	1.33	1.34
5	1.05	1.07	1.09	1.11	1.12	1.14	1.16	1.17	1.19	1.20
6	.92	.94	.96	.98	1.00	1.02	1.03	1.05	1.07	1.08
7	.80	.82	.85	.87	.89	.91	.92	.94	.96	.98
8	.69	.72	.74	.76	.79	.81	.83	.85	.86	.88
9	.60	.62	.65	.67	.69	.72	.73	.75	.77	.79
10	.50	.53	.56	.58	.60	.63	.65	.67	.69	.71
11	.41	.44	.47	.50	.52	.54	.57	.59	.61	.63
12	.33	.36	.39	.41	.44	.47	.49	.51	.54	.56
13	.24	.28	.31	.34	.36	.39	.42	.44	.46	.49
14	.16	.20	.23	.26	.29	.32	.34	.37	.39	.42
15	.08	.12	.15	.18	.22	.24	.27	.30	.33	.35
16	.00	.04	.08	.11	.14	.17	.20	.23	.26	.28
17			.00	.04	.07	.10	.14	.16	.19	.22
18					.00	.03	.07	.10	.13	.16
19							.00	.03	.06	.09
20									.00	.03

Rank order \ n	41	42	43	44	45	46	47	48	49	50
1	2.17	2.18	2.19	2.20	2.21	2.22	2.22	2.23	2.24	2.25
2	1.76	1.78	1.79	1.80	1.81	1.82	1.83	1.84	1.85	1.85
3	1.53	1.54	1.55	1.57	1.58	1.59	1.60	1.61	1.62	1.63
4	1.36	1.37	1.38	1.40	1.41	1.42	1.43	1.44	1.45	1.46
5	1.22	1.23	1.25	1.26	1.27	1.28	1.30	1.31	1.32	1.33
6	1.10	1.11	1.13	1.14	1.16	1.17	1.18	1.19	1.21	1.22
7	.99	1.01	1.02	1.04	1.05	1.07	1.08	1.09	1.11	1.12
8	.90	.91	.93	.95	.96	.98	.99	1.00	1.02	1.03
9	.81	.83	.84	.86	.88	.89	.91	.92	.94	.95
10	.73	.75	.76	.78	.80	.81	.83	.84	.86	.87
11	.65	.67	.69	.71	.72	.74	.76	.77	.79	.80
12	.58	.60	.62	.64	.65	.67	.69	.70	.72	.74
13	.51	.53	.55	.57	.59	.60	.62	.64	.66	.67
14	.44	.46	.48	.50	.52	.54	.56	.58	.59	.61
15	.37	.40	.42	.44	.46	.48	.50	.52	.53	.55
16	.31	.33	.36	.38	.40	.42	.44	.46	.48	.49
17	.25	.27	.29	.32	.34	.36	.38	.40	.42	.44
18	.18	.21	.23	.26	.28	.30	.32	.34	.36	.38
19	.12	.15	.17	.20	.22	.25	.27	.29	.31	.33
20	.06	.09	.12	.14	.17	.19	.21	.24	.26	.28
21	.00	.03	.06	.09	.11	.14	.16	.18	.21	.23
22			.00	.03	.06	.08	.11	.13	.15	.18
23					.00	.03	.05	.08	.10	.13
24							.00	.03	.05	.08
25									.00	.03

TABLE **28** Critical values of the number of runs

This table furnishes the critical number of runs (sequences of one or more like elements preceded and/or followed by unlike elements) in a sequence of $n_1 + n_2$ randomly arranged items, n_1 of one kind, n_2 of another. The critical values of α listed across the top of the table are 0.005, 0.01, 0.025, 0.05, 0.95, 0.975, 0.99, and 0.995. Any number of runs that is *equal to or less than* the desired critical value at the left half of any row in the table or is *equal to or greater than* the desired critical value in the right half of any row leads to a rejection of the hypothesis of random arrangement. The table is divided into two parts: the first permits n_1 and n_2 to vary up to 20 and the sample sizes need not be equal. The second part is for equal sample sizes ($n_1 = n_2$) from 10 to 100.

To look up values significant at 5% in this table, proceed as follows. If in a sequence of 27 items there are $n_1 = 10$ items of type A and $n_2 = 17$ of type B and 20 runs, enter the first part of Table **28** for $n_1 = 10$ and $n_2 = 17$. Always make the smaller sample size equal n_1. The critical values in the table for the percentage points 0.025 and 0.975 are 8 and 19, respectively. Thus as few as 8 runs (or fewer) would indicate a significant departure from randomness, as would 19 or more runs. The above procedure implied a two-tailed significance test in which the alternative hypothesis was either too few or too many runs. For this reason critical percentage points corresponding to proportion $\alpha/2$ and $1 - \alpha/2$ are chosen when one is prepared to accept a type I error of α. For tests of one-tailed hypotheses use critical percentage points corresponding to α or $1 - \alpha$, depending on the tail of the distribution being tested. Minus signs in the left half of a row and asterisks in the right half of a row indicate cases in the first part of the table where no critical values of runs are possible. The second part of the table is looked up similarly. Thus for a sample of 100 items, of which 50 are of type A and 50 are of type B, the 0.01 critical values, 37 and 65 runs, respectively, can be found in the row for $n_1 = n_2 = 50$.

For sample sizes exceeding those listed in this table, employ normal approximations given in Box 18.2 (R. R. Sokal and F. J. Rohlf, *Biometry,* 2nd ed., W. H. Freeman and Company, 1981).

This table is used to evaluate tests of randomness by means of runs tests as discussed and illustrated in Section 18.2.

This table has been copied from tables II and III by F. S. Swed and C. Eisenhart (*Ann. Math. Stat.,* **14**:66–87, 1943), except that 1 has been added to the values in the right half of each row to make the critical values in this table compatible in formulation with those of other statistical tables. Inconsistencies pointed out by these authors in five values of their table III have been corrected.

TABLE 28 Critical values of the number of runs

For sample sizes up to $n_1 \leq n_2 = 20$

n_1	n_2	Cumulative probability 0.005	0.01	0.025	0.05	0.95	0.975	0.99	0.995
2	2	-	-	-	-	*	*	*	*
	3	-	-	-	-	*	*	*	*
	4	-	-	-	-	*	*	*	*
	5	-	-	-	-	*	*	*	*
	6	-	-	-	-	*	*	*	*
	7	-	-	-	-	*	*	*	*
	8	-	-	-	2	*	*	*	*
	9	-	-	-	2	*	*	*	*
	10	-	-	-	2	*	*	*	*
	11	-	-	-	2	*	*	*	*
	12	-	-	2	2	*	*	*	*
	13	-	-	2	2	*	*	*	*
	14	-	-	2	2	*	*	*	*
	15	-	-	2	2	*	*	*	*
	16	-	-	2	2	*	*	*	*
	17	-	-	2	2	*	*	*	*
	18	-	-	2	2	*	*	*	*
	19	-	2	2	2	*	*	*	*
	20	-	2	2	2	*	*	*	*
3	3	-	-	-	-	*	*	*	*
	4	-	-	-	-	7	*	*	*
	5	-	-	-	2	*	*	*	*
	6	-	-	2	2	*	*	*	*
	7	-	-	2	2	*	*	*	*
	8	-	-	2	2	*	*	*	*
	9	-	2	2	2	*	*	*	*
	10	-	2	2	3	*	*	*	*
	11	-	2	2	3	*	*	*	*
	12	2	2	2	3	*	*	*	*
	13	2	2	2	3	*	*	*	*
	14	2	2	2	3	*	*	*	*
	15	2	2	3	3	*	*	*	*
	16	2	2	3	3	*	*	*	*
	17	2	2	3	3	*	*	*	*
	18	2	2	3	3	*	*	*	*
	19	2	2	3	3	*	*	*	*
	20	2	2	3	3	*	*	*	*

TABLE 28 Critical values of the number of runs

For sample sizes up to $n_1 \leq n_2 = 20$

n_1	n_2	0.005	0.01	0.025	0.05	0.95	0.975	0.99	0.995
4	4	-	-	-	2	8	*	*	*
	5	-	-	2	2	9	9	9	*
	6	-	2	2	3	9	9	*	*
	7	-	2	2	3	9	*	*	*
	8	2	2	3	3	*	*	*	*
	9	2	2	3	3	*	*	*	*
	10	2	2	3	3	*	*	*	*
	11	2	2	3	3	*	*	*	*
	12	2	3	3	4	*	*	*	*
	13	2	3	3	4	*	*	*	*
	14	2	3	3	4	*	*	*	*
	15	3	3	3	4	*	*	*	*
	16	3	3	4	4	*	*	*	*
	17	3	3	4	4	*	*	*	*
	18	3	3	4	4	*	*	*	*
	19	3	3	4	4	*	*	*	*
	20	3	3	4	4	*	*	*	*
5	5	-	2	2	3	9	10	10	*
	6	2	2	3	3	10	10	11	11
	7	2	2	3	3	10	11	11	*
	8	2	2	3	3	11	11	*	*
	9	2	3	3	4	11	*	*	*
	10	3	3	3	4	11	*	*	*
	11	3	3	3	4	11	*	*	*
	12	3	3	3	4	11	*	*	*
	13	3	3	4	4	*	*	*	*
	14	3	3	4	5	*	*	*	*
	15	3	4	4	5	*	*	*	*
	16	3	4	4	5	*	*	*	*
	17	3	4	4	5	*	*	*	*
	18	4	4	5	5	*	*	*	*
	19	4	4	5	5	*	*	*	*
	20	4	4	5	5	*	*	*	*

TABLE **28** Critical values of the number of runs

For sample sizes up to $n_1 \leq n_2 = 20$

		Cumulative probability							
n_1	n_2	0.005	0.01	0.025	0.05	0.95	0.975	0.99	0.995
6	6	2	2	3	3	11	11	12	12
	7	2	3	3	4	11	12	12	13
	8	3	3	3	4	12	12	13	13
	9	3	3	4	4	12	13	13	*
	10	3	3	4	5	12	13	*	*
	11	3	4	4	5	13	13	*	*
	12	3	4	4	5	13	13	*	*
	13	3	4	5	5	13	*	*	*
	14	4	4	5	5	13	*	*	*
	15	4	4	5	6	*	*	*	*
	16	4	4	5	6	*	*	*	*
	17	4	5	5	6	*	*	*	*
	18	4	5	5	6	*	*	*	*
	19	4	5	6	6	*	*	*	*
	20	4	5	6	6	*	*	*	*
7	7	3	3	3	4	12	13	13	13
	8	3	3	4	4	13	13	14	14
	9	3	4	4	5	13	14	14	15
	10	3	4	5	5	13	14	15	15
	11	4	4	5	5	14	14	15	15
	12	4	4	5	6	14	14	15	*
	13	4	5	5	6	14	15	*	*
	14	4	5	5	6	14	15	*	*
	15	4	5	6	6	15	15	*	*
	16	5	5	6	6	15	*	*	*
	17	5	5	6	7	15	*	*	*
	18	5	5	6	7	15	*	*	*
	19	5	6	6	7	15	*	*	*
	20	5	6	6	7	15	*	*	*

TABLE 28 Critical values of the number of runs

For sample sizes up to $n_1 \leq n_2 = 20$

		Cumulative probability							
n_1	n_2	0.005	0.01	0.025	0.05	0.95	0.975	0.99	0.995
8	8	3	4	4	5	13	14	14	15
	9	3	4	5	5	14	14	15	15
	10	4	4	5	6	14	15	15	16
	11	4	5	5	6	15	15	16	16
	12	4	5	6	6	15	16	16	17
	13	5	5	6	6	15	16	17	17
	14	5	5	6	7	16	16	17	17
	15	5	5	6	7	16	16	17	*
	16	5	6	6	7	16	17	17	*
	17	5	6	7	7	16	17	*	*
	18	6	6	7	8	16	17	*	*
	19	6	6	7	8	16	17	*	*
	20	6	6	7	8	17	17	*	*
9	9	4	4	5	6	14	15	16	16
	10	4	5	5	6	15	16	16	17
	11	5	5	6	6	15	16	17	17
	12	5	5	6	7	16	16	17	18
	13	5	6	6	7	16	17	18	18
	14	5	6	7	7	17	17	18	18
	15	6	6	7	8	17	18	18	19
	16	6	6	7	8	17	18	18	19
	17	6	7	7	8	17	18	19	19
	18	6	7	8	8	18	18	19	*
	19	6	7	8	8	18	18	19	*
	20	7	7	8	9	18	18	19	*
10	10	5	5	6	6	16	16	17	17
	11	5	5	6	7	16	17	18	18
	12	5	6	7	7	17	17	18	19
	13	5	6	7	8	17	18	19	19
	14	6	6	7	8	17	18	19	19
	15	6	7	7	8	18	18	19	20
	16	6	7	8	8	18	19	20	20
	17	7	7	8	9	18	19	20	20
	18	7	7	8	9	19	19	20	21
	19	7	8	8	9	19	20	20	21
	20	7	8	9	9	19	20	20	21

TABLE 28 Critical values of the number of runs

For sample sizes up to $n_1 \leq n_2 = 20$

n_1	n_2	0.005	0.01	0.025	0.05	0.95	0.975	0.99	0.995
		Cumulative probability							
11	11	5	6	7	7	17	17	18	19
	12	6	6	7	8	17	18	19	19
	13	6	6	7	8	18	19	19	20
	14	6	7	8	8	18	19	20	20
	15	7	7	8	9	19	19	20	21
	16	7	7	8	9	19	20	21	21
	17	7	8	9	9	19	20	21	22
	18	7	8	9	10	20	20	21	22
	19	8	8	9	10	20	21	22	22
	20	8	8	9	10	20	21	22	22
12	12	6	7	7	8	18	19	19	20
	13	6	7	8	9	18	19	20	21
	14	7	7	8	9	19	20	21	21
	15	7	8	8	9	19	20	21	22
	16	7	8	9	10	20	21	22	22
	17	8	8	9	10	20	21	22	22
	18	8	8	9	10	21	21	22	23
	19	8	9	10	10	21	22	23	23
	20	8	9	10	11	21	22	23	23
13	13	7	7	8	9	19	20	21	21
	14	7	8	9	9	20	20	21	22
	15	7	8	9	10	20	21	22	22
	16	8	8	9	10	21	21	22	23
	17	8	9	10	10	21	22	23	23
	18	8	9	10	11	21	22	23	24
	19	9	9	10	11	22	23	24	24
	20	9	10	10	11	22	23	24	24
14	14	7	8	9	10	20	21	22	23
	15	8	8	9	10	21	22	23	23
	16	8	9	10	11	21	22	23	24
	17	8	9	10	11	22	23	24	24
	18	9	9	10	11	22	23	24	25
	19	9	10	11	12	23	23	24	25
	20	9	10	11	12	23	24	25	25

TABLE 28 Critical values of the number of runs

For sample sizes up to $n_1 \leq n_2 = 20$

		Cumulative probability							
n_1	n_2	0.005	0.01	0.025	0.05	0.95	0.975	0.99	0.995
15	15	8	9	10	11	21	22	23	24
	16	9	9	10	11	22	23	24	24
	17	9	10	11	11	22	23	24	25
	18	9	10	11	12	23	24	25	25
	19	10	10	11	12	23	24	25	26
	20	10	11	12	12	24	25	26	26
16	16	9	10	11	11	23	23	24	25
	17	9	10	11	12	23	24	25	26
	18	10	10	11	12	24	25	26	26
	19	10	11	12	13	24	25	26	27
	20	10	11	12	13	25	25	26	27
17	17	10	10	11	12	24	25	26	26
	18	10	11	12	13	24	25	26	27
	19	10	11	12	13	25	26	27	27
	20	11	11	13	13	25	26	27	28
18	18	11	11	12	13	25	26	27	27
	19	11	12	13	14	25	26	27	28
	20	11	12	13	14	26	27	28	29
19	19	11	12	13	14	26	27	28	29
	20	12	12	13	14	27	27	29	29
20	20	12	13	14	15	27	28	29	30

TABLE 28 Critical values of the number of runs

For equal sample sizes

	Cumulative probability							
n	0.005	0.01	0.025	0.05	0.95	0.975	0.99	0.995
10	5	5	6	6	16	16	17	17
11	5	6	7	7	17	17	18	19
12	6	6	7	8	18	19	20	20
13	7	7	8	9	19	20	21	21
14	7	8	9	10	20	21	22	23
15	8	9	10	11	21	22	23	24
16	9	10	11	11	23	23	24	25
17	10	10	11	12	24	25	26	26
18	10	11	12	13	25	26	27	28
19	11	12	13	14	26	27	28	29
20	12	13	14	15	27	28	29	30
21	13	14	15	16	28	29	30	31
22	14	14	16	17	29	30	32	32
23	14	15	16	17	31	32	33	34
24	15	16	17	18	32	33	34	35
25	16	17	18	19	33	34	35	36
26	17	18	19	20	34	35	36	37
27	18	19	20	21	35	36	37	38
28	18	19	21	22	36	37	39	40
29	19	20	22	23	37	38	40	41
30	20	21	22	24	38	40	41	42
31	21	22	23	25	39	41	42	43
32	22	23	24	25	41	42	43	44
33	23	24	25	26	42	43	44	45
34	23	24	26	27	43	44	46	47
35	24	25	27	28	44	45	47	48
36	25	26	28	29	45	46	48	49
37	26	27	29	30	46	47	49	50
38	27	28	30	31	47	48	50	51
39	28	29	30	32	48	50	51	52
40	29	30	31	33	49	51	52	53
41	29	31	32	34	50	52	53	55
42	30	31	33	35	51	53	55	56
43	31	32	34	35	53	54	56	57
44	32	33	35	36	54	55	57	58
45	33	34	36	37	55	56	58	59
46	34	35	37	38	56	57	59	60
47	35	36	38	39	57	58	60	61
48	35	37	38	40	58	60	61	63
49	36	38	39	41	59	61	62	64
50	37	38	40	42	60	62	64	65
51	38	39	41	43	61	63	65	66
52	39	40	42	44	62	64	66	67
53	40	41	43	45	63	65	67	68
54	41	42	44	45	65	66	68	69

TABLE 28 Critical values of the number of runs

For equal sample sizes

	Cumulative probability							
n	0.005	0.01	0.025	0.05	0.95	0.975	0.95	0.995
55	42	43	45	46	66	67	69	70
56	42	44	46	47	67	68	70	72
57	43	45	47	48	68	69	71	73
58	44	46	47	49	69	71	72	74
59	45	46	48	50	70	72	74	75
60	46	47	49	51	71	73	75	76
61	47	48	50	52	72	74	76	77
62	48	49	51	53	73	75	77	78
63	49	50	52	54	74	76	78	79
64	49	51	53	55	75	77	79	81
65	50	52	54	56	76	78	80	82
66	51	53	55	57	77	79	81	83
67	52	54	56	58	78	80	82	84
68	53	54	57	58	80	81	84	85
69	54	55	58	59	81	82	85	86
70	55	56	58	60	82	84	86	87
71	56	57	59	61	83	85	87	88
72	57	58	60	62	84	86	88	89
73	57	59	61	63	85	87	89	91
74	58	60	62	64	86	88	90	92
75	59	61	63	65	87	89	91	93
76	60	62	64	66	88	90	92	94
77	61	63	65	67	89	91	93	95
78	62	64	66	68	90	92	94	96
79	63	64	67	69	91	93	96	97
80	64	65	68	70	92	94	97	98
81	65	66	69	71	93	95	98	99
82	66	67	69	71	95	97	99	100
83	66	68	70	72	96	98	100	102
84	67	69	71	73	97	99	101	103
85	68	70	72	74	98	100	102	104
86	69	71	73	75	99	101	103	105
87	70	72	74	76	100	102	104	106
88	71	73	75	77	101	103	105	107
89	72	74	76	78	102	104	106	108
90	73	74	77	79	103	105	108	109
91	74	75	78	80	104	106	109	110
92	75	76	79	81	105	107	110	111
93	75	77	80	82	106	108	111	113
94	76	78	81	83	107	109	112	114
95	77	79	82	84	108	110	113	115
96	78	80	82	85	109	112	114	116
97	79	81	83	86	110	113	115	117
98	80	82	84	87	111	114	116	118
99	81	83	85	87	113	115	117	119
100	82	84	86	88	114	116	118	120

TABLE **29** Critical values of U, the Mann–Whitney statistic

The quantity U is known as Wilcoxon's two-sample statistic or as the Mann–Whitney statistic. Critical values are tabulated for two samples of sizes n_1 and n_2, where $n_1 \geq n_2$, up to $n_1 = n_2 = 20$. As here presented (and discussed in Section 13.11) the upper bounds of the critical values are furnished so that the sample statistic U_s has to be greater than a given critical value. Some other tables of these critical values give the lower bounds. The probabilities at the heads of the columns are based on a one-tailed test and represent the proportion of the area of the distribution of U in one tail beyond the critical value. The following one-tailed probabilities are furnished: 0.10, 0.05, 0.025, 0.01, 0.005, and 0.001. For a two-tailed test use the same critical values but double the probability at the heads of the columns.

We find the critical value of U ($P = 0.025$, one-tailed) for two samples $n_1 = 14$, $n_2 = 12$ to equal 123. Any value of $U_s > 123$ will be significant at $P < 0.025$. When $n_2 > 20$, or when there are tied values across both samples, the significance of U_s can be tested by a formula given in Box 13.6.

This table is useful for significance testing in the Mann–Whitney U-test and the Wilcoxon two-sample test (see Section 13.11 and Box 13.6), both being nonparametric tests of differences between two samples.

This table was extracted from a more extensive one (table 11.4) in D. B. Owen, *Handbook of Statistical Tables* (Addison-Wesley, Reading, Mass., 1962) with permission of the publishers.

TABLE 29 Critical values of U, the Mann–Whitney statistic

		α					
n_1	n_2	**0.10**	**0.05**	**0.025**	**0.01**	**0.005**	**0.001**
3	2	6					
	3	8	9				
4	2	8					
	3	11	12				
	4	13	15	16			
5	2	9	10				
	3	13	14	15			
	4	16	18	19	20		
	5	20	21	23	24	25	
6	2	11	12				
	3	15	16	17			
	4	19	21	22	23	24	
	5	23	25	27	28	29	
	6	27	29	31	33	34	
7	2	13	14				
	3	17	19	20	21		
	4	22	24	25	27	28	
	5	27	29	30	32	34	
	6	31	34	36	38	39	42
	7	36	38	41	43	45	48
8	2	14	15	16			
	3	19	21	22	24		
	4	25	27	28	30	31	
	5	30	32	34	36	38	40
	6	35	38	40	42	44	47
	7	40	43	46	49	50	54
	8	45	49	51	55	57	60
9	1	9					
	2	16	17	18			
	3	22	23	25	26	27	
	4	27	30	32	33	35	
	5	33	36	38	40	42	44
	6	39	42	44	47	49	52
	7	45	48	51	54	56	60
	8	50	54	57	61	63	67
	9	56	60	64	67	70	74
10	1	10					
	2	17	19	20			
	3	24	26	27	29	30	
	4	30	33	35	37	38	40
	5	37	39	42	44	46	49
	6	43	46	49	52	54	57
	7	49	53	56	59	61	65
	8	56	60	63	67	69	74
	9	62	66	70	74	77	82
	10	68	73	77	81	84	90

TABLE 29 Critical values of U, the Mann–Whitney statistic

n_1	n_2	α 0.10	0.05	0.025	0.01	0.005	0.001
11	1	11					
	2	19	21	22			
	3	26	28	30	32	33	
	4	33	36	38	40	42	44
	5	40	43	46	48	50	53
	6	47	50	53	57	59	62
	7	54	58	61	65	67	71
	8	61	65	69	73	75	80
	9	68	72	76	81	83	89
	10	74	79	84	88	92	98
	11	81	87	91	96	100	106
12	1	12					
	2	20	22	23			
	3	28	31	32	34	35	
	4	36	39	41	42	45	48
	5	43	47	49	52	54	58
	6	51	55	58	61	63	68
	7	58	63	66	70	72	77
	8	66	70	74	79	81	87
	9	73	78	82	87	90	96
	10	81	86	91	96	99	106
	11	88	94	99	104	108	115
	12	95	102	107	113	117	124
13	1	13					
	2	22	24	25	26		
	3	30	33	35	37	38	
	4	39	42	44	47	49	51
	5	47	50	53	56	58	62
	6	55	59	62	66	68	73
	7	63	67	71	75	78	83
	8	71	76	80	84	87	93
	9	79	84	89	94	97	103
	10	87	93	97	103	106	113
	11	95	101	106	112	116	123
	12	103	109	115	121	125	133
	13	111	118	124	130	135	143
14	1	14					
	2	24	25	27	28		
	3	32	35	37	40	41	
	4	41	45	47	50	52	55
	5	50	54	57	60	63	67
	6	59	63	67	71	73	78
	7	67	72	76	81	83	89
	8	76	81	86	90	94	100
	9	85	90	95	100	104	111
	10	93	99	104	110	114	121
	11	102	108	114	120	124	132
	12	110	117	123	130	134	143
	13	119	126	132	139	144	153
	14	127	135	141	149	154	164

TABLE 29 Critical values of U, the Mann–Whitney statistic

n_1	n_2	α = 0.10	0.05	0.025	0.01	0.005	0.001
15	1	15					
	2	25	27	29	30		
	3	35	38	40	42	43	
	4	44	48	50	53	55	59
	5	53	57	61	64	67	71
	6	63	67	71	75	78	83
	7	72	77	81	86	89	95
	8	81	87	91	96	100	106
	9	90	96	101	107	111	118
	10	99	106	111	117	121	129
	11	108	115	121	128	132	141
	12	117	125	131	138	143	152
	13	127	134	141	148	153	163
	14	136	144	151	159	164	174
	15	145	153	161	169	174	185
16	1	16					
	2	27	29	31	32		
	3	37	40	42	45	46	
	4	47	50	53	57	59	62
	5	57	61	65	68	71	75
	6	67	71	75	80	83	88
	7	76	82	86	91	94	101
	8	86	92	97	102	106	113
	9	96	102	107	113	117	125
	10	106	112	118	124	129	137
	11	115	122	129	135	140	149
	12	125	132	139	146	151	161
	13	134	143	149	157	163	173
	14	144	153	160	168	174	185
	15	154	163	170	179	185	197
	16	163	173	181	190	196	208
17	1	17					
	2	28	31	32	34		
	3	39	42	45	47	49	51
	4	50	53	57	60	62	66
	5	60	65	68	72	75	80
	6	71	76	80	84	87	93
	7	81	86	91	96	100	106
	8	91	97	102	108	112	119
	9	101	108	114	120	124	132
	10	112	119	125	132	136	145
	11	122	130	136	143	148	158
	12	132	140	147	155	160	170
	13	142	151	158	166	172	183
	14	153	161	169	178	184	195
	15	163	172	180	189	195	208
	16	173	183	191	201	207	220
	17	183	193	202	212	219	232

TABLE 29 Critical values of U, the Mann–Whitney statistic

n_1	n_2	α = 0.10	0.05	0.025	0.01	0.005	0.001
18	1	18					
	2	30	32	34	36		
	3	41	45	47	50	52	54
	4	52	56	60	63	66	69
	5	63	68	72	76	79	84
	6	74	80	84	89	92	98
	7	85	91	96	102	105	112
	8	96	103	108	114	118	126
	9	107	114	120	126	131	139
	10	118	125	132	139	143	153
	11	129	137	143	151	156	166
	12	139	148	155	163	169	179
	13	150	159	167	175	181	192
	14	161	170	178	187	194	206
	15	172	182	190	200	206	219
	16	182	193	202	212	218	232
	17	193	204	213	224	231	245
	18	204	215	225	236	243	258
19	1	18	19				
	2	31	34	36	37	38	
	3	43	47	50	53	54	57
	4	55	59	63	67	69	73
	5	67	72	76	80	83	88
	6	78	84	89	94	97	103
	7	90	96	101	107	111	118
	8	101	108	114	120	124	132
	9	113	120	126	133	138	146
	10	124	132	138	146	151	161
	11	136	144	151	159	164	175
	12	147	156	163	172	177	188
	13	158	167	175	184	190	202
	14	169	179	188	197	203	216
	15	181	191	200	210	216	230
	16	192	203	212	222	230	244
	17	203	214	224	235	242	257
	18	214	226	236	248	255	271
	19	226	238	248	260	268	284

TABLE 29 Critical values of U, the Mann–Whitney statistic

		α					
n_1	n_2	0.10	0.05	0.025	0.01	0.005	0.001
20	1	19	20				
	2	33	36	38	39	40	
	3	45	49	52	55	57	60
	4	58	62	66	70	72	77
	5	70	75	80	84	87	93
	6	82	88	93	98	102	108
	7	94	101	106	112	116	124
	8	106	113	119	126	130	139
	9	118	126	132	140	144	154
	10	130	138	145	153	158	168
	11	142	151	158	167	172	183
	12	154	163	171	180	186	198
	13	166	176	184	193	200	212
	14	178	188	197	207	213	226
	15	190	200	210	220	227	241
	16	201	213	222	233	241	255
	17	213	225	235	247	254	270
	18	225	237	248	260	268	284
	19	237	250	261	273	281	298
	20	249	262	273	286	295	312

TABLE **30** Critical values of the Wilcoxon rank sum

This table furnishes critical values for the one-tailed test of significance of the rank sum T_s obtained in Wilcoxon's matched-pairs signed-ranks test. Critical values are given for one-tailed probabilities 0.05, 0.025, 0.01, and 0.005 and for sample sizes from $n = 5$ to $n = 50$. Since the exact probability level desired cannot be obtained with integral critical values of T, two such values and their attendant probabilities bracketing the desired significance level are furnished. Thus, to find the significant 1% values for $n = 19$ we note the two critical values of T, 37 and 38, in the table. The probabilities corresponding to these two values of T are 0.0090 and 0.0102. Clearly a rank sum of $T_s = 37$ would have a probability of less than 0.01 and would be considered significant by the stated criterion. For two-tailed tests in which the alternative hypothesis is that the pairs could differ in either direction, double the probabilities stated at the head of the table. For sample sizes $n > 50$ compute

$$t_{\alpha[\infty]} = \left[T_s - \frac{n(n+1)}{4}\right] \Big/ \sqrt{\frac{n(n+1)(2n+1)}{24}}$$

for a two-tailed test.

This table is used for Wilcoxon's matched-pairs signed-ranks test, described in Section 13.12.

The table was prepared on a computer using a recursion equation given in D. B. Owen, *Handbook of Statistical Tables* (Addison-Wesley, Reading, Mass., 1962, p. 325).

TABLE 30 Critical values of the Wilcoxon rank sum.

Nominal α	0.05		0.025		0.01		0.005	
n	T	α	T	α	T	α	T	α
5	0	.0312						
	1	.0625						
6	2	.0469	0	.0156				
	3	.0781	1	.0312				
7	3	.0391	2	.0234	0	.0078		
	4	.0547	3	.0391	1	.0156		
8	5	.0391	3	.0195	1	.0078	0	.0039
	6	.0547	4	.0273	2	.0117	1	.0078
9	8	.0488	5	.0195	3	.0098	1	.0039
	9	.0645	6	.0273	4	.0137	2	.0059
10	10	.0420	8	.0244	5	.0098	3	.0049
	11	.0527	9	.0322	6	.0137	4	.0068
11	13	.0415	10	.0210	7	.0093	5	.0049
	14	.0508	11	.0269	8	.0122	6	.0068
12	17	.0461	13	.0212	9	.0081	7	.0046
	18	.0549	14	.0261	10	.0105	8	.0061
13	21	.0471	17	.0239	12	.0085	9	.0040
	22	.0549	18	.0287	13	.0107	10	.0052
14	25	.0453	21	.0247	15	.0083	12	.0043
	26	.0520	22	.0290	16	.0101	13	.0054
15	30	.0473	25	.0240	19	.0090	15	.0042
	31	.0535	26	.0277	20	.0108	16	.0051
16	35	.0467	29	.0222	23	.0091	19	.0046
	36	.0523	30	.0253	24	.0107	20	.0055
17	41	.0492	34	.0224	27	.0087	23	.0047
	42	.0544	35	.0253	28	.0101	24	.0055
18	47	.0494	40	.0241	32	.0091	27	.0045
	48	.0542	41	.0269	33	.0104	28	.0052
19	53	.0478	46	.0247	37	.0090	32	.0047
	54	.0521	47	.0273	38	.0102	33	.0054
20	60	.0487	52	.0242	43	.0096	37	.0047
	61	.0527	53	.0266	44	.0107	38	.0053

TABLE 30 Critical values of the Wilcoxon rank sum.

Nominal α	0.05		0.025		0.01		0.005	
n	T	α	T	α	T	α	T	α
21	67	.0479	58	.0230	49	.0097	42	.0045
	68	.0516	59	.0251	50	.0108	43	.0051
22	75	.0492	65	.0231	55	.0095	48	.0046
	76	.0527	66	.0250	56	.0104	49	.0052
23	83	.0490	73	.0242	62	.0098	54	.0046
	84	.0523	74	.0261	63	.0107	55	.0051
24	91	.0475	81	.0245	69	.0097	61	.0048
	92	.0505	82	.0263	70	.0106	62	.0053
25	100	.0479	89	.0241	76	.0094	68	.0048
	101	.0507	90	.0258	77	.0101	69	.0053
26	110	.0497	98	.0247	84	.0095	75	.0047
	111	.0524	99	.0263	85	.0102	76	.0051
27	119	.0477	107	.0246	92	.0093	83	.0048
	120	.0502	108	.0260	93	.0100	84	.0052
28	130	.0496	116	.0239	101	.0096	91	.0048
	131	.0521	117	.0252	102	.0102	92	.0051
29	140	.0482	126	.0240	110	.0095	100	.0049
	141	.0504	127	.0253	111	.0101	101	.0053
30	151	.0481	137	.0249	120	.0098	109	.0050
	152	.0502	138	.0261	121	.0104	110	.0053
31	163	.0491	147	.0239	130	.0099	118	.0049
	164	.0512	148	.0251	131	.0105	119	.0052
32	175	.0492	159	.0249	140	.0097	128	.0050
	176	.0512	160	.0260	141	.0103	129	.0053
33	187	.0485	170	.0242	151	.0099	138	.0049
	188	.0503	171	.0253	152	.0104	139	.0052
34	200	.0488	182	.0242	162	.0098	148	.0048
	201	.0506	183	.0252	163	.0103	149	.0051
35	213	.0484	195	.0247	173	.0096	159	.0048
	214	.0501	196	.0257	174	.0100	160	.0051

TABLE 30 Critical values of the Wilcoxon rank sum.

Nominal α	0.05		0.025		0.01		0.005	
n	T	α	T	α	T	α	T	α
36	227	.0489	208	.0248	185	.0096	171	.0050
	228	.0505	209	.0258	186	.0100	172	.0052
37	241	.0487	221	.0245	198	.0099	182	.0048
	242	.0503	222	.0254	199	.0103	183	.0050
38	256	.0493	235	.0247	211	.0099	194	.0048
	257	.0509	236	.0256	212	.0104	195	.0050
39	271	.0493	249	.0246	224	.0099	207	.0049
	272	.0507	250	.0254	225	.0103	208	.0051
40	286	.0486	264	.0249	238	.0100	220	.0049
	287	.0500	265	.0257	239	.0104	221	.0051
41	302	.0488	279	.0248	252	.0100	233	.0048
	303	.0501	280	.0256	253	.0103	234	.0050
42	319	.0496	294	.0245	266	.0098	247	.0049
	320	.0509	295	.0252	267	.0102	248	.0051
43	336	.0498	310	.0245	281	.0098	261	.0048
	337	.0511	311	.0252	282	.0102	262	.0050
44	353	.0495	327	.0250	296	.0097	276	.0049
	354	.0507	328	.0257	297	.0101	277	.0051
45	371	.0498	343	.0244	312	.0098	291	.0049
	372	.0510	344	.0251	313	.0101	292	.0051
46	389	.0497	361	.0249	328	.0098	307	.0050
	390	.0508	362	.0256	329	.0101	308	.0052
47	407	.0490	378	.0245	345	.0099	322	.0048
	408	.0501	379	.0251	346	.0102	323	.0050
48	426	.0490	396	.0244	362	.0099	339	.0050
	427	.0500	397	.0251	363	.0102	340	.0051
49	446	.0495	415	.0247	379	.0098	355	.0049
	447	.0505	416	.0253	380	.0100	356	.0050
50	466	.0495	434	.0247	397	.0098	373	.0050
	467	.0506	435	.0253	398	.0101	374	.0051

TABLE 31 Critical values of the two-sample Kolmogorov–Smirnov statistic

This table furnishes *upper* critical values of $n_1 n_2 D$, the Kolmogorov–Smirnov test statistic D multiplied by the two sample sizes n_1 and n_2. In the two-sample case D is defined as the largest unsigned difference between the two relative cumulative frequency distributions representing the observed frequencies of the two samples. Critical values are tabulated for two samples of sizes n_1 and n_2 up to $n_1 = n_2 = 25$. Sample sizes n_1 are given at the left margin of the table, while sample sizes n_2 are given across its top at the heads of the columns. The six values furnished at the intersection of two sample sizes represent the following six two-tailed probabilities: 0.10, 0.05, 0.025, 0.01, 0.005, and 0.001.

For the sample statistic $n_1 n_2 D$ to be significant, it has to equal or exceed a tabled critical value.

For two samples with $n_1 = 16$ and $n_2 = 10$, the 5% critical value of $n_1 n_2 D$ is 84. Any value of $n_1 n_2 D \geq 84$ will be significant at $P \leq 0.05$. When either n_1 or n_2 are > 25, approximate tests can be carried out as shown in Box 13.8.

This table is used for significance testing in the two-sample Kolmogorov–Smirnov test (see Section 13.11 and Box 13.8) in nonparametric tests for the difference between two sample distributions.

When a one-sided test is desired, approximate probabilities can be obtained from this table by doubling the nominal α values. However, these are not exact since the distribution of cumulative frequencies is discrete. A one-sided table is furnished by M. H. Gail and S. B. Green (*J. Am. Stat. Assoc.*, **71**:757–760, 1976).

This table was copied from table 55 in E. S. Pearson and H. O. Hartley, *Biometrika Tables for Statisticians*, Volume II (Cambridge University Press, 1972) with permission of the publishers.

TABLE 31 Critical values of the two-sample Kolmogorov–Smirnov statistic

n_1	α	n_2: 1	2	3	4	5	6	7	8	9	10	11	12	13	14	15	16	17	18	19	20	21	22	23	24	25
1	.1	-	-	-	-	-	-	-	-	-	-	-	-	-	-	-	-	-	-	19	20	21	22	23	24	25
	.05	-	-	-	-	-	-	-	-	-	-	-	-	-	-	-	-	-	-	-	-	-	-	-	-	-
	.025	-	-	-	-	-	-	-	-	-	-	-	-	-	-	-	-	-	-	-	-	-	-	-	-	-
	.01	-	-	-	-	-	-	-	-	-	-	-	-	-	-	-	-	-	-	-	-	-	-	-	-	-
	.005	-	-	-	-	-	-	-	-	-	-	-	-	-	-	-	-	-	-	-	-	-	-	-	-	-
	.001	-	-	-	-	-	-	-	-	-	-	-	-	-	-	-	-	-	-	-	-	-	-	-	-	-
2	.1	-	-	-	-	10	12	14	16	18	18	20	22	24	24	26	28	30	32	32	34	36	38	38	40	42
	.05	-	-	-	-	-	-	-	16	18	20	22	24	26	26	28	30	32	34	36	38	38	40	42	44	46
	.025	-	-	-	-	-	-	-	-	-	-	-	24	26	28	30	32	34	36	38	40	40	42	44	46	48
	.01	-	-	-	-	-	-	-	-	-	-	-	-	-	-	-	-	-	-	38	40	42	44	46	48	50
	.005	-	-	-	-	-	-	-	-	-	-	-	-	-	-	-	-	-	-	-	-	-	-	-	-	-
	.001	-	-	-	-	-	-	-	-	-	-	-	-	-	-	-	-	-	-	-	-	-	-	-	-	-
3	.1	-	-	9	12	15	15	18	21	21	24	27	27	30	33	33	36	36	39	42	42	45	48	48	51	54
	.05	-	-	-	-	15	18	21	21	24	27	30	30	33	36	36	39	42	45	45	48	51	51	54	57	60
	.025	-	-	-	-	-	18	21	24	27	30	30	33	36	39	39	42	45	48	51	51	54	57	60	60	63
	.01	-	-	-	-	-	-	-	-	27	30	33	36	39	42	42	45	48	51	54	57	57	60	63	66	69
	.005	-	-	-	-	-	-	-	-	-	-	-	36	39	42	45	48	51	54	57	57	60	63	66	69	72
	.001	-	-	-	-	-	-	-	-	-	-	-	-	-	-	-	-	-	-	-	-	63	66	69	72	75
4	.1	-	-	12	16	16	18	21	24	27	28	29	36	35	38	40	44	44	46	49	52	52	56	57	60	63
	.05	-	-	-	16	20	20	24	28	28	30	33	36	39	42	44	48	48	50	53	60	59	62	64	68	68
	.025	-	-	-	-	20	24	28	28	32	36	36	40	44	44	45	52	52	54	57	64	63	66	69	72	75
	.01	-	-	-	-	-	24	28	32	36	36	40	44	48	48	52	56	60	60	64	68	72	72	76	80	84
	.005	-	-	-	-	-	-	-	32	36	40	44	48	48	52	56	60	64	64	68	72	76	76	80	84	88
	.001	-	-	-	-	-	-	-	-	-	-	-	-	52	56	60	64	68	72	76	76	80	84	88	92	96

n_1	α	1	2	3	4	5	6	7	8	9	10	11	12	13	14	15	16	17	18	19	20	21	22	23	24	25
5	**.1**	-	10	15	16	20	24	25	27	30	35	35	36	40	42	50	48	50	52	56	60	60	63	65	67	75
	.05	-	-	15	20	25	24	28	30	35	40	39	43	45	46	55	54	55	60	61	65	69	70	72	76	80
	.025	-	-	-	20	25	30	30	32	36	40	44	45	47	51	55	59	60	65	66	75	74	78	80	81	90
	.01	-	-	-	-	25	30	35	35	40	45	45	50	52	56	60	64	68	70	71	80	80	83	87	90	95
	.005	-	-	-	-	-	30	35	40	45	45	50	55	55	60	65	70	70	72	76	85	84	88	92	95	100
	.001	-	-	-	-	-	-	-	-	45	50	55	60	65	70	70	75	80	85	85	90	95	100	105	105	110
6	**.1**	-	12	15	18	24	30	28	30	33	36	38	48	46	48	51	54	56	66	64	66	69	70	73	78	78
	.05	-	-	18	20	24	30	30	34	39	40	43	48	52	54	57	60	62	72	70	72	75	78	80	90	88
	.025	-	-	18	24	30	36	35	36	42	44	48	54	54	58	63	64	67	78	76	78	81	86	86	96	96
	.01	-	-	-	24	30	36	36	40	45	48	54	60	60	64	69	72	73	84	83	88	90	92	97	102	107
	.005	-	-	-	-	30	36	42	42	48	50	55	60	65	66	72	74	79	84	89	90	96	98	103	108	113
	.001	-	-	-	-	-	-	-	48	54	60	66	66	72	78	84	84	85	96	96	100	105	110	114	120	125
7	**.1**	-	14	18	21	25	28	35	34	36	40	44	46	50	56	56	59	61	65	69	72	77	77	80	84	86
	.05	-	-	21	24	28	30	42	40	42	46	48	53	56	63	62	64	68	72	76	79	91	84	89	92	97
	.025	-	-	21	28	30	35	42	41	45	49	52	56	58	70	68	73	77	80	84	86	98	96	98	102	105
	.01	-	-	-	28	35	36	42	48	49	53	59	60	65	77	75	77	84	87	91	93	105	103	108	112	115
	.005	-	-	-	-	35	42	49	48	54	56	63	65	70	77	77	84	85	91	95	99	112	110	112	119	122
	.001	-	-	-	-	-	-	49	56	63	63	70	72	78	84	90	96	98	101	107	112	119	125	126	133	136
8	**.1**	-	16	21	24	27	30	34	40	40	44	48	52	54	58	60	72	68	72	74	80	81	84	89	96	95
	.05	-	16	21	28	30	34	40	48	46	48	53	60	62	64	67	80	77	80	82	88	89	94	98	104	104
	.025	-	-	24	28	32	36	41	48	48	54	58	64	65	70	74	80	80	86	90	96	97	102	106	112	112
	.01	-	-	-	32	35	40	48	56	55	60	64	68	72	76	81	88	88	94	98	104	107	112	115	128	125
	.005	-	-	-	32	40	42	48	56	56	62	66	72	78	82	88	96	96	100	104	112	115	120	122	136	134
	.001	-	-	-	-	-	48	56	64	64	70	77	80	88	90	97	104	111	112	117	124	126	132	137	152	150

Column header: n_2

TABLE 31 Critical values of the two-sample Kolmogorov–Smirnov statistic

n_1	α	n_2: 1	2	3	4	5	6	7	8	9	10	11	12	13	14	15	16	17	18	19	20	21	22	23	24	25
9	.1	-	18	21	27	30	33	36	40	54	50	52	57	59	63	69	69	74	81	80	84	90	91	94	99	101
	.05	-	18	24	28	35	39	42	46	54	53	59	63	65	70	75	78	82	90	89	93	99	101	106	111	114
	.025	-	-	27	32	36	42	45	48	63	60	63	69	72	76	81	85	90	99	98	100	108	110	115	120	123
	.01	-	-	27	36	40	45	49	55	63	63	70	75	78	84	90	94	99	108	107	111	117	122	126	132	135
	.005	-	-	-	36	45	48	54	56	72	70	72	78	82	89	93	99	102	117	114	117	123	127	134	138	144
	.001	-	-	-	-	45	54	63	64	72	80	81	87	91	98	105	110	117	126	126	133	138	144	152	156	162
10	.1	-	18	24	28	35	36	40	44	50	60	57	60	64	68	75	76	79	82	85	100	95	98	101	106	110
	.05	-	20	27	30	40	40	46	48	53	70	60	66	70	74	80	84	89	92	94	110	105	108	114	118	125
	.025	-	-	30	36	40	44	49	54	60	70	68	72	77	82	90	90	96	100	103	120	116	118	124	128	135
	.01	-	-	30	36	45	48	53	60	63	80	77	80	84	90	100	100	106	108	113	130	126	130	137	140	150
	.005	-	-	-	40	45	50	56	62	70	80	79	84	90	96	105	108	110	116	122	130	130	138	144	148	155
	.001	-	-	-	-	50	60	63	70	80	90	89	96	100	106	115	118	126	132	133	150	149	154	160	166	175
11	.1	-	20	27	29	35	38	44	48	52	57	66	64	67	73	76	80	85	88	92	96	101	110	108	111	117
	.05	-	22	30	33	39	43	48	53	59	60	77	72	75	82	84	89	93	97	102	107	112	121	119	124	129
	.025	-	-	30	36	44	48	52	58	63	68	77	76	84	87	94	96	102	107	111	116	123	132	131	137	140
	.01	-	-	33	40	45	54	59	64	70	77	88	86	91	96	102	106	110	118	122	127	134	143	142	150	154
	.005	-	-	-	44	50	55	63	66	72	79	88	88	97	101	109	112	119	125	130	136	143	154	153	159	164
	.001	-	-	-	-	55	66	70	77	81	89	99	99	108	115	120	127	132	140	146	154	157	176	173	176	184
12	.1	-	22	27	36	36	48	46	52	57	60	64	72	71	78	84	88	90	96	99	104	108	110	113	132	120
	.05	-	24	30	36	43	48	53	60	63	66	72	84	81	86	93	96	100	108	108	116	120	124	125	144	138
	.025	-	24	33	40	45	54	56	64	69	72	76	96	84	94	99	104	108	120	120	124	129	134	137	156	150
	.01	-	-	36	44	50	60	60	68	75	80	86	96	95	104	108	116	119	126	130	140	141	148	149	168	165
	.005	-	-	36	48	55	60	65	72	78	84	88	108	104	108	117	124	127	138	140	148	150	154	160	180	175
	.001	-	-	-	-	60	66	72	80	87	96	99	120	117	120	129	136	141	150	156	164	168	174	182	192	192

n_1	α	1	2	3	4	5	6	7	8	9	10	11	12	13	14	15	16	17	18	19	20	21	22	23	24	25
13	**.1**	-	24	30	35	40	46	50	54	59	64	67	71	91	78	87	91	96	99	104	108	113	117	120	125	131
	.05	-	26	33	39	45	52	56	62	65	70	75	81	91	89	96	101	105	110	114	120	126	130	135	140	145
	.025	-	26	36	44	47	54	58	65	72	77	84	84	104	100	104	111	114	120	126	130	137	141	146	151	158
	.01	-	-	39	48	52	60	65	72	78	84	91	95	117	104	115	121	127	131	138	143	150	156	161	166	172
	.005	-	-	39	48	55	65	70	78	82	90	97	104	117	115	122	128	135	141	145	154	161	168	171	177	184
	.001	-	-	-	52	65	72	78	88	91	100	108	117	130	129	137	143	152	156	164	169	179	185	191	199	200
14	**.1**	-	24	33	38	42	48	56	58	63	68	73	78	78	98	92	96	100	104	110	114	126	124	127	132	136
	.05	-	26	36	42	46	54	63	64	70	74	82	86	89	112	98	106	111	116	121	126	140	138	142	146	150
	.025	-	28	39	44	51	58	70	70	76	82	87	94	100	112	110	116	122	126	133	138	147	148	154	160	166
	.01	-	-	42	48	56	64	77	76	84	90	96	104	104	126	123	126	134	140	148	152	161	164	170	176	182
	.005	-	-	42	52	60	66	77	82	89	96	101	108	115	126	125	136	140	148	154	160	175	174	179	186	194
	.001	-	-	-	56	70	78	84	90	98	106	115	120	129	154	140	152	159	166	176	180	189	196	202	210	219
15	**.1**	-	26	33	40	50	51	56	60	69	75	76	84	87	92	105	101	105	111	114	125	126	130	134	141	145
	.05	-	28	36	44	55	57	62	67	75	80	84	93	96	98	120	114	116	123	127	135	138	144	149	156	160
	.025	-	30	39	45	55	63	68	74	81	90	94	99	104	110	135	119	129	135	141	150	153	154	163	168	175
	.01	-	-	42	52	60	69	75	81	90	100	102	108	115	123	135	133	142	147	152	160	168	173	179	186	195
	.005	-	-	45	56	65	72	77	88	93	105	109	117	122	125	150	144	148	156	161	170	177	182	187	198	205
	.001	-	-	-	60	70	84	90	97	105	115	120	129	137	140	165	162	165	174	180	195	198	205	210	222	230
16	**.1**	-	28	36	44	48	54	59	72	69	76	80	88	91	96	101	112	109	116	120	128	130	136	141	152	149
	.05	-	30	39	48	54	60	64	80	78	84	89	96	101	106	114	128	124	128	133	140	145	150	157	168	167
	.025	-	32	42	52	59	64	73	80	85	90	96	104	111	116	119	144	136	140	145	156	157	164	169	184	181
	.01	-	-	45	56	64	72	77	88	94	100	106	116	121	126	133	160	143	154	160	168	173	180	187	200	199
	.005	-	-	48	60	70	74	84	96	99	108	112	124	128	136	144	160	157	162	170	180	183	192	198	208	213
	.001	-	-	-	64	75	84	96	104	110	118	127	136	143	152	162	176	174	186	192	200	208	216	221	232	238

TABLE 31 Critical values of the two-sample Kolmogorov–Smirnov statistic

n_1	α	1	2	3	4	5	6	7	8	9	10	11	12	13	14	15	16	17	18	19	20	21	22	23	24	25
17	**.1**	-	30	36	44	50	56	61	68	74	79	85	90	96	100	105	109	136	118	126	132	136	142	146	151	156
	.05	-	32	42	48	55	62	68	77	82	89	93	100	105	111	116	124	136	133	141	146	151	157	163	168	173
	.025	-	34	45	52	60	67	77	80	90	96	102	108	114	122	129	136	153	148	151	160	166	170	179	183	190
	.01	-	-	48	60	68	73	84	88	99	106	110	119	127	134	142	143	170	164	166	175	180	187	196	203	207
	.005	-	-	51	64	70	79	85	96	102	110	119	127	135	140	148	157	170	168	179	186	193	199	207	214	222
	.001	-	-	-	68	80	85	98	111	117	126	132	141	152	159	165	174	204	187	200	209	217	225	232	240	249
18	**.1**	-	32	39	46	52	66	65	72	81	82	88	96	99	104	111	116	118	144	133	136	144	148	152	162	162
	.05	-	34	45	50	60	72	72	80	90	92	97	108	110	116	123	128	133	162	142	152	159	164	170	180	180
	.025	-	36	48	54	65	78	80	86	99	100	107	120	120	126	135	140	148	162	159	166	174	178	184	198	196
	.01	-	-	51	60	70	84	87	94	108	108	118	126	131	140	147	154	164	180	176	182	189	196	204	216	216
	.005	-	-	54	64	72	84	91	100	117	116	125	138	141	148	156	162	168	198	180	194	201	208	216	228	231
	.001	-	-	-	72	85	96	101	112	126	132	140	150	156	166	174	186	187	216	212	214	225	234	242	252	257
19	**.1**	19	32	42	49	56	64	69	74	80	85	92	99	104	110	114	120	126	133	152	144	147	152	159	164	168
	.05	-	36	45	53	61	70	76	82	89	94	102	108	114	121	127	133	141	142	171	160	163	169	177	183	187
	.025	-	38	51	57	66	76	84	90	98	103	111	120	126	133	141	145	151	159	190	169	180	185	190	199	205
	.01	-	38	54	64	71	83	91	98	107	113	122	130	138	148	152	160	166	176	190	187	199	204	209	218	224
	.005	-	-	57	68	76	89	95	104	114	122	130	140	145	154	161	170	179	180	209	204	207	219	224	232	241
	.001	-	-	-	76	85	96	107	117	126	133	146	156	164	176	180	192	200	212	228	225	237	242	253	261	268
20	**.1**	20	34	42	52	60	66	72	80	84	100	96	104	108	114	125	128	132	136	144	160	154	160	164	172	180
	.05	-	38	48	60	65	72	79	88	93	110	107	116	120	126	135	140	146	152	160	180	173	176	184	192	200
	.025	-	40	51	64	75	78	86	96	100	120	116	124	130	138	150	156	160	166	169	200	180	192	199	208	215
	.01	-	40	57	68	80	88	93	104	111	130	127	140	143	152	160	168	175	182	187	220	199	212	219	228	235
	.005	-	-	57	72	85	90	99	112	117	130	136	148	154	160	170	180	186	194	204	220	217	226	233	244	250
	.001	-	-	-	76	90	100	112	124	133	150	154	164	169	180	195	200	209	214	225	260	239	254	262	272	280

n_1	α	1	2	3	4	5	6	7	8	9	10	11	12	13	14	15	16	17	18	19	20	21	22	23	24	25
21	**.1**	21	36	45	52	60	69	77	81	90	95	101	108	113	126	126	130	136	144	147	154	168	163	171	177	182
	.05	-	38	51	59	69	75	91	89	99	105	112	120	126	140	138	145	151	159	163	173	189	183	189	198	202
	.025	-	40	54	63	74	81	98	97	108	116	123	129	137	147	153	157	166	174	180	180	210	203	206	213	220
	.01	-	42	57	72	80	90	105	107	117	126	134	141	150	161	168	173	180	189	199	199	231	223	227	237	244
	.005	-	-	60	76	84	96	112	115	123	130	143	150	161	175	177	183	193	201	207	217	252	229	242	252	258
	.001	-	-	63	80	95	105	119	126	138	149	157	168	179	189	198	208	217	225	237	239	273	267	269	282	290
22	**.1**	22	38	48	56	63	70	77	84	91	98	110	110	117	124	130	136	142	148	152	160	163	198	173	182	189
	.05	-	40	51	62	70	78	84	94	101	108	121	124	130	138	144	150	157	164	169	176	183	198	194	204	209
	.025	-	42	57	66	78	86	96	102	110	118	132	134	141	148	154	164	170	178	185	192	203	220	214	222	228
	.01	-	44	60	72	83	92	103	112	122	130	143	148	156	164	173	180	187	196	204	212	223	242	237	242	250
	.005	-	-	63	76	88	98	110	120	127	138	154	154	168	174	182	192	199	208	219	226	229	264	253	258	268
	.001	-	-	66	84	100	110	125	132	144	154	176	174	185	196	205	216	225	234	242	254	267	286	282	292	299
23	**.1**	23	38	48	57	65	73	80	89	94	101	108	113	120	127	134	141	146	152	159	164	171	173	207	183	195
	.05	-	42	54	64	72	80	89	98	106	114	119	125	135	142	149	157	163	170	177	184	189	194	230	205	216
	.025	-	44	60	69	80	86	98	106	115	124	131	137	146	154	163	169	179	184	190	199	206	214	230	226	237
	.01	-	46	63	76	87	97	108	115	126	137	142	149	161	170	179	187	196	204	209	219	227	237	253	249	262
	.005	-	-	66	80	92	103	112	122	134	144	153	160	171	179	187	198	207	216	224	233	242	253	276	270	274
	.001	-	-	69	88	105	114	126	137	152	160	173	182	191	202	210	221	232	242	253	262	269	282	299	296	312
24	**.1**	24	40	51	60	67	78	84	96	99	106	111	132	125	132	141	152	151	162	164	172	177	182	183	216	204
	.05	-	44	57	68	76	90	92	104	111	118	124	144	140	146	156	168	168	180	183	192	198	204	205	240	225
	.025	-	46	60	72	81	96	102	112	120	128	137	156	151	160	168	184	183	198	199	208	213	222	226	264	238
	.01	-	48	66	80	90	102	112	128	132	140	150	168	166	176	186	200	203	216	218	228	237	242	249	288	262
	.005	-	-	69	84	95	108	119	136	138	148	159	180	177	186	198	208	214	228	232	244	252	258	270	288	283
	.001	-	-	72	92	105	120	133	152	156	166	176	192	199	210	222	232	240	252	261	272	282	292	296	336	312

Column header: n_2

TABLE 31 Critical values of the two-sample Kolmogorov–Smirnov statistic

n_1	α	n_2: 1	2	3	4	5	6	7	8	9	10	11	12	13	14	15	16	17	18	19	20	21	22	23	24	25
25	.1	25	42	54	63	75	78	86	95	101	110	117	120	131	136	145	149	156	162	168	180	182	189	195	204	225
	.05	-	46	60	68	80	88	97	104	114	125	129	138	145	150	160	167	173	180	187	200	202	209	216	225	250
	.025	-	48	63	75	90	96	105	112	123	135	140	150	158	166	175	181	190	196	205	215	220	228	237	238	275
	.01	-	50	69	84	95	107	115	125	135	150	154	165	172	182	195	199	207	216	224	235	244	250	262	262	300
	.005	-	-	72	88	100	113	122	134	144	155	164	175	184	194	205	213	222	231	241	250	258	268	274	283	325
	.001	-	-	75	96	110	125	136	150	162	175	184	192	200	219	230	238	249	257	268	280	290	299	312	312	350

TABLE **32** Critical values of the one-sample Kolmogorov–Smirnov statistic

This table furnishes critical values of the Kolmogorov–Smirnov test statistic D, which is defined as the maximum unsigned difference between two relative cumulative frequency distributions, one expected and the other observed. The probabilities 0.20, 0.10, 0.05, 0.02, 0.01 given at the heads of the columns of the table are the probabilities α that a given value of D equals to or exceeds the critical value shown when the observed frequency distribution was sampled from the expected one. The table furnishes critical values for every sample size between 1 and 100.

For a sample of $n = 42$ items, the 5% critical value of D is found to be 0.20517. Observed values of D greater than this critical value would be considered significant. For sample sizes $n > 100$ the two-tailed critical value D_α can be found as $D_\alpha = \sqrt{-\ln(\frac{1}{2}\alpha)/2n}$. For $\alpha = 0.05$, this is $1.358/\sqrt{n}$, and for $\alpha = 0.01$, it is $1.628/\sqrt{n}$.

The Kolmogorov–Smirnov test is used to test the deviations of observed continuous frequency distributions from expected ones and for setting confidence limits to a cumulative frequency distribution. Both methods are illustrated in Section 17.2. The parameters of the expected distribution are *not* obtained from the observed distribution; that is, the observations are fitted to an extrinsic hypothesis. In the more common case of fitting an intrinsic hypothesis, critical values of the one-sample Kolmogorov–Smirnov statistic can be looked up in Table **33.**

This table has been extracted from L. H. Miller (*J. Am. Stat. Assoc.*, **51**:111–121, 1956) with permission of the author and publisher.

TABLE 32 Critical values of the one-sample Kolmogorov–Smirnov statistic

$n \backslash \alpha$	0.2	0.1	0.05	0.02	0.01
1	.90000	.95000	.97500	.99000	.99500
2	.68377	.77639	.84189	.90000	.92929
3	.56481	.63604	.70760	.78456	.82900
4	.49265	.56522	.62394	.68887	.73424
5	.44698	.50945	.56328	.62718	.66853
6	.41037	.46799	.51926	.57741	.61661
7	.38148	.43607	.48342	.53844	.57581
8	.35831	.40962	.45427	.50654	.54179
9	.33910	.38746	.43001	.47960	.51332
10	.32260	.36866	.40925	.45662	.48893
11	.30829	.35242	.39122	.43670	.46770
12	.29577	.33815	.37543	.41918	.44905
13	.28470	.32549	.36143	.40362	.43247
14	.27481	.31417	.34890	.38970	.41762
15	.26588	.30397	.33760	.37713	.40420
16	.25778	.29472	.32733	.36571	.39201
17	.25039	.28627	.31796	.35528	.38086
18	.24360	.27851	.30936	.34569	.37062
19	.23735	.27136	.30143	.33685	.36117
20	.23156	.26473	.29408	.32866	.35241
21	.22617	.25858	.28724	.32104	.34427
22	.22115	.25283	.28087	.31394	.33666
23	.21645	.24746	.27490	.30728	.32954
24	.21205	.24242	.26931	.30104	.32286
25	.20790	.23768	.26404	.29516	.31657
26	.20399	.23320	.25907	.28962	.31064
27	.20030	.22898	.25438	.28438	.30502
28	.19680	.22497	.24993	.27942	.29971
29	.19348	.22117	.24571	.27471	.29466
30	.19032	.21756	.24170	.27023	.28987
31	.18732	.21412	.23788	.26596	.28530
32	.18445	.21085	.23424	.26189	.28094
33	.18171	.20771	.23076	.25801	.27677
34	.17909	.20472	.22743	.25429	.27279
35	.17659	.20185	.22425	.25073	.26897
36	.17418	.19910	.22119	.24732	.26532
37	.17188	.19646	.21826	.24404	.26180
38	.16966	.19392	.21544	.24089	.25843
39	.16753	.19148	.21273	.23786	.25518
40	.16547	.18913	.21012	.23494	.25205
41	.16349	.18687	.20760	.23213	.24904
42	.16158	.18468	.20517	.22941	.24613
43	.15974	.18257	.20283	.22679	.24332
44	.15796	.18053	.20056	.22426	.24060
45	.15623	.17856	.19837	.22181	.23798
46	.15457	.17665	.19625	.21944	.23544
47	.15295	.17481	.19420	.21715	.23298
48	.15139	.17302	.19221	.21493	.23059
49	.14987	.17128	.19028	.21277	.22828
50	.14840	.16959	.18841	.21068	.22604
51	.14697	.16796	.18659	.20864	.22386
52	.14558	.16637	.18482	.20667	.22174
53	.14423	.16483	.18311	.20475	.21968
54	.14292	.16332	.18144	.20289	.21768
55	.14164	.16186	.17981	.20107	.21574
56	.14040	.16044	.17823	.19930	.21384
57	.13919	.15906	.17669	.19758	.21199
58	.13801	.15771	.17519	.19590	.21019
59	.13686	.15639	.17373	.19427	.20844
60	.13573	.15511	.17231	.19267	.20673
61	.13464	.15385	.17091	.19112	.20506
62	.13357	.15263	.16956	.18960	.20343
63	.13253	.15144	.16823	.18812	.20184
64	.13151	.15027	.16693	.18667	.20029
65	.13052	.14913	.16567	.18525	.19877
66	.12954	.14802	.16443	.18387	.19729
67	.12859	.14693	.16322	.18252	.19584
68	.12766	.14587	.16204	.18119	.19442
69	.12675	.14483	.16088	.17990	.19303
70	.12586	.14381	.15975	.17863	.19167
71	.12499	.14281	.15864	.17739	.19034
72	.12413	.14183	.15755	.17618	.18903
73	.12329	.14087	.15649	.17498	.18776
74	.12247	.13993	.15544	.17382	.18650
75	.12167	.13901	.15442	.17268	.18528
76	.12088	.13811	.15342	.17155	.18408
77	.12011	.13723	.15244	.17045	.18290
78	.11935	.13636	.15147	.16938	.18174
79	.11860	.13551	.15052	.16832	.18060
80	.11787	.13467	.14960	.16728	.17949
81	.11716	.13385	.14868	.16626	.17840
82	.11645	.13305	.14779	.16526	.17732
83	.11576	.13226	.14691	.16428	.17627
84	.11508	.13148	.14605	.16331	.17523
85	.11442	.13072	.14520	.16236	.17421
86	.11376	.12997	.14437	.16143	.17321
87	.11311	.12923	.14355	.16051	.17223
88	.11248	.12850	.14274	.15961	.17126
89	.11186	.12779	.14195	.15873	.17031
90	.11125	.12709	.14117	.15786	.16938
91	.11064	.12640	.14040	.15700	.16846
92	.11005	.12572	.13965	.15616	.16755
93	.10947	.12506	.13891	.15533	.16666
94	.10889	.12440	.13818	.15451	.16579
95	.10833	.12375	.13746	.15371	.16493
96	.10777	.12312	.13675	.15291	.16408
97	.10722	.12249	.13606	.15214	.16324
98	.10668	.12187	.13537	.15137	.16242
99	.10615	.12126	.13469	.15061	.16161
100	.10563	.12067	.13403	.14987	.16081

TABLE **33** Critical values of the one-sample Kolmogorov–Smirnov statistic for intrinsic hypotheses

This table furnishes critical values of the Kolmogorov–Smirnov test-statistic D, which is defined as the maximum unsigned difference between two relative cumulative frequency distributions, one expected and the other observed. The values in this table are appropriate for the case in which the parameters of the expected distribution were obtained from the observed frequency distribution (i.e., the observations are fitted to an intrinsic hypothesis). The probabilities 0.2, 0.15, 0.10, 0.05, and 0.01 given at the heads of the columns of the table are the probabilities α that a given value of D equals or exceeds the critical value shown when the observed frequency distribution was sampled from the expected one. The table furnishes critical values for every sample size between 4 and 20 and for sample sizes 25 and 30.

For a sample of 17 items, the 5% critical value of D is found to be 0.206. Observed values of D equal to or greater than this critical value would be considered significant. For sample sizes > 30, critical values can be approximated by the terms given in the last line of the table.

This table is used to test the deviations of observed continuous frequency distributions from expected ones and for setting confidence limits to a cumulative frequency distribution. Both methods are illustrated in Section 17.2.

The values in this table were obtained from a Monte Carlo calculation by H. W. Lilliefors (*J. Am. Stat. Assoc.*, **62**:399–402, 1967) and are reproduced here with permission of the author and publisher.

TABLE 33 Critical values of the one-sample Kolmogorov–Smirnov statistic for intrinsic hypotheses

n \ α	.20	.15	.10	.05	.01
4	.300	.319	.352	.381	.417
5	.285	.299	.315	.337	.405
6	.265	.277	.294	.319	.364
7	.247	.258	.276	.300	.348
8	.233	.244	.261	.285	.331
9	.223	.233	.249	.271	.311
10	.215	.224	.239	.258	.294
11	.206	.217	.230	.249	.284
12	.199	.212	.223	.242	.275
13	.190	.202	.214	.234	.268
14	.183	.194	.207	.227	.261
15	.177	.187	.201	.220	.257
16	.173	.182	.195	.213	.250
17	.169	.177	.189	.206	.245
18	.166	.173	.184	.200	.239
19	.163	.169	.179	.195	.235
20	.160	.166	.174	.190	.231
25	.142	.147	.158	.173	.200
30	.131	.136	.144	.161	.187
>30	$\frac{.736}{\sqrt{n}}$	$\frac{.768}{\sqrt{n}}$	$\frac{.805}{\sqrt{n}}$	$\frac{.886}{\sqrt{n}}$	$\frac{1.031}{\sqrt{n}}$

TABLE **34** Critical values for Page's test

This table furnishes critical values for Page's L-test for ordered alternatives. Critical values are furnished for a, the number of treatments, ranging from 2 to 10 and for b, the number of blocks, ranging from 2 to 24. At each intersection of a given treatment and block number three numbers are furnished that, from the top down, are the 0.05, 0.01, and 0.001 critical values of L, the statistic for Page's test.

We find the critical value of L for 10 treatments and 4 blocks to be 1382 at $\alpha = 0.001$. If an observed value of L is greater than this tabled critical value we can reject the null hypothesis that the rank sums are unrelated to the tabulated order of the treatments. For critical values beyond these arguments, employ the following expression:

$$L_{\alpha} = \frac{b(a^3 - a)}{12}\left[\frac{t_{(2\alpha)[\infty]}}{\sqrt{b(a-1)}} + \frac{3(a+1)}{a-1}\right]$$

This table is used for Page's L-test for ordered alternatives (see Section 14.12), a nonparametric test for linear relationships of treatment means in a randomized complete blocks design.

Values in this table for $a \leq 8$ and $b \leq 12$ and $a = 3$ for $13 \leq b \leq 20$ were taken from E. B. Page (*J. Am. Stat. Assoc.*, **58**:216–230, 1963) with permission of the author and publisher. Other values were computed using the normal approximation given above.

TABLE 34 Critical values for Page's test

b	nominal α	α 3	4	5	6	7	8	9	10
	.05	28	58	103	166	252	362	500	669
2	**.01**	-	60	106	173	261	376	520	696
	.001	-	-	109	178	269	388	544	726
	.05	41	84	150	244	370	532	736	986
3	**.01**	42	87	155	252	382	549	761	1019
	.001	-	89	160	260	394	567	790	1056
	.05	54	111	197	321	487	701	970	1301
4	**.01**	55	114	204	331	501	722	999	1339
	.001	56	117	210	341	516	743	1032	1382
	.05	66	137	244	397	603	869	1204	1614
5	**.01**	68	141	251	409	620	893	1236	1656
	.001	70	145	259	420	637	917	1273	1704
	.05	79	163	291	474	719	1037	1436	1926
6	**.01**	81	167	299	486	737	1063	1472	1972
	.001	83	172	307	499	757	1090	1512	2025
	.05	91	189	338	550	835	1204	1668	2238
7	**.01**	93	193	346	563	855	1232	1706	2288
	.001	96	198	355	577	876	1262	1750	2344
	.05	104	214	384	625	950	1371	1899	2548
8	**.01**	106	220	393	640	972	1401	1940	2602
	.001	109	225	403	655	994	1433	1987	2662
	.05	116	240	431	701	1065	1537	2130	2859
9	**.01**	119	246	441	717	1088	1569	2174	2915
	.001	121	252	451	733	1113	1603	2223	2980
	.05	128	266	477	777	1180	1703	2361	3169
10	**.01**	131	272	487	793	1205	1736	2407	3228
	.001	134	278	499	811	1230	1773	2459	3296
	.05	141	292	523	852	1295	1868	2591	3478
11	**.01**	144	298	534	869	1321	1905	2639	3541
	.001	147	305	546	888	1348	1943	2694	3612
	.05	153	317	570	928	1410	2035	2821	3787
12	**.01**	156	324	581	946	1437	2072	2872	3852
	.001	160	331	593	965	1465	2112	2929	3927

TABLE 34 Critical values for Page's test

b	nominal α	α 3	4	5	6	7	8	9	10
	.05	165	343	615	1002	1524	2201	3051	4096
13	.01	169	350	628	1022	1553	2240	3104	4164
	.001	172	358	642	1044	1585	2285	3163	4241
	.05	178	368	661	1078	1639	2366	3281	4405
14	.01	181	376	674	1098	1668	2407	3335	4475
	.001	185	384	689	1121	1702	2453	3397	4556
	.05	190	394	707	1153	1753	2532	3511	4713
15	.01	194	402	721	1174	1784	2574	3567	4786
	.001	197	410	736	1197	1818	2622	3631	4869
	.05	202	419	753	1228	1868	2697	3740	5021
16	.01	206	427	767	1249	1899	2740	3798	5097
	.001	210	436	783	1274	1935	2790	3864	5183
	.05	215	445	799	1303	1982	2862	3969	5330
17	.01	218	453	814	1325	2014	2907	4029	5407
	.001	223	463	830	1350	2051	2958	4098	5496
	.05	227	471	845	1378	2097	3028	4199	5637
18	.01	231	479	860	1401	2129	3073	4260	5717
	.001	235	489	876	1427	2167	3126	4330	5808
	.05	239	496	891	1453	2210	3193	4428	5945
19	.01	243	505	906	1476	2245	3240	4491	6027
	.001	248	515	923	1503	2283	3294	4563	6121
	.05	251	522	937	1528	2325	3358	4657	6253
20	.01	256	531	953	1552	2360	3406	4722	6337
	.001	260	541	970	1579	2399	3461	4796	6433
	.05	263	547	983	1603	2439	3523	4885	6560
21	.01	268	556	999	1628	2475	3572	4952	6647
	.001	273	567	1017	1656	2515	3629	5028	6745
	.05	275	573	1029	1678	2553	3687	5114	6868
22	.01	280	582	1045	1703	2589	3738	5182	6956
	.001	285	593	1063	1732	2631	3796	5260	7057
	.05	288	598	1075	1753	2667	3852	5343	7175
23	.01	292	608	1091	1778	2704	3904	5413	7265
	.001	298	619	1110	1808	2747	3963	5492	7368
	.05	300	624	1121	1828	2781	4017	5571	7482
24	.01	305	633	1138	1854	2819	4070	5643	7574
	.001	310	644	1157	1884	2863	4130	5724	7679

TABLE **35** Critical values of Olmstead and Tukey's test criterion

This table furnishes conservative critical values for the absolute value of the quadrant sum employed by Olmstead and Tukey as a test criterion for their corner test of association. The technique for obtaining a quadrant sum is explained in Section 15.8 and the significance level of a given sum is furnished in this table. Where two magnitudes of the quadrant sums are furnished for any significance level, the smaller magnitude applies to larger sample sizes, the larger magnitude to small sample sizes. Magnitudes of the quadrant sum $\geq (2n - 6)$ should not be employed in the test.

This table is taken from P. S. Olmstead and J. W. Tukey (*Ann. Math. Stat.*, **18**:496–513, 1947) with permission of the publisher.

α	Absolute value of quadrant sum
.10	9
.05	11
.02	13
.01	14 - 15
.005	15 - 17
.002	17 - 19
.001	18 - 21

TABLE **36** Critical values for testing outliers (according to Dixon)

This table features critical values for Dixon's test statistic for outliers. Critical values are furnished for sample sizes from 3 to 25 and for α values of 0.10, 0.05, and 0.01. The table is divided into four parts for four different test statistics shown in the right hand column. The probabilities given are one-tailed. For a two-tailed test use the same critical values but double the probability at the heads of the columns.

To find the 5% critical value for a sample of 12, employ test statistic r_{21} where Y_1 is the suspected outlier and the other subscripted values of Y are the respective variates in an ordered array of the variates with the first being the suspected outlier. Thus, the 5% critical value of r_{21} is 0.546. If the observed test statistic equals or exceeds that value we would consider it an outlier.

This table is used when employing Dixon's test for outliers in normal populations as discussed in Section 13.4.

The table was extracted from a more extensive one (table A-8e) in W. J. Dixon and F. J. Massey, Jr., *Introduction to Statistical Analysis,* 3rd Edition (McGraw-Hill, New York, 1969) with permission of the publishers.

TABLE 36 Critical values for testing outliers (according to Dixon)

$n \backslash \alpha$	**0.10**	**0.05**	**0.01**	
3	0.886	0.941	0.988	
4	0.679	0.765	0.889	
5	0.557	0.642	0.780	$r_{10} = \dfrac{Y_2 - Y_1}{Y_n - Y_1}$
6	0.482	0.560	0.698	
7	0.434	0.507	0.637	
8	0.479	0.554	0.683	
9	0.441	0.512	0.635	$r_{11} = \dfrac{Y_2 - Y_1}{Y_{n-1} - Y_1}$
10	0.409	0.477	0.597	
11	0.517	0.576	0.679	
12	0.490	0.546	0.642	$r_{21} = \dfrac{Y_3 - Y_1}{Y_{n-1} - Y_1}$
13	0.467	0.521	0.615	
14	0.492	0.546	0.641	
15	0.472	0.525	0.616	
16	0.454	0.507	0.595	
17	0.438	0.490	0.577	
18	0.424	0.475	0.561	
19	0.412	0.462	0.547	$r_{22} = \dfrac{Y_3 - Y_1}{Y_{n-2} - Y_1}$
20	0.401	0.450	0.535	
21	0.391	0.440	0.524	
22	0.382	0.430	0.514	
23	0.374	0.421	0.505	
24	0.367	0.413	0.497	
25	0.360	0.406	0.489	

TABLE **37** Critical values for testing outliers (according to Grubbs)

This table features critical values for Grubbs' test statistic for outliers. Critical values are tabled for sample sizes $n = 3$ to $n = 40$ in increments of 1, and for sample sizes $n = 40$ to $n = 140$ in increments of 10. The one-tailed probabilities furnished are 0.10, 0.05, 0.025, 0.01, and 0.005. For a two-tailed test use the same critical values but double the probabilities α.

Grubbs' test statistic is $(Y_1 - \overline{Y})/s$ where Y_1 is the suspected outlier, $\overline{Y}$ is the sample mean, and s is the sample standard deviation. Whenever the outlier is below the mean, consider the ratio as positive before comparing it with critical values in the table. For a sample of 37 items we would find a 5% critical value of 2.835.

This table is useful when carrying out Grubbs' test for outliers in normal samples, discussed in Section 13.4.

This table has been extracted from a more extensive one (table I) in F. E. Grubbs and G. Beck (*Technometrics,* **14**:847–854, 1972) with permission of the authors and publisher.

TABLE 37 Critical values for testing outliers (according to Grubbs)

$n \setminus \alpha$	**0.10**	**0.05**	**0.025**	**0.01**	**0.005**
3	1.148	1.153	1.155	1.155	1.155
4	1.425	1.463	1.481	1.492	1.496
5	1.602	1.672	1.715	1.749	1.764
6	1.729	1.822	1.887	1.944	1.973
7	1.828	1.938	2.020	2.097	2.139
8	1.909	2.032	2.126	2.221	2.274
9	1.977	2.110	2.215	2.323	2.387
10	2.036	2.176	2.290	2.410	2.482
11	2.088	2.234	2.355	2.485	2.564
12	2.134	2.285	2.412	2.550	2.636
13	2.175	2.331	2.462	2.607	2.699
14	2.213	2.371	2.507	2.659	2.755
15	2.247	2.409	2.549	2.705	2.806
16	2.279	2.443	2.585	2.747	2.852
17	2.309	2.475	2.620	2.785	2.894
18	2.335	2.504	2.651	2.821	2.932
19	2.361	2.532	2.681	2.854	2.968
20	2.385	2.557	2.709	2.884	3.001
21	2.408	2.580	2.733	2.912	3.031
22	2.429	2.603	2.758	2.939	3.060
23	2.448	2.624	2.781	2.963	3.087
24	2.467	2.644	2.802	2.987	3.112
25	2.486	2.663	2.822	3.009	3.135
26	2.502	2.681	2.841	3.029	3.157
27	2.519	2.698	2.859	3.049	3.178
28	2.534	2.714	2.876	3.068	3.199
29	2.549	2.730	2.893	3.085	3.218
30	2.563	2.745	2.908	3.103	3.236
31	2.577	2.759	2.924	3.119	3.253
32	2.591	2.773	2.938	3.135	3.270
33	2.604	2.786	2.952	3.150	3.286
34	2.616	2.799	2.965	3.164	3.301
35	2.628	2.811	2.979	3.178	3.316
36	2.639	2.823	2.991	3.191	3.330
37	2.650	2.835	3.003	3.204	3.343
38	2.661	2.846	3.014	3.216	3.356
39	2.671	2.857	3.025	3.228	3.369
40	2.682	2.866	3.036	3.240	3.381
50	2.768	2.956	3.128	3.336	3.483
60	2.837	3.025	3.199	3.411	3.560
70	2.893	3.082	3.257	3.471	3.622
80	2.940	3.130	3.305	3.521	3.673
90	2.981	3.171	3.347	3.563	3.716
100	3.017	3.207	3.383	3.600	3.754
110	3.049	3.239	3.415	3.632	3.787
120	3.078	3.267	3.444	3.662	3.817
130	3.104	3.294	3.470	3.688	3.843
140	3.129	3.318	3.493	3.712	3.867

TABLE **38** Critical values for Kendall's rank correlation coefficient τ

This table furnishes 0.10, 0.05, and 0.01 critical values for Kendall's rank correlation coefficient τ. These values are given for every sample size between $n = 4$ and $n = 40$. The probabilities are based on a two-tailed test. When a one-tailed test is desired, halve the probabilities at the heads of the columns.

The table is used to test the null hypothesis that the variates in two samples are arrayed at random with respect to each other, that is, that the parametric rank correlation coefficient is zero.

To test the significance of a correlation coefficient, enter the table with the appropriate sample size and find the appropriate critical value. For example, for a sample size of 15 the 5% critical value of $\tau = 0.390$. Thus, an observed value of 0.498 would be considered significant at the 5% but not at the 1% level. Negative correlations are considered as positive for purposes of this test. For sample sizes $n > 40$ use an asymptotic approximation given in Box 15.6.

This table is used to test the statistical significance of Kendall's rank correlation coefficients, as illustrated in Section 15.8.

The values in this table have been derived from those furnished in table XI of J. V. Bradley, *Distribution-Free Statistical Tests* (Prentice-Hall, New Jersey, 1968).

TABLE 38 Critical values for Kendall's rank correlation coefficient τ

$n \backslash \alpha$	**0.10**	**0.05**	**0.01**
4	1.000	-	-
5	0.800	1.000	-
6	0.733	0.867	1.000
7	0.619	0.714	0.905
8	0.571	0.643	0.786
9	0.500	0.556	0.722
10	0.467	0.511	0.644
11	0.418	0.491	0.600
12	0.394	0.455	0.576
13	0.359	0.436	0.564
14	0.363	0.407	0.516
15	0.333	0.390	0.505
16	0.317	0.383	0.483
17	0.309	0.368	0.471
18	0.294	0.346	0.451
19	0.287	0.333	0.439
20	0.274	0.326	0.421
21	0.267	0.314	0.410
22	0.264	0.307	0.394
23	0.257	0.296	0.391
24	0.246	0.290	0.377
25	0.240	0.287	0.367
26	0.237	0.280	0.360
27	0.231	0.271	0.356
28	0.228	0.265	0.344
29	0.222	0.261	0.340
30	0.218	0.255	0.333
31	0.213	0.252	0.325
32	0.210	0.246	0.323
33	0.205	0.242	0.314
34	0.201	0.237	0.312
35	0.197	0.234	0.304
36	0.194	0.232	0.302
37	0.192	0.228	0.297
38	0.189	0.223	0.292
39	0.188	0.220	0.287
40	0.185	0.218	0.285

TABLE **39** Critical values for runs up and down

This table furnishes the critical number of runs up and down (sequences of successive differences of like sign) in a sequence of n randomly arranged items. The values of α listed across the top of the table are 0.01, 0.025, 0.05, 0.95, 0.975, and 0.99. The sample size n ranges from 5 to 25. Any number of runs that is equal to or less than the critical value at the left half of any row in the table or is equal to or greater than the critical value at the right half of any row leads to a rejection of the hypothesis of random arrangement of successive differences with a one-tail type I error rate of α or $1-\alpha$.

For a two-tailed significance test in which the alternative hypothesis is either too few or too many runs, one uses percentage points corresponding to the proportions $\alpha/2$ and $1 - \alpha/2$ when one is prepared to accept a type I error of α. To look up critical values significant at the 5% level, proceed as follows. For a sequence of 20 observations enter the table under sample size $n = 20$ and note the values 8 and 16 in the columns labeled 0.025 and 0.975, respectively. Thus as few as 8 runs or as many as 16 runs up and down indicate a significant departure from randomness at the 5% level in a sample of 20 items. For tests of one-tailed hypotheses use the percentage points corresponding to α or $1 - \alpha$, depending on the tail of the distribution being tested. Minus signs in the left half of a row and asterisks in the right half of a row indicate cases in the first part of the table where no significant numbers of runs are possible. For sample sizes $n > 40$, employ the asymptotic approximate formula given in Box 18.3.

This table is used to evaluate the randomness of trends by the runs up and down test discussed and illustrated in Section 18.2. It has been constructed from information furnished in table X of J. V. Bradley, *Distribution-Free Statistical Tests* (Prentice-Hall, New Jersey, 1968).

TABLE 39 Critical values for runs up and down

n \ nominal α	**0.01**	**0.025**	**0.05**	**0.95**	**0.975**	**0.99**
5	-	1	1	*	*	*
6	1	1	1	*	*	*
7	1	2	2	*	*	*
8	2	2	2	*	*	*
9	2	3	3	7	*	*
10	3	3	3	8	*	*
11	3	4	4	9	9	*
12	4	4	4	10	10	*
13	4	5	5	11	11	11
14	5	6	6	11	12	12
15	5	6	6	12	13	13
16	6	7	7	13	13	14
17	6	7	7	14	14	15
18	7	7	8	14	15	15
19	7	8	8	15	16	16
20	8	8	9	16	16	17
21	8	9	10	17	17	18
22	9	10	10	17	18	19
23	10	10	11	18	18	19
24	10	11	11	19	19	20
25	11	11	12	20	20	21

TABLE **40** Some mathematical constants

The following mathematical constants are useful in a variety of statistical applications.

	Value	Reciprocal	$\log_{10}$
π	3.14159 26535 89793	0.31830 98861 83791	0.49714 98726 94134
2π	6.28318 53071 79586	0.15915 49430 91895	0.79817 98683 58115
$\pi/2$	1.57079 63267 94897	0.63661 97723 67581	0.19611 98770 30153
π^2	9.86960 44010 89359	0.10132 11836 42338	0.99429 97453 88268
$\sqrt{\pi}$	1.77245 38509 05516	0.56418 95835 47756	0.24857 49363 47067
$\sqrt{2\pi}$	2.50662 82746 31001	0.39894 22804 01433	0.39908 99341 79058
e	2.71828 18284 59045	0.36787 94411 71442	0.43429 44819 03252
e^2	7.38905 60989 30650	0.13533 52832 36613	0.86858 89638 06504
$\sqrt{e}$	1.64872 12707 00128	0.60653 06597 12633	0.21714 72409 51626
$\sqrt{2}$	1.41421 35623 73095	0.70710 67811 86548	0.15051 49978 31991
$\sqrt{3}$	1.73205 08075 68877	0.57735 02691 89626	0.23856 06273 59831
$\sqrt{10}$	3.16227 76601 68379	0.31622 77660 16838	0.50000 00000 00000
1 radian	57.29577 95130 82321°	0.01745 32925 19943	1.75812 26324 09172
ln 10	2.30258 50929 94046